D0867096

The Ecology of Urban Habitats

The Ecology of Urban Habitats

O.L. Gilbert

Reader in Landscape Architecture
The University of Sheffield

London New York
Chapman and Hall

First published in 1989 by
Chapman and Hall Ltd
11 New Fetter Lane, London EC4P 4EE
Published in the USA by
Chapman and Hall
29 West 35th Street, New York NY 10001

Typeset in 10½/11½pt Bembo by
Photoprint, Torquay, Devon
Printed in Great Britain at the
University Press, Cambridge

ISBN 0 412 28270 4

British Library Cataloguing in Publication Data

Gilbert, O.L. (Oliver L), *1936–*
The ecology of urban habitats
1. Great Britain. Urban regions. Natural history
I. Title
508.41

ISBN 0–412–28270–4

Library of Congress Cataloging in Publication Data

Gilbert, O.L.
The ecology of urban habitats / O.L. Gilbert.
p. cm.
Bibliography: p.
Includes index.
ISBN 0–412–28270–4
1. Urban ecology (Biology) I. Title.
QH541.5.C6G53 1989
574.5′268––dc 19 89–548
CIP

CONTENTS

PREFACE

This book is about the plants and animals of urban areas, not the urban fringe, not encapsulated countryside but those parts of towns where man's impact is greatest. The powerful anthropogenic influences that operate in cities have, until recently, rendered them unattractive to ecologists who find the high proportion of exotics and mixtures of planted and spontaneous vegetation bewildering. They are also unused to considering fashion, taste, mowing machines and the behaviour of dog owners as habitat factors. I have always maintained, however, and I hope this book demonstrates, that there are as many interrelationships to be uncovered in a flower bed as in a field, in a cemetery as on a sand dune; and due to the well documented history of urban sites, together with the strong effects of management, they are frequently easier to interpret than those operating in more natural areas.

The potential of these communities as rewarding areas for study is revealed in the literature on the pests of stored products, urban foxes and birds. The journals of local natural history societies have also provided a rich source of material as amateurs have never been averse to following the fortunes of their favourite groups into the heart of our cities. It is predictable that among the few professionals to specialize in this discipline have been those enclosed in West Berlin, who must be regarded as among the leading exponents of urban ecology.

My working life, spent first among the well studied communities of the Pennines, then as an academic botanist investigating air pollution effects on Tyneside, and finally as a landscape ecologist at Sheffield University, gradually forced to my attention the acute lack of knowledge about the ecology of urban areas. My students needed knowledge of the wildlife capital present in parks, allotments, precincts, canals, railways, roadsides, disturbed sycamore woods, cemeteries, gardens and industrial land. This was not available, so ten years ago I started to make my own observations, to scan the literature and to interrogate anyone likely to possess relevant information. Coverage has been uneven with urban wasteland, here named 'urban commons', receiving sufficient attention to illustrate successional relationships, regional variation, edaphic influences and to provide a reasonably detailed account of most plant and animal groups. Some other habitats need more detailed study to uncover those subtle and sophisticated interrelationships that ecologists love to read about.

Much recent urban ecology has been involved with habitat creation and how to set about greening a city. This book hardly touches on these subjects; it concentrates on what is already there. Too often mature, attractive, spontaneously developed, almost self-maintaining communities which provide that precious commodity, local character, are destroyed, to be replaced by tended open space or sometimes imitation countryside. One aim of this book is to reveal, through an understanding of its ecology, the interest and attraction of the wildlife already present in our towns. At the human and planning level the case is made for fostering links between man and nature within the urban environment. While the quality of human life is merely improved by contact with nature, people are essential if urban wildlife is to survive.

Progress on this book and during the preceding years of research has not been accomplished single-handed. The wide spectrum of plant and animal groups encountered has at times necessitated the use of referees to determine unfamiliar or critical material. A list of names is not an adequate acknowledgement of their help and enthusiasm but regrettably will have to suffice: A. Brackenbury (hoverflies), K. Clarkson (birds), R. Clinging (molluscs), B.J. Coppins (lichens), W.A. Ely (millipedes and centipedes), C.A. Collingwood (ants), S. Garland (lepidoptera), E.C.M. Haes (grasshoppers), G.Hanson (seeds), F. Harrison (lepidoptera), C.A. Howes (mammals), A. Lazenby (beetles), A.J.E. Lyon (fungi), D.J. McCosh (hawkweeds), A. Newton (brambles), M.J. Roberts (spiders), H.L.G. Stroyan (aphids), D. Whiteley (mammals). Others who have helped are G.C. Ainsworth, J.G. Doney, J.P. Grime and J.R. Palmer.

Space, not ingratitude, precludes the naming of every person who has made a contribution of value. These include cemetery managers, allotment officers, waterboard officials, pest control officers, railway engineers, park keepers, arboriculturalists, planning officers and many others. Most of the photographs and line drawings have been prepared specially for this book; I am indebted to Mike Birkhead, Tim Birkhead, John Gay, Bryan Jones, William Purvis and particularly Ian Smith for taking or supplying plates, also to Mike Lindley for his effective artwork.

Finally I wish to thank Margretha Pearman who, with patience and skill, kept a critical eye on the text, typed most of it, helped check the proofs and contributed directly to the work on urban rivers.

O.L. Gilbert
Sheffield

Chapter 1

INTRODUCTION

The view that the maintenance of civilization should include a battle against nature still has many adherents. Most planners and landscape architects take as a basic tenet that people inherently like the picturesque and given the choice would decide to live in a setting not dissimilar to eighteenth century parkland. This represents a controlled and improved aspect of nature catering for artistic and horticultural interests; indeed it has been suggested that Britain's major contribution to the visual arts is perfecting this type of landscape. Our entire cultural background, reinforced by modern advertising, encourages us to appreciate the gardenesque approach. It is currently the accepted landscape treatment for most public open space in towns, along road corridors, in parks, cemeteries, the environs of local authority housing, industrial estates and buffer zones. Being primarily aesthetic rather than for active use there is a big design and maintenance input which includes mowing extensive areas of grass, tending shrubs, planting and replanting trees, weeding, removing dead material, sweeping leaves and giving a lot of attention to edges. The design concept is to produce a static controlled sequence of aesthetic events.

An alternative to the gardenesque fashion is the technological landscape style which dominates in city centres where the density of people requires soft surfaces to be largely replaced by hard ones. These high-cost areas are mostly created by machinery. Due to cut and fill the landforms are rarely natural, and alien artificial materials such as concrete, tarmac or fibre glass are used in their construction. For these reasons they tend to be uniform, lacking local character on both a national and continental scale. Rather a narrow range of plant material is used. This high-cost hard-wearing method of treating city centres has earned them the nickname 'concrete jungles'. In common with the first type this is a static, highly designed, costly landscape solution, functional rather than aesthetic, but when well conceived both are eminently acceptable to their users.

Recently a third approach to urban landscape has emerged which could be called ecological. Most towns in Britain have a few examples of this style – Warrington New Town has many – while the Dutch have been involved with them for many decades (Ruff, 1979). It is a low-cost low-input type of landscape in which man coexists with nature and is not too dominant over it; such areas may arise in three ways. First, encapsulated countryside such as Ruislip Woods or Wimbledon Common in London, which has persisted by chance. Most towns include a few such sites particularly towards the urban fringe. Secondly, local authorities and others are now seeking to create such areas believing them to be a useful addition to the land-use mosaic in towns with the result that a wave of rather piecemeal projects involving habitat creation is sweeping urban areas. The third and in many ways most satisfying type of ecological area in cities are the unofficial wild spaces where nature has gained a foothold. These may vary in size from a railway bank to a crack in a pavement; they permeate the entire built up area. All tend to be informal, exhibit varying degrees of local character, require only a minimal or no design input and one of their great attractions is the low level of management required which enables plant succession to take its course introducing a dynamic element. The presence of wildlife provides the unexpected, abrupt edges are replaced by ecotones, native and naturalized species dominate, dead material is allowed to accumulate and above all it is a habitat for people to interact with, actively, passively, even destructively as it is composed of those species best suited to the site, well able to perpetuate themselves.

These three types of landscape have been defined as:

1. technological, where the biological landscape has been substantially replaced by artificial substitutes
2. gardenesque, where biological elements can function only under continuous management
3. ecological, where natural elements are allowed to function in a natural manner.

The landscape pattern of cities is determined by physical, historical and social conditions. In Britain, more than almost any other country, our attitude to landscape is dominated by culture-bound aesthetics which lead us to expect and to appreciate the gardenesque. However, in thickly populated areas, the technological approach is the only workable solution to high-density living imposed by land shortages, so this style dominates in city centres. The ecological approach is so new it still frequently requires justification. It is championed by the 'Green Movement', a growing grass-roots awareness of the beauty of nature, and increasingly the low capital and maintenance costs are attractive to land managers. It is clear that for the foreseeable future cities will be composed of all three landscape types. It should be borne in mind, however, that the ecological style is at an experimental stage; in the same way that gardening and engineering practices have evolved with time, far-reaching changes in the manner it is implemented can be expected.

There is a widely held assumption that the ecological approach is more favourable to wildlife than the gardenesque, which in turn is better than the technological. This is not entirely true, the three broad habitats being complimentary to each other and providing far richer variety than if only two were present. City centres for example have been likened to a cliff/organic detritus zone which is justified with regard to birds, a number of which have forsaken their rock faces and adapted to breeding on tall buildings. These include feral pigeon, kestrel, black redstart, starling, house sparrow, several species of gull, and, in America, night hawk and chimney swift. Avian density can be exceptionally high with Campbell and Dagg (1976) reporting 796 pairs per square kilometre from a town centre in Ontario. Other birds such as pied wagtails use city centres for roosting. Several well-defined communities of spontaneous vegetation penetrate the heart of cities; they include ruderals, garden escapes and unusual aliens (Chapter 10). For all these species technological landscapes have furnished opportunities for range extension not equalled elsewhere in the urban environment. Similarly, gardenesque and ecological areas have their own special fauna and flora. It is as invidious to infer that one part of the urban mosaic is better for widlife than another as it is to claim that, for example, salt marsh is superior to sand dune; they are both integral parts of the maritime zone.

What pass for the leading ecologically based landscapes in cities (see Smyth, 1987) are frequently fragments of encapsulated countryside little influenced by urban habitat factors. They have much in common with rural communities and the publicity which sites like Moseley Bog, Birmingham ('A beautiful woodland and bog area supporting a wide variety of wildlife including many rare plants, butterflies and birds', Smyth, 1987) receive have tended to direct attention away from mainstream urban communities. Relic countryside surrounded by housing is well worth campaigning to save but it should be recognized that few specifically urban species are likely to be present. This highlights an unresolved dilemma concerning the evaluation of ecological sites in towns. Should the traditional biological criteria used for rural areas (Ratcliffe, 1977), which put a high value on rarity, large size, non-recreateability, richness, diversity and historical continuity, be employed or is it more appropriate to give prominence to social factors like ease of public access, aesthetic appeal, proximity to the town centre, ability to withstand disturbance and occurrence in areas of local deficiency? During the last few years most towns in Britain have had an inventory made of their wildlife habitats. The surveys vary from cities such as Leicester and London where they took several years, to others like Sheffield, and Tyneside where the primary work was accomplished during a single summer. An almost universal finding is that most of the best sites lie near the city limits, being meadow, marsh, bluebell woodland, heathland, hedgerow, etc. When traditional criteria are used to evaluate data collected within the city boundary this is inevitable. Their apparent high status is an artifact produced by the contrived limits of the survey area; in the countryside beyond they are less remarkable. The idea of urban wildlife groups and local-authority ecologists devoting scarce resources to promoting slivers of countryside at the edge of towns is alarming. They would do better to concentrate, as a priority, on the special culture-favoured communities that occur in heavily built-up areas; no-one else will. Attention has already been drawn to some of these habitats through books like *Wild in London* (Goode, 1986) which is particularly informative about birds, but interesting wildlife invades every corner of the built-up area, much of which cannot be seen elsewhere or, like foxes and magpies, behaves differently in towns.

A familiar example of such a community is urban wasteland where a plant succession occurs, every stage of which suffers disturbance and is rich in attractive alien plants. These ‘urban commons’ or ‘urban fields’ as they are sometimes known start with a pioneer community in which Oxford ragwort is prominent, then move on to a tall herb association where garden escapes such as Michaelmas daisy, golden-rod, wormwood, goats rue, Japanese knotweed, evening primrose, lupin, mints and rosebay willowherb form the bulk of vegetation. If the site remains undeveloped, succession may proceed to an open scrub stage in which native species mingle with buddleia, laburnum, sycamore, Swedish whitebeam and domestic apple to form a further assemblage the like of which is not known in rural areas. The animal communities of these sites are less well studied, but synanthropic, often introduced species of slug, snail, woodlice, millipede, centipede and spider are frequent together with a selection of opportunistic native species. Further examples of highly urban habitats each providing their own environment and each exploited by specialized groups of plants and animals include cemeteries, industrial areas, railway land, town parks and allotments.

Gardens are an enigma. They may cover 50% of a town but, with the exception of work by the Owens (see references) and a few studies carried out in America, they are usually ignored by ecologists. This is partly due to the difficulties of studying land in multiple ownership but chiefly because they have never been considered to fall within the remit of pukka ecologists. However, using reciprocal averaging techniques Whitney and Adams (1980) have shown that distinct communities occur within garden vegetation. Further by overlaying socioeconomic variables they identified the primary features responsible for ordering the plants as fashion, taste, species availability, property value and age of house. This is an extreme case of a phenomenon frequently encountered by urban ecologists – the communities are as much a product of the cultural environment as they are a part of the physical landscape. The ecology of gardens is an exciting field where pioneer studies have still to be made. Their complexity, however, is daunting; about a third of British insects are expected to visit a 0.6 ha garden in Leicester, and interwar housing is considered to support a higher density of breeding birds than the richest deciduous woodland.

The urban ecologist finds less of interest in purpose-designed ecological parks and recently created wildlife areas. These have been extremely valuable for education and raising awareness concerning plants and animals but usually few urban species are present. Often attempts are made to imitate the countryside by producing a hay meadow, calcareous grassland, thicket, small patch of woodland, a marsh and a pond (Fig. 1.1). People relate well to these familiar scenes with their connotations of the countryside but it is time to move on and to learn to appreciate urban wildlife. It will always be expensive to maintain the countryside in the town as it is out of context, but partnerships of man and urban nature can be almost self-maintaining. Too often naturalistic wildlife areas in towns are created by destroying perfectly good existing communities of self-sown native and alien plants. In addition communities that have arisen spontaneously provide local character; it will be demonstrated in Chapter 6 how each town has a distinctive urban flora often as a result of trade and commerce.

The next generation of ecological parks is likely to devote substantial areas to initiating natural successions on a range of substrates, e.g. brick rubble, concrete, ash, broken tarmac; perhaps re-creating in an area of old sidings the formerly widespread rail-side community of annuals now almost eliminated by spray trains; managing disused cemeteries for wildlife as at Highgate, or possibly siting the parks so they include some older urban plant associations. Owing to pressures on land, it is unusual for sites to lie dormant for more than a few decades. Occasionally, however, the backwaters of a canal, a steep bank, an island, a fragment of railway property, neglected allotments or the corner of a cemetery

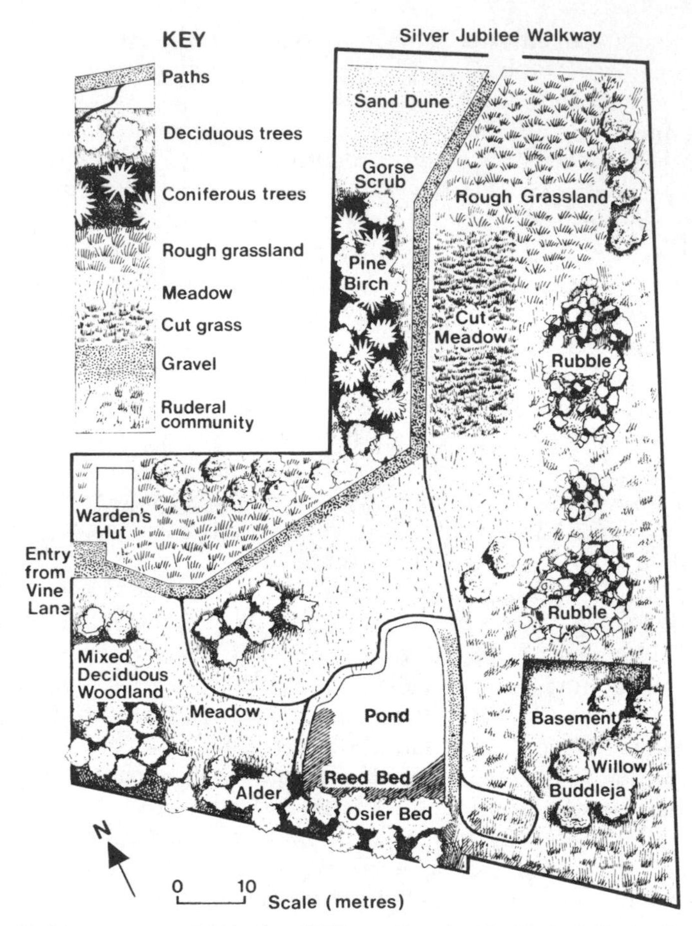

Fig. 1.1. Habitat areas within the William Curtis Ecological Park, London. Though today the layout would be considered overcontrived, this site demonstrated that attractive wildlife could exist in city centres (Ecological Parks Trust, 1982).

may be found where succession has proceeded uninterrupted for many years. These are likely to be particularly rich in synanthropic wildlife, robust, self-maintaning and visually acceptable. The Gunnersbury Railway Triangle in West London is an example of such a site, a public inquiry was fought over it, the inspector refusing to allow development because of the value of ordinary wildlife to townspeople in the place where they live. In addition to being a local amenity and useful for education, a study of such habitats provides ecologists with an insight into the middle stages of urban plant succession.

A feature of wildlife in towns that has been almost overlooked is that the majority of species are so well adapted to the habitat that they will flourish there whether consciously provided for or not. Many attractive urban species are so robust, so well suited to the conditions, invade so efficiently and on such a broad front (no need for green corridors) that it would be impossible to keep them out. Rather than working habitats up through design and management it is perhaps more appropriate to direct attention to increasing appreciation of what is already present. Encounters with mammals and birds have an immense appeal (Fig. 1.2) and it is only a small step to move from an enjoyment of butterflies visiting flowers to an awareness of other visitors such as diptera. In Victorian parks it is possible to enjoy the ornamental horticulture but at the same time to be aware of field woodrush and birdsfoot trefoil picking out the older areas of turf, to walk along a pavement but still take delight in the plant of millet growing outside the pet food shop or notice limestone-loving ferns on the mortared wall of a railway bridge.

The case for fostering links between man and nature within the urban environment and providing the opportunity for city dwellers to experience this nature first hand has already been forcibly made elsewhere (Nicholson, 1987). It will by now be clear that the most appropriate method of accommodating and presenting wildlife is still

Fig. 1.2. Grey squirrel in a London park. Unexpected encounters with animals have an immense appeal (Bryan James).

undecided. This requires skills that planners, landscape architects and ecologists do not have; social scientists, human geographers and environmental psychologists need to be consulted. This new and exciting field is just starting to be opened up in the UK by people like Paul Fitzpatrick, Jacqui Burgess, Carolyn Harrison, Alison Millward and Barbara Mostyn, whose work is discussed in the final chapter. They have discovered that the urban green is appreciated at many levels: simply walking along a road, from the top of a bus, a view across a valley as well as more intimately from walking through it. Ideally it should be an integral part of the urban scene; if site-based, a special expedition has to be made to experience it. The role of urban wildlife in people's lives appears to be multifaceted. Often it is the pleasure of contact with nature and the natural world, the sensuous pleasures of touch, sight, smell and sound. There is also an element of escapism, a release from the predictable urban environment into a more spontaneous one with the chance of an unexpected encounter with living things be it a bird, butterfly, ladybird or unknown wildflower. People do not need instruction in how to enjoy nature. On the negative side the scruffy or untidy vegetation that is particularly good for wildlife often makes people feel uneasy so they prefer to visit such areas with a dog or a companion. These problems can usually be resolved by a small design input.

The benefits of urban wildlife to the city population have so far turned out to be wider than expected and include physical, social, emotional and intellectual components. A satisfying aspect is that the relationship is mutual; while people need nature, urban wildlife needs people if it is to thrive. Not so much a helping hand through habitat creation, just a high density of people pursuing their work, rest and play.

Chapter 2

CHARACTERISTICS OF THE URBAN FLORA AND FAUNA

This chapter and the next two examine how ecological relationships in urban areas differ from those in the surrounding countryside. Most accounts highlight the urban climate, including air pollution, and the soil as aspects which particularly distinguish cities. These, two of the classic trio, climatic, edaphic and biotic factors are treated later, as in my experience they are less significant than certain other forces at work. If the founding fathers of ecology had studied cities rather than the most natural areas available to them they would have given greater prominence to anthropogenic factors. The full extent of man's involvement in determining the present pattern of vegetation in seminatural woodland, grassland and moorland is only now emerging (Rackham, 1986) so it is not surprising that in urban ecosystems it is all-pervading. The shortage of detailed ecological studies carried out in highly urbanized areas means that only incomplete evidence and scattered examples are available to illustrate anthropogenic influences at work. This lack is particularly severe with regard to the fauna. Ensuing chapters provide detail; this one identifies some unifying characteristics of urban ecosystems.

2.1 PROLIFERATION OF INTRODUCED SPECIES

The large numbers of foreign organisms that are intentionally and unintentionally introduced into cities each year impose a considerable inoculation pressure on urban habitats with the result that many of the species eventually find a vacant niche and become established. In the case of plants, horticultural species are a most important source of exotics as they are to some extent matched to the local climate. Quite a small provincial botanical garden such as that in Sheffield contains around 5 500 different plants and in one study a medium-sized domestic garden in Leicester was reported to have held 183 alien species over an 8-year period (Owen, 1983). One of the best known examples of a plant that escaped from a botanical garden is Oxford ragwort (*Senecio squalidus*) first noticed on the walls of a garden at Oxford in 1794. After an initial slow spread it now pioneers the urban succession in many English towns. Most garden plants that become naturalized have multiple points of introduction which pass unrecorded. Examples are Michaelmas daisy, golden-rod, Japanese knotweed, spotted dead-nettle (*Lamium maculatum*), balm (*Melissa officinalis*), horse-radish, evening primrose, *Oxalis corymbosa, Potentilla norvegica*. Each species has its own ecological preference and method of dispersal so their distribution in towns is seldom random.

Industrial areas and railway yards nearly match the suburbs as points of introduction. For example 267 species of grain alien were found round a brewery at Burton-on-Trent (Curtis, 1931; Burgess, 1946), while in Germany up to the mid-1930s, 814 citrus weed species and 692 hay seed aliens had been recorded. Many of the species involved are annuals originating from much warmer climates so they are unable to establish persistent populations. In Britain we call them casuals and on the continent they are labelled ephemerophytes; they rely on repeated introduction for their continuous presence. A limited number introduced through these channels become thoroughly naturalized. Foreign arthropods may build up colonies in railway yards, with those from more southern climes establishing themselves in warehouses and other

Table 2.1. Floristic characteristics of zones with different levels of urbanization, Berlin

Zone	*Built-up*	*Partly built-up*	*Inner suburb*	*Outer suburb*
Vegetation cover (%)	32	55	75	95
No. of spp. of vascular plant (per km^2)	380	424	415	357
Aliens (%)	50	47	43	28
Archaeophytes (%)*	15	14	14	10
Neophytes (%)†	24	23	21	16
No. of rare spp. (per km^2)‡	17	23	35	58

After Sukopp *et al.* (1979).
* Archaeophytes, species introduced up till the year 1500.
† Neophytes, species introduced after 1500.
‡ Rare species, plants in Berlin with only 10 records.

buildings. Every urban land use has its spectrum of introduced species which is described in the relevant chapter.

The net effect of this influx on the composition of the urban flora has been described for Berlin (Sukopp *et al.*, 1979). Table 2.1 illustrates how with increasing urbanization the percentage of introduced species rises to 50%, with neophytes (plants introduced since AD 1500) particularly prominent at 24%. The manner in which the build up of neophyte ruderal species has paralleled the increase in the human population is shown in Fig. 2.1. The alien component of the urban flora is reported to rise to 70% in certain Polish cities (Falinski, 1971). The overall result is that urbanized areas, despite a large reduction in the total vegetation cover, support a higher number of species than the surrounding countryside. This has been demonstrated for Berlin, Hanover (Haeupler, 1974) and Cambridge (Walters, 1970) and could equally well be shown for London if the data in Burton's *Flora of the London Area* (1983) was fully analysed. It has been suggested (Haeupler, 1974) that this enhanced diversity is only shown when the urban population exceeds 50 000. The majority of the new species in the above cities had migrated from southern Europe.

The extent to which towns have become invaded by introduced taxa is often underestimated. For example, most dandelions in city centres are

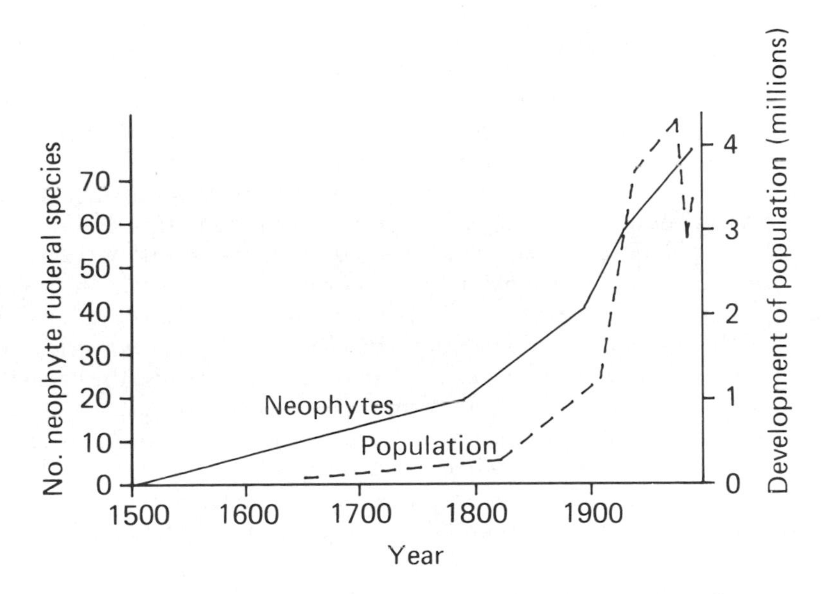

Fig. 2.1. Population development of Berlin related to the number of ruderal species introduced since the year AD 1500 (Sukopp *et al.*, 1979).

introduced species, e.g. *Taraxacum atactum, T. exacutum* and *T. proceris-quameum*, and similarly many of the hawkweeds and brambles. What appears to be tufted vetch (*Viccia cracca*) may well be one of its continental lookalikes (*V. tenuifolia, V. cassubica, V. villosa*). More confusing is that native British species such as Bermuda-grass (*Cynodon dactylon*), tansy, soapwort, wormwood (*Artemisia absinthum*), rock stone crop (*Sedum forsterianum*), columbine and lesser meadow rue (*Thalictrum minus*) are present in towns as populations that have arisen through accidental introduction or as garden rejects. Most of the above examples can, with patience, be assigned to either 'native' or 'alien' status but this is far harder when common and widespread taxa are involved. The very long lists of species currently being introduced as components of bird seed mixes (Hanson and Mason, 1985), wildflower mixes, 3562 kg of which were sold in the UK during 1982 (Wells, 1987), and in the past as grain seed and wool shoddy aliens means that foreign genotypes of many of our native species are now present in towns. I am unaware of any studies that have attempted to assess the scale of this phenomenon, but it suggests that the 50% figure for alien plants in the built-up area of Berlin will be an underestimate.

The typical pattern of spread of introduced taxa, such as *Epilobium angustifolium* (Rackham, 1986), *Senecio squalidus* (Kent, 1955, 1956, 1960, 1964a,b,c). *Veronica filiformis* (Bangerter and Kent, 1957), *Galinsoga* spp. (Salisbury, 1953), sycamore (Jones, 1945), and *Reynoutria japonica* (Conolly, 1977), is a slow and uncertain start which eventually speeds up, the trigger being land use changes, the introduction of new methods of man-aided dispersal, a more vigorous genotype evolving or the population passing what is loosely termed a threshold inoculum pressure. This concept is borrowed from plant pathologists who use it to explain epidemics. It is likely that the ingress of seminatural communities in lowland Britain by introduced species, e.g. sycamore, sweet chestnut and rhododendron, is only just starting. Alien trees which are currently expanding into wild vegetation in towns include Norway maple, Swedish whitebeam, laburnum, domestic apple, Turkey oak and *Robinia pseudoacacia* which together with shrubs such as *Buddleia davidii, Cotoneaster* spp., *Pyracantha coccinea, Ribes nigrum* and *Prunus lusitanica* possess the capacity to invade disturbed seminatural communities once strong populations have built up. Already certain communities are entirely urbanized such as the Sisymbrion and Tanaceo-artemisietum in Central Europe (Sukopp and Werner, 1983). Many of the species involved belong to what is becoming a cosmopolitan urban flora that is continually being added to.

2.2 SEVERE DISTURBANCE

Grime (1979) has defined disturbance as 'the mechanisms which limit plant biomass by causing partial or total destruction'. In cities disturbance through human impact is widespread the destruction being caused by bulldozers, tipping, the creation of over steep slopes, foot traffic, vehicles, horticultural operations such as herbicides, digging, weeding and mowing, occasionally by war damage and regularly by storm water run-off leading to enhanced erosion along water courses. These high levels of disturbance facilitate the survival of aliens through reducing competition. This effect is well documented in naturally disturbed communities such as sand dunes which are readily invaded by North American *Oenothera* spp. and stream sides by *Epilobium brunnescens, Montia sibirica* and *Mimulus* spp., but towns provide a better example. Here erratic severe disturbance is continually initiating successions in which aliens reach their peak densities during the first ten years after which many start to decline due to competition from native grasses. In residential areas irregular spraying and clearing at the base of walls enables garden escapes, which often overhang the structure anyway, to build up considerable populations at the back edge of pavements. If left too long, however, succession proceeds and they are eliminated.

A further consequence of disturbance and its concomitant reduction in competition is that a rather diverse assemblage of native species can be found in towns growing outside their normal context. A 6-year-old demolition site may support hedgerow and wood margin species (*Geranium robertianum, Scrophularia nodosa* and *Stachys sylvatica*), grassland plants (*Centaurea nigra, Hypochoeris radicata,* and *Leontodon autumnalis*), damp ground species (*Epilobium hirsutum* and *Phalaris arundinacea*), arable weeds of fertile soils and stress tolcraters (*Arabidopsis thaliana* and *Sedum acre*) growing alongside the expected ruderals. The point here is that native species display a wider ecological amplitude in the disturbed conditions of towns than they do in the closed vegetation of the surrounding countryside.

The effects of disturbance on insect populations has been studied at Rothamsted Experimental Station in Hertfordshire (Taylor *et al.*, 1978). In a series of cycles an allotment site was disturbed, allowed to revert to dense vegetation, again disturbed and so on. As the vegetation grew up moth productivity (total number of individuals in the population) increased but not diversity. This was attributed to an increase in plant biomass but no profound change in the general composition of the flora. When the area was again cleared moth productivity declined drastically but, if anything, the diversity index increased (Fig. 2.2), probably because a number of abundant opportunistic species were most severely affected by the disturbance. The conclusion was that the species structure of a moth population is more stable with regard to disturbance than population density. The opposite is true of bird communities in urban woodland (Tilghman, 1987).

Fig. 2.2. The effect of cycles of disturbance on the moth population of an allotment site at Rothamsted. N, total number of individuals in the population. D, an index of diversity (Taylor *et al.*, 1978).

2.3 INCREASED OPPORTUNITIES FOR DISPERSAL

Many successful urban species have small wind dispersed seeds that saturate the area, with the result that patches of disturbed ground are rapidly colonized by plants such as *Aster novi-belgii, Epilobium angustifolium, E. ciliatum, Linaria vulgaris, Senecio squalidus, Solidago canadensis, Taraxacum officinale, Tussilago fafara* and small-seeded grasses. This strategy is well documented for the North American prairies where it is associated particularly with the initial phase of colonization of earth mounds produced by burrowing animals (Grime, 1979). It seems reasonable to conclude that this form of regeneration is adapted to exploit unpredictable forms of disturbance. Visitors to the prairies (A. Ruff, personal communication) have commented on the similarities between the prairie flora and that of urban wasteland in Britain. Many of these species appear to have a dispersal method pre-adapted to urban conditions, wind gusting down streets and the slipstream of traffic ensuring the necessary wide dissemination of their seeds.

The increased opportunities for dispersal in towns rely largely on novel methods provided by man's activities. The best synoptic account of dispersal in an urban area is Allen's (1965) observations on the flora of Hyde Park and Kensington Gardens made over five summers. On a citywide scale the transport of topsoil and rubble is particularly important, especially for rhizomatous species like *Reynoutria japonica* and *R. sachalinensis* which are not known to reproduce by seed in this country. The success of *Aegopodium podagraria, Calystegia sepium,*

Elymus repens and *Tussilago farfara* is also often due to the presence of vegetative fragments transported in soil which enable them to dominate a site quickly. Other successful urban plants accumulate a persistent bank of buried seeds in the soil and these can be dispersed in mud adhering to footwear (Clifford, 1956) or stuck in the tread of vehicle tyres. Darlington (1969) found that one car he examined was carrying the seeds of 13 angiosperms; the dominants *Matricaria matricarioides* (220 seeds), *Poa annua* (387) and *Stellaria media* (274) all accumulate seed banks.

Horticultural practice is responsible for spreading many species by deliberately planting them. At the establishment stage weeds of rich soil may also inadvertently be disseminated through being associated with container-grown nursery stock, e.g. *Galinsoga* spp., *Mercuralis annua, Oxalis* spp. and *Urtica urens*. A relatively new dispersal route connected with landscape work is the use of bark or peat mulches; these ameliorants contain the seeds of species such as *Juncus* spp., *Rumex acetosella* and foxglove which establish readily. Another horticultural activity is the dumping of refuse. The margins of wasteland in towns are regularly colonized by garden rejects such as potato, mint, montbretia, nasturtium, *Bergenia, Galeobdolon argentatum, Geranium* × *magnificum* and many others. Most persist for a few years only but are constantly replaced by new arrivals.

Almost every activity that involves transport provides an opportunity for the dispersal of organisms, and towns are nodes on transport systems. Evolution has equipped plant propagules to travel by air, water, attached to animals, inside animals, even to be carried by ants, but in towns these ingenious methods are considerably supplemented and often replaced by anthropogenic agencies. Urban areas can be described as showing enhanced permeability to immigrants due to the increased opportunities for dispersal coupled with the open nature of the habitat.

2.4 HABITAT DIVERSITY

The higher number of plant species found in towns compared to an equal area of countryside can only partly be explained by the influx of aliens. Equally important is the varied small-scale habitat mosaic imposed by man. In more natural areas habitat diversity is controlled by factors such as topography and geology interacting with current and past land uses; these often remain uniform over considerable areas. Cities are different. Owing to anthropogenic influences a very wide spectrum of environments may be present in a small area. If Grime's (1979) triangular model representing the range of plant strategies is translated into a habitat diagram it can be seen that almost the entire spectrum may be present in a town (Fig. 2.3). The small empty area in the right-hand side of the model is where advanced successional stages are normally found.

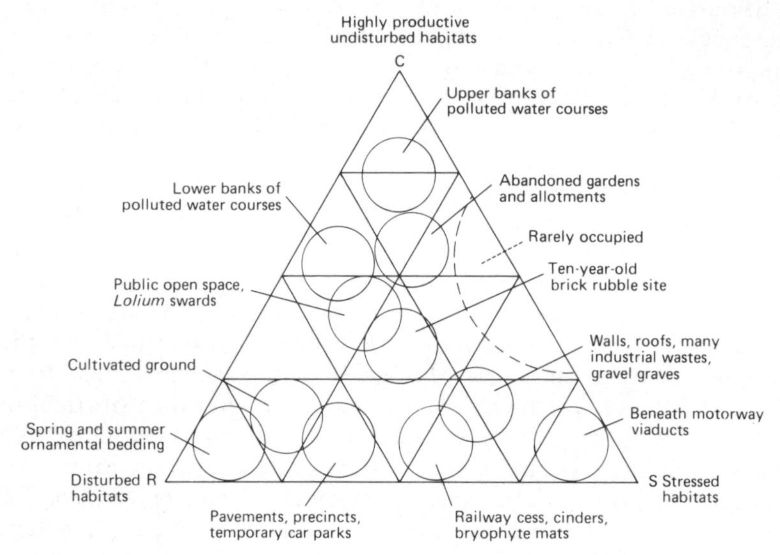

Fig. 2.3. Grime's C–R–S model converted to illustrate the range of habitats present in cities.

The largest areas are occupied by disturbed and productive habitats which is not unexpected due to the high level of biotic activity and the way towns are used as receiving sites for large amounts of materials including nutrients. The stressed examples are mostly connected with hard landscape features such as walls and roofs which provide conditions similar to those on cliff or scree. An extreme example of a stressed habitat occurs under the viaducts of motorway intersections such as Spaghetti Junction, Birmingham where one of the few organisms present in the dry shady environment is the introduced spider *Tegenaria agrestis* (Fig. 2.4). The older urban structures such as Victorian parks, cemeteries, railways and canals provide habitats suitable for stress-tolerant ruderals. In the interests of efficient use of space the division between land uses is normally sharp so ecotones are infrequent but can sometimes be found beside linear features such as rivers, canals, railways and roads.

Species, and by implication communities, that are negatively correlated with urbanization can be identified by examining the maps in Burton's (1983) *Flora of London*, which covers the area within a 32 km radius of St Pauls' (Fig. 2.5). Assemblages which are deficient in and immediately around the capital include the ground flora of deciduous woodland (*Dryopteris dilatata, Oxalis acetosella, Mercuralis perennis* and

Fig. 2.4. An extremely stressed habitat, Spaghetti Junction motorway viaduct, Birmingham. The only species of note is an introduced spider (*Tegenaria agrestis*).

Zerna ramosa) and hedgerow and wood margin species (*Arum maculatum*, *Bryonia dioica*, *Geum urbanum* and *Stellaria holostea*). Many of the native species that make up scrub are also more or less absent which leaves a vacant niche in our cities that is being filled by buddleia, cotoneasters and the Duke of Argyll's tea-plant (*Lycium halimifolium*). Urban London is also deficient in such normally widespread meadow and rough

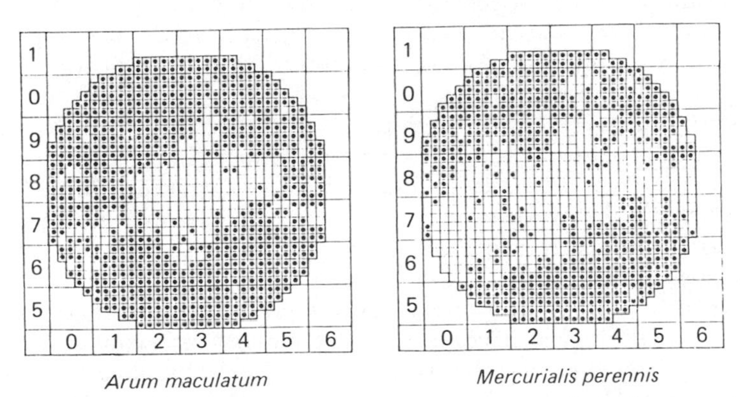

Fig. 2.5. Dot maps showing the distribution of *Arum maculatum* (cuckoo-pint) and *Mercurialis perennis* (dog's mercury) within a 32 km radius of St. Pauls, London (Burton, 1983, courtesy of the London Natural History Society).

grassland plants as *Stellaria graminea, Veronica chamaedrys* and *Vicia sepium*. These four communities together with oligotrophic sites are under-represented in urban areas because they belong to more advanced successional stages and most urban habitats are recent. Often they are only present in islands of encapsulated countryside so are not particularly relevant to urban ecology.

2.5 TAXONOMIC AND EVOLUTIONARY ASPECTS

In a series of papers on commonness and rarity in plants, Hodgson (1986) has pointed out that families of angiosperms are not equally equipped to exploit the habitats created by modern methods of land use. An extensive vegetation survey of the Sheffield region for example revealed that the most successful families include a high proportion of species which combine the ability to exploit fertile habitats with prolific regeneration by seed. He linked this to the controversial evolutionary history of the families as indicated by Sporne (1980), suggesting that primitive families tend to be represented by rare species confined to infertile and relatively undisturbed habitats while the most advanced families include many species of widespread occurrence in man-made sites.

The most abundant families, with regard to polycarpic perennials, in the landscape of this heavily modified region are the Compositae, Gramineae and Labiatae. They possess seeds of intermediate weight capable of immediate germination and rapid growth. This combination of characters is considered to equip them well to exploit modern man-made habitats. In addition, established plants of many of the species in these three families have an ability for extensive lateral spread so they possess unusually flexible regeneration strategies. Families with a greater number of rare species are more stress tolerant and their large, or in the case of Orchidaceae, exceptionally minute seeds, it is suggested, restrict opportunities for regeneration in the conditions now prevailing in the Sheffield landscape. The 11 most abundant families in the region have been tentatively grouped in terms of their capacity to exploit the fertile and disturbed conditions created by man (Table 2.2). The data imply that many ecologically important taxonomically linked adaptations predate the major impact of man upon the world and must have arisen as evolutionary responses to conditions in the past which have similiarities to those found in urban sites today.

For reasons already outlined, species normally separated by geographical or ecological barriers frequently come into contact in urban areas. The cross breeding that results may eventually have far-reaching consequences. Domestic dogs (*Canis familiaris*), for example, can mate successfully with several wild canids, and many of the coyotes living in

Table 2.2. A tentative classification of families in terms of their capacity to exploit fertile disturbed man-made habitats in the Sheffield area. Only polycarpic perennials are considered

Family grouping		*Index of abundance*	*Characteristics*
Group 1	Compositae Gramineae Labiatae	High	Many species in fertile habitats. Often high capacity for lateral spread. Seeds intermediate in size, good dispersal, immediate germination
Group 2	Caryophyllaceae Cruciferae Scrophulariaceae	Intermediate	As in Group 1 but smaller seeds
Group 3	Leguminosae Ranunculaceae Rosaceae Umbelliferae	Low	Fewer species found in fertile habitats. Only limited capacity for vegetative regeneration. Large seeds. Germination often strongly seasonal
Group 4	Orchidaceae	Very low	Less fertile habitats. Little capacity for lateral spread. Minute seeds

Adapted from Hodgson (1986c).

American cities are the progeny of hybrid matings; they are called coy–dogs. The few surviving golden jackals (*Canis aureus*) and wolves (*C. lupis*) in Greece are now largely hybrid animals (Harris, 1986) so, despite nature reserves and legal protection the wild, pure-bred wolves and jackals of Europe may still disappear and be replaced by populations of dog hybrids living as scavengers in the urban fringe. Most of the plants that naturalize readily in towns were already pre-adapted to the conditions but there is evidence that special urban races are evolving. Sukopp *et al.* (1979) give as an example the evening primrose (*Oenothera*) in Europe, where today there are more than 15 different species. With two exceptions (*O. chicagoensis, O. syrticola*) they are not identical with the North American plants from which they descended. These

European taxa have developed since their American parent species were introduced 350 years ago. Michaelmas daisies (*Aster novae-angliae, A. novae-belgii, A. lanceolatus, A. laevis* and hybrids) in British cities appear to be becoming increasingly variable both morphologically and in their ecological amplitude which suggests that new urban taxa may be evolving. Many dandelions particularly in the sections Hamata and Ruderalia are centred on disturbed urban habitats (Graham, 1988), and their local, sometimes endemic, occurrence indicates that they must have originated only a short time ago.

New taxa arise through man-made selection, particularly important here being the numerous grass cultivars of species such as *Lolium perenne, Festuca rubra* and *Poa pratensis*. Breeding programmes select for features like disease resistance, ability to withstand trampling, and persistence, so races of 'super grass' well adapted to urban conditions are continually being introduced into cities. The type of features selected for when breeding ornamental plants, however, often reduces their chances of naturalizing successfully, e.g. yellow, variegated or purple foliage, multiplication of floral parts, gigantism, dwarfism.

Hybridization resulting from a breakdown of geographical isolation produces a fair number of plant hybrids in disturbed urban environments. One of the best known is the cross between Oxford ragwort (*Senecio squalidus*) and sticky groundsel (*S. viscosus*) which was first noticed in London (1944) and named accordingly *S.* × *londinensis*. It is a plant of railway property and waste ground. A further series of hybrids characteristic of railway tracks and sidings involves the toadflaxes *Linaria purpurea, L. repens* and *L. vulgaris*. London pride (*Saxifraga* × *urbinum*) is a sterile hybrid not known in the wild which can frequently be seen in towns in a seminaturalized condition. It is probably totally dependent on urbanization in the same way that our only endemic mammal, the St. Kilda house mouse (*Mus muralis*) was; it became extinct 2 or 3 years after the remote island where it lived was evacuated.

Severe environmental stress in the form of pollution exerts very strong selection pressures on organisms. When polluted environments have been examined in detail, adapted populations have usually been discovered whether the toxicity was salt, heavy metals, sulphur dioxide or smoke. Examples are described in the relevant parts of the book.

2.6 VERTEBRATES

Urbanization has created a number of new ecological niches which, after remaining empty for some time, are increasingly being occupied by vertebrates, particularly birds and mammals. Summers-Smith (1963), when attempting to account for the phenomenal success of the house-sparrow, suggested that its secret was that while becoming adapted to living close to man it had not become so highly specialized that it was restricted to a very limited niche. It is a general feeder, flexible in choice of nest sites, has few successful enemies, and is very tolerant of disturbance. A further reason for the house sparrows success may lie in its high intelligence which maze tests have shown to be comparable to rats and monkeys (Porter in Summers-Smith, 1963). Other birds that are opportunists characterized by a wide ecological amplitude and are also moving into towns include magpie, crow, jay and herring gull. There are also signs that the lesser black-backed gull is adapting to city life. Since the major urban food sources are intentional feeding by man and from refuse dumps, non-territorial seed eaters and omnivores which gather food on the ground are at an advantage. Certain species are to some extent pre-adapted to the new niches, for example the cliff nesting kestrel, jackdaw, town pigeon and black redstart have extended their range to include tall buildings while blackbird and certain finches of the forest edge favour gardens.

Ornithologists have suggested that the behavioural changes required for the exploitation of a new habitat are sometimes the result of unusual weather conditions. For example Hudson (1898) considered that a series of hard winters around 1890 prompted the first inland invasion of the London area by gulls. Other advantageous habits may be learnt by chance and passed on by copying, for example, bullfinches adding the buds of fruit trees to their diet, blue-tits and house-sparrows pecking their way into milk bottles and magpies learning to extract the contents of egg cartons. Safe nest sites, safe roosting sites and an absence of natural enemies or persecution by farmers are additional reasons why towns are favourable to birds. It must be stressed that ordinary habitat factors such as these remain the chief determinant of bird success in towns. In a 5-year study of urban/suburban breeding birds in Massachusetts, Goldstein *et al.* (1986) found that woody vegetation volume alone accounted for 50% of the variations observed.

Successful urban mammals show the same traits as the birds, particularly important being omnivorous feeding coupled, in this group, with a nocturnal or crepuscular habit. In the United States of America, the racoon, opossum, chipmunk, skunk and coyote are now regarded as members of the urban fauna and there is evidence that the Texas armadillo or peba, which has a diet somewhat similar to a hedgehog is starting to invade suburbia. In Britain the fox, grey squirrel, rat, hedgehog and pipistrelle bat have adapted especially well to towns. Behavioural adaptations to city living include an increased proportion of the diet obtained by scavenging, a reduction of home range and, in foxes at least, the high road traffic casualties have implications for the social structure of family groups.

Vertebrates poorly adapted to towns are those with specialized habitat or feeding requirements. Grasshopper warbler, wood warbler, shrike and nightingale which require thick scrub have gradually decreased round London, as have ground nesting birds. Rooks are prepared to fly considerable distances from their rookeries to find open farmland on which to feed but the central areas of large cities are now devoid of this species as the necessary journeys are too far (Goode, 1986). Vertebrates are highly versatile and with badger, fox, Canada goose, mandarin duck and ring-necked parakeet having adapted to various towns in Britain over the last 40 years, gains are expected to exceed losses for the foreseeable future.

2.7 INVERTEBRATES

Davis (1978, 1979) surveyed the ground arthropods occurring in London gardens by employing pitfall traps. His results showed that the best predictor of faunal diversity was the proportion of land occupied by gardens, parks etc. within 1 km radius of a site. This was interpreted in terms of the increasing opportunities which open space affords for different species to sustain populations and the consequence of this for dispersal. The correlation coefficient was 73%. There was also a clear relationship between species number and distance from the city centre. This followed a curved path (Fig. 2.6) with a rapid initial increase moving away from the centre and a more gradual increase in the outer

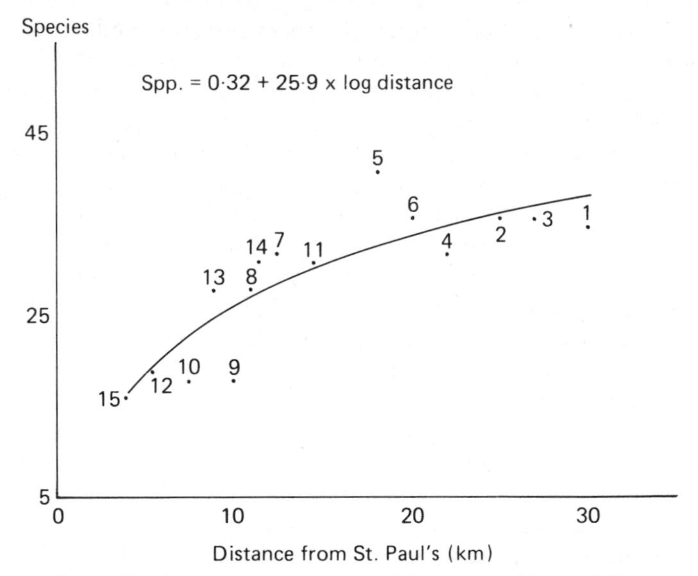

Fig. 2.6. Relationship between total arthropod species, identified from 15 London gardens, and distance from the city centre (Davis, 1979).

suburbs. A combination of both variables improved the correlation coefficient to 82%.

This work highlights a number of factors relevant to invertebrates. The fragmentation of green space presents problems for certain groups. Less mobile ones in particular have difficulty filling their potential distribution as sites become increasingly isolated. This factor would be less important to birds, flying insects and spiders belonging to the family Linyphiidae which broadcast by aeronauting; all these groups disperse on broad fronts over great distances. The second point to note is that distance from the city centre is a gradient associated with a multitude of factors and it is unrealistic to imagine that a synoptic explanation can apply to a diverse collection of arthropods. Autecological studies on the individual species and competitive interactions would need to be investigated to clarify the factors involved. Habitat simplification and particularly levels of disturbance would be promising lines of enquiry to follow. From Fig. 2.6 it can be seen that even at the most central site, a considerable number of ground arthropods are still present. These include species with a wide ecological tolerance such as several centipedes and woodlice, and others which are urban specialists being synanthropic and almost certainly introduced, e.g. the millipede *Choneiulus palmatus*. A feature of the urban environment that is unfavourable to ground arthropods is the generally poor development of a soil litter layer. More needs to be known about the biology of individual species before generalizations can be made, yet studies like this one in which uniform habitats are examined along a gradient of increasing urbanization can be very rewarding. Environments that would lend themselves to such a study include nettle beds, thistles, the canopy of sycamore trees, tree trunks, ornamental planting and walls.

2.8 PLANTED VEGETATION

Up until this point, the characteristics of the urban wild flora and fauna have been considered, but 60–70% of the urban green is planted vegetation. Studies of suburban gardens (Whitney and Adams, 1980) have shown that this type of vegetation can be separated into distinct communities controlled by socioeconomic variables such as taste, fashion, prestige and income. The fact that the primary factors responsible for ordering them are cultural sets them well apart from most other vegetation. Little is known about their structure and functioning so they represent a major challenge. One of the most important aspects of suburban gardens is that they provide the characteristic facies of the town fauna, their fine-grained pattern attracting a wide range of synanthropic animals. Gardens are perhaps the ultimate urban ecosystem embodying man's goals, life styles and value systems. They are considered further in Chapter 14.

2.9 HISTORICAL FACTORS

Pollen analysis has established that many of the native plant genera and species that are abundant in towns today were widespread in the open park–tundra that covered Britain at the end of the Pleistocene. What is less clear is their status during the ensuing forest period and just when they first became associated with man. Clapham (1953) has suggested that many could have colonized drift lines and banks by streams, rivers, lakes and the seashore so would have been available for and capable of spread into forest clearings and along tracks made by early man, e.g. tansy, mugwort, coltsfoot, docks, *Dactylis glomerata, Lolium perenne* and *Elymus repens*. Many are vigorously growing perennials with a marked capacity for vegetative spread and respond to nitrogen; these would have become prominent in the vicinity of human settlements. There, some would attract attention as potential foods or medicines and might be introduced more widely. A history along these lines may be postulated for many native components of the urban flora.

The history of alien species, particularly those introduced since about 1750, is known with more certainty. They present a rich field for investigation by historical ecologists as a recent study has shown that each city has its own spectrum of naturalized exotics (Gilbert, 1983a). Their patterns of distribution can be explained partly in terms of the local industries, the presence of botanical gardens, the nature of the horticultural trade, or sometimes land use and custom. For example, the abundance of alexanders (*Smyrnium olusatrum*) in Norwich is believed to be a result of its early introduction as a pot herb by local monks. Historical factors alone cannot always be invoked; for example the profusion of hoary mullein (*Verbascum pulverulentum*) on waste ground in the same city is probably a climatic phenomenon, while the frequency of evening primroses (*Oenothera* spp.) in Liverpool can be linked to the extensive populations on nearby sand dunes. The discipline of historical ecology which uses the past activities of man to interpret the current situation has much to offer the even younger field of urban ecology.

Chapter 3

THE URBAN CLIMATE AND AIR POLLUTION

Most of the climatic changes brought about by urbanization have been well documented and repeatedly confirmed. Much less well known are their consequences. Even classic studies like Chandler's (1965) *The Climate of London* contain only very general references to its effects on the economic and social life of the city. Direct or indirect mention of the urban climate influencing the ecology of built-up areas is equally scarce, and with the exception of work in West Germany usually anecdotal. By contrast a great deal of scientific information exists on the biological consequences of air pollution. This is discussed separately. Tables expressing average urban climatic differences as a percentage of rural conditions can be found in a number of books (Changnon, 1976; Landsberg, 1981; Spirn, 1984). They show reasonably close agreement; that of Horbert (1978) is reproduced as Table 3.1. While it is appreciated that the climate acts as a whole, the various factors and their ecological consequences will, as far as possible, be discussed separately.

3.1 CLIMATE

Temperature

Annual mean temperatures in the centre of large towns are 0.5–1.5° C higher than those of the surrounding rural area. Chandler (1965) found that this effect was most pronounced on still clear nights when a well-developed heat island of the order of 5° C or more may develop over those parts of London with a high building density. By dawn, most of

Table 3.1. Average change of climatic parameters in built-up areas

Climatic parameters	*Characteristics*	*In comparison to the surrounding area*
Air pollution	Gaseous pollution	5 – 25 times more
Solar radiation	Global solar radiation	15 – 20% less
	Ultraviolet radiation	15 – 20% less
	Duration of bright sunshine	5 – 15% less
Air temperature	Annual mean average	0.5 – 1.5° C higher
	On clear days	2 – 6° C higher
Wind speed	Annual mean average	10 – 20% less
	Calm days	5 – 20% more
Relative humidity	Winter	2% less
	Summer	8 – 10% less
Clouds	Overcast	5 – 10% more
Precipitation	Total rainfall	5 – 10% more

From Horbert (1978).

the heat stored in buildings and road fabrics has been released and the heat island largely dissipated only to slowly build up again during the day. In addition to cities being built of materials with a high thermal capacity the scarcity of vegetation, soil and other damp surfaces means that the heat required for evaporation and powering transpiration is small, so all the more is left for heating the air and buildings. City centres are usually warmer than the countryside on 2 days in 3 and about 4 nights in 5. At their maximum the difference in narrow streets, courtyards and the small open spaces which characterize town centres can be 10° C compared with adjacent rural areas. Wind speeds over 8–12 m s^{-1} (12–18 mph) destroy heat islands; they are also severely weakened by cloudy weather. A factor that helps to maintain them is the burning of fuel; this is so effective that it is said when the Sheffield steel industry was at its peak snow never settled in the area of the foundries.

The main biological result of these temperature anomalies is that the active growing season for plants, defined as the period during which the mean temperature exceeds 5.6° C (42° F), is almost 3 weeks longer in central London and other large cities than in nearby open areas. The frost-free period is extended by 10 weeks in the heart of London, there being in particular a delay in the first air frosts as the heat island is stronger in autumn than spring. As might be expected these fairly substantial, if intermittent, temperature differences have an effect on the plants and animals inhabiting towns though they are not very obvious and need a considerable input of time to demonstrate. All the best work has been done in West Germany where several different approaches have been taken.

Phenological observations, especially of blossom development (Ellenberg, 1954; Schreiber, 1977), can be used to show detailed thermal patterns in early summer. A study of *Forsythia* was made in Hamburg during 1955 (Franken, 1955). Through an appeal in the papers, citizens who regularly travelled the same route to work were asked to note the day on which they first observed at least ten blossoms on *Forsythia* bushes at different locations. From the results a detailed map was produced showing the city divided into four phenological or thermal zones (Fig. 3.1). The pattern was not entirely concentric, being complicated by relief, areas of bomb damage (cooler) and a narrow

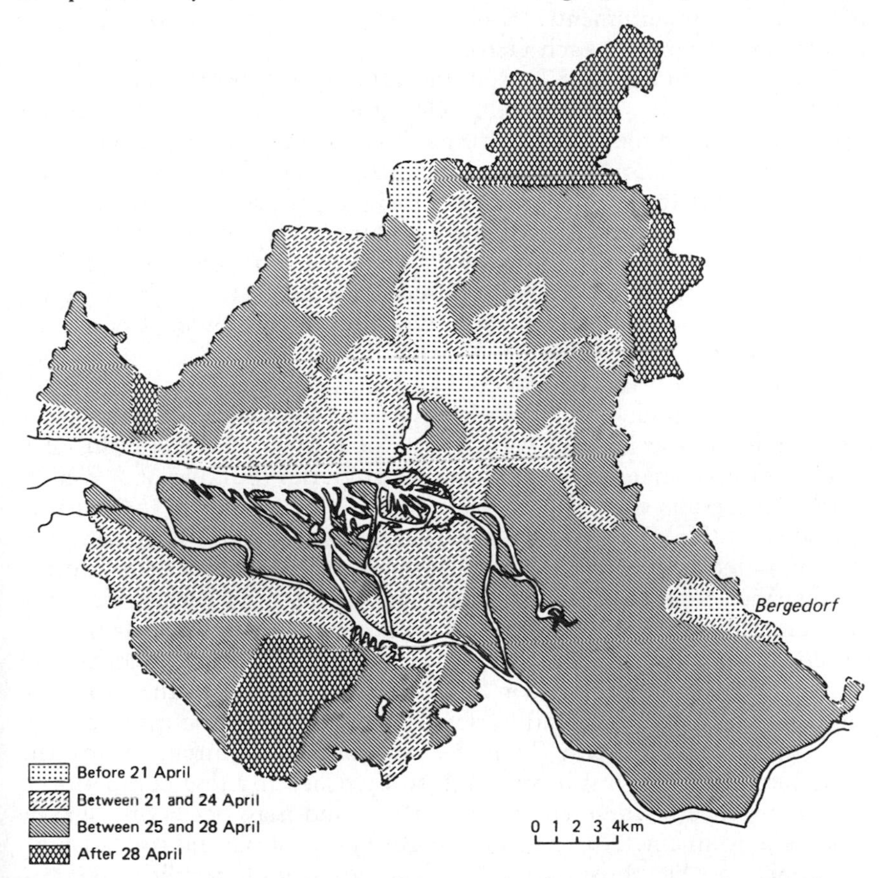

Fig. 3.1. Map showing the onset of blossom development in *Forsythia*, Hamburg, April 1955 (Franken, 1955).

warmer strip beside the river Elbe. A central heat island, with an outlier round the town of Bergedorf, is, however, clearly visible. The year 1955 was particularly suitable for such a study, the cold spring extending blossom development over 17 days. Other species that, owing to their sensitivity to temperature, have been successfully used for phenological mapping are flowering in roadside lime trees (Zacharias, 1972), the flowering of apple, though dates need correcting for cultivar using Cox's Orange Pippin as the standard (Bultitude, 1984), and various parameters of beech (*Fagus sylvatica*) such as the first sign of green, leaves unfolding or shoots reaching 4 cm in length.

A second method employs Ellenberg's (1974) table of the indicator value for vascular plants in Central Europe. This table, which is not well known in Britain, provides information on the ecology of 2000 plants with regard to light, temperature, continentality, moisture, soil reaction and nitrogen requirements. Each species is assigned to one of nine degrees of tolerance to each factor.

Wittig and Durwen (1982) used the table to compare the spontaneous floras in the towns of Bielefeld, Cologne, Dortmund and Munster against the flora of their surroundings. It was conclusively demonstrated that all four cities contained a greater proportion of species with high indicator values for temperature than their respective environs. In other words the mean expression of the spontaneous urban vegetation was more thermophilous. This is a result of elevated annual average daily temperatures, so it reflects a different parameter to a phenological inquiry. Godde and Wittig (1983), working in the town of Munster, West Germany, have refined the use of Ellenberg's table to the point where zoning, heat islands and cold air bands can be clearly pinpointed.

A third method involves plotting the frequency of individual thermophilous species such as the tree of heaven (*Ailanthus altissima*) which has been mapped throughout West Berlin (Sukopp and Weiler, 1986). The results show a high concentration of records in the warmer central areas (Fig. 3.2). The map would be improved if the distribution of seed and sucker parents were also indicated. Durwen (1978) plotted the distribution of wall barley (*Hordeum murinum*) in Munster claiming that the results showed a restriction to the warmer central areas though it is equally possible that the pattern was controlled by factors other than climate. For example, Davison (1970, 1971) working in the North of England concluded that wall barley was restricted to the more central areas of Newcastle upon Tyne by its unusual requirement for the combination of a high soil pH, high soil fertility and low competition. The occurrence of these conditions is rare and may occur only in the urban environment. For example, in Britain most natural soils of high pH appear to be phosphorus deficient while high fertility and low competition is a fleeting if not mutually exclusive combination.

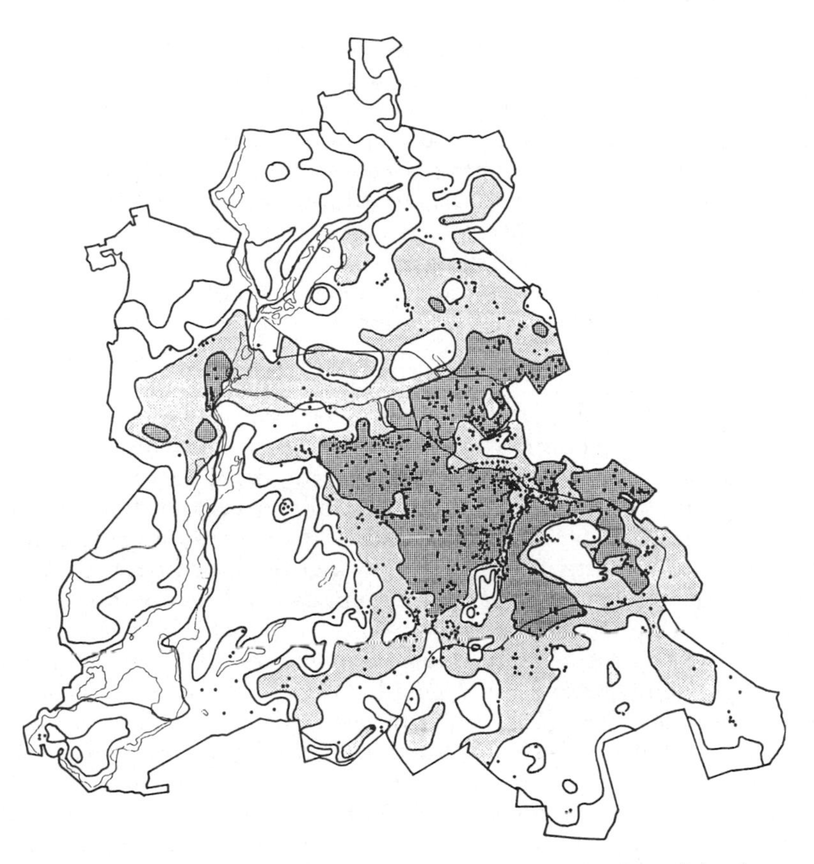

Fig. 3.2. Map of West Berlin showing spontaneous regeneration of the thermophilous Tree of Heaven (*Ailanthus altissima*). The heaviest shading indicates the warmest temperature zones (Sukopp and Weiler, 1986).

Davison (1977) took his studies a stage further by observing the growth of *Hordeum murinum* populations sown into an urban site and two with adverse climates. This revealed that the failure of the species in adverse climates is due to a reduction in the rate of growth and development which has a detrimental effect on grain production and, therefore, on the potential population size the next year. Additionally, late or incomplete ripening of the spike meant there was little or no shattering of the rachis and consequently, poor dispersal. *Hordeum murinum* is sensitive to competition and, as a result, populations are in a state of flux. In many situations, lack of dispersal would therefore lead to rapid elimination of populations by invading competitors. Failure of this autumn germinating annual, growing at the edge of its range, would be gradual and erratic depending on annual variations in the climate.

Davison's detailed analysis of *Hordeum murinum* in Britain is a major contribution to an understanding of its ecology. The cautious conclusion was that the species appears to penetrate into cooler and wetter regions when it is growing in urban rather than rural sites, and this could be due to an ameliorating effect on the general climate, but it is more probably due to the occurrence of microhabitats such as the bases of buildings, where conditions are locally warmer and drier.

It is clear that careful and detailed work is needed to demonstrate the influence of the urban heat island on vegetation. Towns are not generally recognized as sites where thermophilous plant species reach their northern limit. When possible examples of this effect have been investigated many turn out to have alternative explanations; for example, the Breckland rarity *Vulpia ciliata* ssp. *ambigua* was reported from Sheffield, well north of its normal range but the site is beyond the city limits and relates to an unusual substrate, dark, very freely draining steel slag, not the urban heat island. Possibly heat island effects are easier to demonstrate under continental rather than windy atlantic climates. Even horticulturalists, a group very aware of microclimatic differences, can provide few examples of general temperature amelioration allowing tender species to survive in towns. This may partly be due to the fact that the most intensive temperature anomalies occur in sites where hard landscape dominates and any vegetation present is chosen for robustness and known survival qualities. However, one outstanding example is known, the Chelsea Physic Garden in Central London. This 2 ha site contains what is believed to be the most northerly olive tree (*Olea europaea*) in the world; not only is it 120–150 years old and 10 m high, it even ripens fruit, 3 kg being harvested in December 1976. Nearby is a flourishing cork oak (*Quercus suber*) and a kermes oak (*Q. coccifera*) all being linked by further shrubby and herbaceous plants from southern Europe including a pomegranate (*Punica granatum*). Elsewhere in this walled garden the national collection of *Citisus* including the comfry-leaved citisus (*C. symphytifolium*) from the Canary Isles can be found together with a wide range of Salvias and beds of South American, South African and Australian plants. The curator of the garden reports that the plants are regularly 2–3 weeks ahead of those on the edge of London which corresponds closely to Chandler's prediction of spring growth starting 3 weeks earlier in Central London. Though a number of the Chelsea Physic Garden species can be grown in milder parts of the country such as Cornwall, Devon and West Wales, their occurrence in London is remarkable.

Zoologists have little to add apart from casual observations that birds may start nesting earlier in town centres and that certain species may roost in towns because of the warmth. There are alternative explanations to 'temperature triggers' for these phenomena. Davis (1982) has pointed out that for invertebrates very local factors such as the occurrence of greenhouses can obscure simple linear climatic gradients. The case of melanism in the two-spot ladybird is considered later.

Wind, humidity and precipitation

Few other climatic parameters cited in Table 3.1 are of sufficient magnitude to have detectable effects on urban ecosystems, apart of course, from those related to air pollution. Wind speeds are important in the general sense that when they are strong there is so much mixing there is no such thing as an urban climate. The mean 10–20% reduction in wind speed is due to the frictional drag of buildings on the air moving round them. High buildings produce violent eddies and can cause accelerated wind speeds at ground level. Isolated trees near the base of tall buildings can be shaped by this effect and the exposure-linked phenomenon of bearing smaller leaves on the windward side of the canopy is sometimes discernible. The place to see pronounced wind shaping of trees is close to air pollution sources. The aluminium-producing town of Ardal in Norway has many street trees showing unequal canopies where eddies of air carrying fluoride pollution have locally restricted their growth.

The slightly lower air humidity in towns has not been shown to have any effect on the plants and animals living there but has been the subject of a famous controversy. Throughout the 1960s, when most lichenologists were moving towards the idea that sulphur dioxide was the main factor limiting the lichen flora of urban areas, Jan Rydzak in Poland dissented from this view, putting forward his 'Drought Hypothesis' to explain the zonation patterns seen in urban lichen floras. Though workers showed errors in Rydzak's interpretation of various data (Coppins, 1973), he never altered his view. His theory died with him, but he is remembered as an eccentric who caused lichenologists to examine the 'Toxic Gas Hypothesis' more rigorously than they might otherwise have done.

The increase in cloud and precipitation over cities is one aspect which has taken some time to prove. Recent work, however, has shown conclusively that increases in precipitation of the order of 5–10% are normal. This is not a significant factor in the water regime of vegetation, however, because most occurs in the form of heavy rain, often thunderstorms, which runs off rapidly into underground drains. An indirect result of this is that foul water services, which are designed to accommodate only two to six times the normal flow, frequently become overloaded with the consequence that raw sewage enters local water courses to the detriment of the stream fauna. German work on the indicator value of spontaneous urban floras (Wittig and Durwen, 1982) has shown that biologically the small increase in total rainfall is more than cancelled out by its rapid run-off; so the spectrum of the urban flora indicates somewhat dryer conditions than in the surrounding countryside.

3.2 AIR POLLUTION

Most town dwellers perceive air pollution as visible emissions from chimneys or obnoxious smells and are reminded by smoke-blackened stonework that it was once far worse. They mostly pass through life unaware of any of its numerous biological effects apart from perhaps noticing an absence of lichens and having heard somewhere that it promotes melanism in the peppered moth. This chapter is designed to alert urban ecologists to the multifarious but subtle ways in which the major pollutants sulphur dioxide (SO_2) and smoke have affected and continue to affect urban ecosystems. Most of these effects can still be detected in our towns and cities but are much better developed in the capitals of Eastern Europe. Vehicle exhaust is considered in Chapter 9.

Higher plants

A large literature records the effects of air pollution on higher plants; it reveals that all the most serious examples of damage involve emissions from industrial point sources. Accounts of the generally lower urban pollution levels causing severe damage are quite rare and in Britain date from before 1965. At that time annual average smoke concentrations over considerable areas of the London Basin, Tyneside, Birmingham, West Yorkshire, Lancashire, and the South Wales and Scottish coalfields were in the range 250–350 $\mu g\ m^{-3}$ while the mean for sulphur dioxide (SO_2) was 200–250 $\mu g\ m^{-3}$. It must be remembered that these average values mask much higher monthly peaks and during the London smogs of December 1952 and 1962 daily values for smoke crept up to 4000 $\mu g\ m^{-3}$ and SO_2 to around 2500 $\mu g\ m^{-3}$.

In a series of almost forgotten papers Salisbury (1954) and Metcalf (1953) recorded the effects of air pollution at Kew while McMillan (1954) documented the effects of extremely heavy smoke and SO_2 in Manchester. At the same time Scurfield (1955, 1960) was collating the observations of parks superintendents in the North of England. Despite their intimate knowledge of plants these botanists/horticulturalists found it difficult to attribute specific damage to air pollution other than a general stunting and consequent requirement for more frequent replacement planting. All, however, reported a severe effect on conifers. Smoke was thought to be responsible for golden varieties of privet and elder, and the variegated foliage of other species reverting to a normal green colour. A number of evergreen species such as the privet (*Ligustrum lucidum*) and even holly tended to become deciduous in the more heavily polluted areas. As a consequence in Burnley holly was replaced by laburnum. A general phenomenon was that deciduous trees shed their leaves earlier. Autumn bedding of *Myosotis,* wallflowers, *Bellis* and *Primula* did not succeed and the practice was adopted of raising these species outside the town then transplanting them to their flowering position in the spring. It was impossible to grow sun-loving species such as cinerarias, *Aubretia* and *Salvia harbenger* until smokeless zones were introduced.

All these phenomena are associated with a reduced light intensity caused by the smoke cloud. Light reduction has been quantified for London by comparing the hours of bright sunshine received with that in the surrounding countryside. Bilham (1938) showed that during the interwar years winter sunshine in central London was cut by half, though by mid-summer the reduction was only 10%. Losses in other British cities over the 5 months November–March ranged from 25 to 55% (Chandler, 1965). Reductions of this order are quite sufficient to explain the observed phenomena and also the lack of persistence of vernal plants which are active early in the year. For example, regular replacement planting was required for spring bulbs, crocus, winter aconites, etc. which could not manufacture sufficient food reserves to maintain themselves and flower for more than 1 or 2 years.

The commonly expressed opinion that soot also acted directly on plants by interfering with gas exchange through blocking stomata is now known to be entirely wrong. Particulate matter jams stomata open thus facilitating gas exchange (Williams *et al.*, 1971). Today urban smoke levels in Britain have fallen on average by 90% and arc expected to stay low; never again will so much coal be burnt so inefficiently. In countries such as Poland and Czechoslovakia, urban smoke levels remain very high and the types of damage outlined are still widespread.

Sulphur dioxide is an invisible gas produced whenever coal or oil is burnt, but not by the combustion of natural gas or petrol. It acts on plants through a toxic effect attacking cells inside the leaf. It also causes a general acidification of the environment including a lowering of soil pH. Conifers are particularly sensitive to SO_2 and can be eliminated by the levels once found in urban areas. In 1924 the National Pinetum was moved from Kew, London to the much cleaner air of Bedgebury, Kent because the trees were not thriving and replacement planting was becoming difficult. Not all conifers are equally affected. Most *Pinus, Picea, Pseudotsuga* and *Abies* spp. are highly sensitive and had been eliminated from the centre of all our larger towns by 1950. Among the more resistant are yew (*Taxus baccata*) and Austrian pine (*Pinus nigra*) but even these are no longer present in town centres.

Although urban SO_2 levels are now much reduced the distribution of conifers still shows a concentric zoning in our larger cities. Gardens in the outer suburbs, particularly on the windward side, support fairly diverse collections of conifers, while the only mature ones present in the inner suburban ring are *Pinus nigra* and *Taxus baccata*. City centres and old industrial areas contain either none or a much younger age class planted after pollution levels fell dramatically in the early 1970s. The same pattern occurs throughout the developed world, a recent census of trees in the heart of Tokyo revealed that *Abies firma* had already disappeared by 1950 and since then *Cryptomeria japonica, Pinus densiflora* and *Castanea crenata* had decreased by 95% (Numata, 1982). Today urban SO_2 levels in Britain have fallen by 80% compared with 30 years ago, and the European Economic Community guideline value for housing areas of 60 $\mu g\ m^{-3}$ as an annual average is rarely exceeded.

At its height, pollution in manufacturing towns seems to have suppressed all tree growth with the exception of fast growing species such as the Manchester poplar (*Populus nigra,* var *betulifolia,* Fig. 3.3), hybrid black poplar (*P.* × *canadensis*), crack and white willows (*Salix*

Fig. 3.3. Being resistant to smoky air, the Manchester poplar (*Populus nigra* var. *betulifolia*) has long been planted in northern towns, particularly those on Merseyside. The heavily burred trunk is characteristic.

fragilis, *S. alba*) and plane (*Platanus* × *acerifolia*), together with occasional ash and sycamore. Beech and oak are rather sensitive; it is almost unknown to see large trees of these species near the centre of industrial towns. Even in the suburbs of Sheffield they exhibit dense, compact, flat topped canopies composed of contorted zig-zagging branches. Such deformities in pine have been used to map zones of severe and moderately intense pollution in the Polish city of Cracow (Grodzinska, 1982). Since the early 1970s when pollution levels in Sheffield plummeted due to the introduction of natural gas, these reduced canopies have expanded by sending isolated, strongly growing leaders into the cleaner atmosphere (Fig. 3.4). This production of a 'double canopy' has been accompanied by increases in the width of annual growth rings. The air pollution history of a site and the relative sensitivity of trees can be discovered from investigating tree ring widths using an incremental borer. Figure 3.5 shows a growth ring analysis of

Fig. 3.4. Beech (*Fagus sylvatica*) sending out strongly growing branches following a reduction in air pollution 15 years ago. Photograph 2 km from the centre of Sheffield.

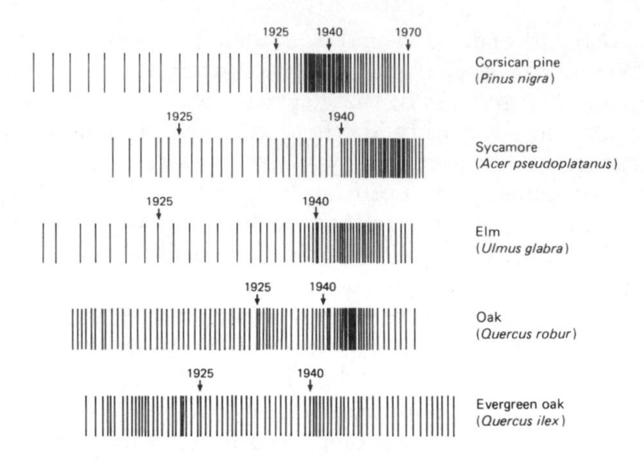

Fig. 3.5. Growth-ring analysis of five tree species occurring downwind of a steel works in Rotherham. The air-pollution history of the site has influenced the width of the rings. See the text for further details.

five species occurring together on a ridge downwind of a steel works in Rotherham. The important events were an expansion of the works in 1925, a massive and sustained increase in steel production during World War II, then a conversion of furnaces from open hearth to electric arc in the mid 1960s which enabled the flue gases (SO_2,F) to be scrubbed. The order of tolerance of the five trees to this combination of pollutants is seen to be *Quercus ilex, Ulmus glabra, Acer pseudoplatanus, Quercus robur* and *Pinus nigra* (most sensitive).

Lichens

It is common knowledge that lichens show a wide range of sensitivity to SO_2 and have been used to construct scales (Gilbert, 1970b, 1974; Hawksworth and Rose, 1970) that allow even schoolchildren to make semiquantitative maps of SO_2, within the range 35–170 $\mu g\ m^{-3}$. While these scales worked well enough in the 1960s and early 1970s when SO_2 levels were fairly stable they are unable to detect rapidly changing levels of air pollution due to a time lag in response. For this reason they are out of favour. A study more relevant to the current situation has been carried out by Seaward (1982) who investigated the reinvasion of *Lecanora muralis* on asbestos roofs throughout the West Yorkshire conurbation. Under conditions of ameliorating SO_2 this lichen exhibited a 5-year time lag before reinvasion became discernible as a spatter of tiny colonies. After establishment the annual increase in thallus diameter was reasonably constant falling within the range 2.29–2.67 mm (av. 2.50), though there was evidence of an urban super race with a faster growth rate (av. 3.87). Over the same period, the ecological amplitude of this species was seen to expand as SO_2 levels fell. An initial restriction to asbestos-cement roofing widened to take in cement-tile roofs and later successively, sandstone, tarmacadam and worked wood. The lesson to be learnt from Seaward's work is that studies that attempt to investigate dynamic situations currently have more relevance than the construction of one-off presence or absence maps which leave the following important questions to be answered. Are the boundaries distinguished static or moving? To what extent is the map out of date due to time lag phenomena? Which parameter of pollution are the lichens responding to?

The lichen deserts of Western Europe are the best studied in the world. In them it has repeatedly been confirmed that the communities respond to pollution by simplification; there is a sharp fall in diversity, biomass and structural complexity leading to the selection of new dominants often of exotic origin. Tolerant species are characterized by a high reproductive capacity and frequently display a broader ecological amplitude than when growing in their native habitats. Additionally the city conditions tend to select for nitrotolerance. There are very few introduced lichens in the flora of any region but it seems likely that *Lecanora conizaeoides*, the most toxitolerant epiphytic lichen known, has such a history; this is absolutely typical of urban dominants.

Bryophytes

Air pollution exercises considerable control over bryophytes in towns. Though they show as a wide a range of sensitivity to SO_2 as lichens, they have never been popular for investigating its distribution. This is partly because under pollution stress many retreat into sheltered niches where they do not experience the regional SO_2 levels. Also resistant species maintain a high biomass into city centres, so the reduced diversity is noticed only by bryologists.

The leading community in highly polluted areas is dominated by the mosses *Bryum argenteum* and *Ceratodon purpureus*; it is not known outside towns. Widespread on wall tops, asbestos roofs, in cracks between paving stones, on tarmac, soil, horizontal timber surfaces, ledges, slates, tiles, in flower pots, gutters and on trodden paths, it is present in most sites where bryophytes can grow. This range of habitats in a small village would support over 30 species exhibiting a wide range of growth forms so the extreme simplification to two species sharing the short turf habit can be appreciated. In less eutrophic sites *Ceratodon purpureus* is usually dominant with the silvery *Bryum argenteum* increasing from 1% to complete dominance where there is strong nitrogen enrichment associated with dust, perching birds or dog faeces.

A conspectus of urban byrophyte communities can be found in Gilbert (1968a, 1970a). They are typified by a low species diversity imposed by SO_2. All the characteristic species *Bryum argenteum, B. capillare, Ceratodon purpureus, Funaria hygrometrica, Leptobryum pyriforme, Lunularia cruciata, Pohlia annotina, Tortula muralis,* etc. are international weeds possessing a high reproductive capacity, exhibiting a fast growth rate, and showing resistance to SO_2. Most are nitrophilous or at least nitrotolerant, but *Dicranella heteromalla* is a calcifuge; it occurs on stable surface-leached soils, pH range 3.2–5.1, under shrubberies in parks and churchyards. Owing to their constant association with man, a number are suspected of being introduced, but others such as *B. argenteum* have spread from specialized niches such as bird colonies.

Fungi

For a long time mycologists have been aware that certain leaf parasites and inflorescence diseases are rare or absent in urban areas, e.g. oak mildew (*Microsphaera alphitoides*), black-spot of roses (*Diplocarpon rosae*), tar-spot of sycamore (*Rhytisma acerinum*), several common rusts, ergot (*Claviceps purpurea*) and choke (*Epichloe typhina*). Detailed studies of black-spot (Saunders, 1966) and tar-spot (Vick and Bevan, 1976) have determined the limiting concentrations of SO_2 which controls their distribution to be about 100 $\mu g\ m^{-3}$, and about 90 $\mu g\ m^{-3}$, respectively. As sycamore is widespread and tar-spot easy to identify, this disease has been integrated into lichen scales and used to help map SO_2 in South Yorkshire and Merseyside. It should be noted that these fungi are depressed over a much smaller area than most lichens and bryophytes but as they rely on annual infection, their response to changing levels of air pollution is far more rapid. For this reason they deserve further study.

The suppression of oak mildew by SO_2 may be responsible for the healthy state of many urban oakwoods. Ecologists are aware that regeneration within British oakwoods has been almost non-existent this century, but urban woods uncharacteristically contain a wide range of age classes. Rackham (1976) has suggested that this may be correlated with the virtual absence of the mildew which, since it spread rapidly in this country after its introduction around 1900, has taken a heavy toll by weakening oak saplings growing in the humid interior of rural woods.

Pollution levels are now too low in Britain for a meaningful survey of their effects on macro-fungi to be undertaken, but a recent study in Poland (Lawrynowicz, 1982) revealed consequences that have gone largely unnoticed elsewhere. His 10-year project in Lodz, the second largest town, broke new ground by showing that macro-fungi occur throughout the city on lawns, in parks, roadsides, ruderal sites, etc., the total number of species, at 476, exceeding the number of wild higher plants. The distribution of certain ecological groupings, however, appeared to be determined by air pollution. A central zone lacked epiphytes with woody perennial fruit bodies, such as *Fomes fomentarius, Piptoporus betulinus,* and contained only a limited number of mycorrhizal symbionts. The middle urban zone contained many more mycorrhizal symbionts such as *Amanita* spp. and *Boletus* spp., and epiphytes with long-lived woody fruit bodies were also present, e.g. *Ganoderma applanatum, Daedalea quercina* and *P. betulinus* but parasites were reduced compared with the next zone. The outer suburban area supported a normal macro-fungus flora which owing to habitat diversity was richer than that found in some undisturbed forests. To what extent these changes are the result of air pollution remains conjectural; lower humidity could also be important but it is significant that tree trunks are the most sensitive habitat, and Grodzinska (1982) working a pollution gradient in the Polish city of Cracow is more definite about correlating similar changes with industrial emissions.

Invertebrates

One of the more publicized aspects of urban ecology is the indirect effect of air pollution in promoting melanism among moths. As tree trunks lose their lichen cover (SO_2) and become blackened (soot), species such as the light coloured peppered moth (*Biston betularia*), which normally spend the day resting on the bark relying on cryptic coloration to avoid detection, suddenly become conspicuous. Under this selection pressure the rare black morph (var. *carbonaria*) begins to increase in number relative to the usual pale form as they are now well camouflaged on the dark tree trunks. Since melanism is inherited, by the 1950s the great majority of peppered moths in urban areas were black, though in country areas the pale form remained commonest. Though the classic study used *Biston betularia* (Kettlewell, 1973), the effect is quite widespread among nocturnal moths which rest on tree trunks during the day, e.g. pale brindled beauty, scalloped hazel, waved umber, grey dagger, mottled beauty, etc. Problems with this theory are that var. *carbonaria* is fairly frequent in East Anglia where trees are still lichen clad, typical forms are never quite eliminated in towns, and on a transect from North Wales to Manchester there has been an inward shift in the distribution of var *typica* over the 1973–1986 period but no significant shift in the abundance of grey foliose lichen species to form a cryptic background. However, the abundance of the lichen *Lecanora conizaeoides* has increased greatly at the urban end of the transect and this does not form a protective background for either morph so there the predation pattern on var. *carbonaria* has changed for the worse. A topic that needs urgent investigation is the location of resting sites of the moth on trees as it is possible they prefer sites in the canopy rather than on the trunks which is where the lichen cover is usually assessed.

Industrial melanism has been reported in other invertebrates such as spiders and psocids where it is assumed to result from the same mechanism but this is not always the case. The high melanic frequency in many urban populations of the two-spot ladybird (*Adalia bipunctata*) was interpreted by Creed (1974) as being due to intense local air pollution. A reanalysis of his data by Muggleton *et al.* (1975) showed that it also supported the opinion of Lusis (1961) that sunshine level is the principal factor determining melanic frequency in the two-spot ladybird. Thermal melanism, by enabling the black morphs to absorb solar radiation faster, gives them a considerable competitive advantage over normal forms in the search for food as well as in mating activity in areas where sunshine levels are reduced, for example by smoke. Over the country as a whole a highly significant negative correlation was shown between annual hours of bright sunshine and melanic frequency so Muggleton *et al.* (1975) suggest that it is a geographic rather than an industrial melanic. The last word on this subject probably still has to be written.

Apart from some inconclusive work on oribatid mites (Andre and Lebrun, 1982), urban levels of air pollution have not been shown to affect any invertebrate directly. The consequences of the disappearance of a lichen cover for the invertebrate component of the tree trunk ecosystem has been studied on a transect into Newcastle-upon-Tyne. It showed that the effects of pollution were greatest on epiphytes which were eliminated, the repercussions of this spreading to the herbivores, mainly psocids and lichenophagous caterpillars, which also became extinct. The predatory fauna, however, survived undiminished by feeding on the large number of aphids which fell out of the canopy. It appeared that a small amount of pollution favoured certain groups, i.e. the psocids *Elipsocus* spp., *Mesopsocus unipunctatus* and *Reuterella helvimacula* which preferentially feed on *Pleurococcus* and *Lecanora conizaeoides,* by producing conditions in the suburbs where their staple food reached maximum abundance (Gilbert 1971).

An attractive theory, which it will be hard to disprove, concerns the spread of breeding colonies of house martins and swifts into Central London during the 1960s. It was suggested by Cramp and Gooders (1967) that their increase was linked with the declining smoke levels and a consequent increase in flying insects.

Chapter 4

SOILS IN URBAN AREAS

4.1 THE CLASSIFICATION OF URBAN SOILS

Engineers, landscape architects, horticulturalists and others concerned with urban soil usually recognize three types, topsoil, subsoil and no soil. The extremes of texture, heavy clay and sand may also be acknowledged. This simplistic classification is a result of the almost total lack of interest shown by the Soil Survey of England and Wales who spend most of their time collecting information on agricultural land. True, one of the 10 major groups they recognize is 'man-made soils' defined as 'soils with a thick man-made A horizon, a disturbed subsurface layer, or both,' but they possess little information on this category. Despite this, the soil survey's publications are always worth consulting for background information especially in cases where urbanization has spread over agricultural land already mapped.

Avery (1980) has made a start to a classification of urban soils by recognizing two broad divisions. The first is 'man-made humus soils' which have an abnormally thick A horizon of dark well-structured topsoil. This type is associated with allotments and old gardens where they result from particularly deep cultivation (double digging) coupled with heavy organic manuring. Another activity which produces them, night soiling, is the tipping for a hundred years or more of sewage and domestic refuse. Profiles tend to contain quantities of debris such as claypipe stems, broken china, cinders and glass. At depth the remains of buildings may occur. The second division recognized by Avery is 'disturbed soils', these are described as predominantly mineral soils over 40 cm deep, consisting of artificially dispaced material. They can have any composition depending on the source of the material that is derived at least in part from the A_0, A, B or BC horizons of pre-existing soils. An additional category 'disturbed rock waste' is C horizon material. This heterogeneous group needs further subdivision before it can be of use to ecologists.

A provisional conspectus of soils occurring in urban areas is given below. It attempts to match them with categories recognized by the soil survey so they can be intercalated with an existing system. It would be unnecessary and wrong to regard them as a totally separate branch of pedology. It should be appreciated, however, that all kinds of intermediate conditions occur so that, for example, an ecologist attempting to decide whether a primary or secondary plant succession has been initiated will often remain in doubt.

4.2 CONSPECTUS

1. Man-made humus soil. A thick (>40 cm) man-made A horizon resulting from bulky amendments of manure, mineral matter or domestic rubbish. Allotments, old gardens, small holdings, rubbish dumps.
2. Topsoiled sites. A/C or A/BC soils created by spreading topsoil of variable quality over raw disturbed mineral soil. May be compacted at one or more levels. The evolution of these rankers to brown soils is helped by cultivation. Recently landscaped sites.
3. Raw–lithomorphic soils. Initially these consist of little altered raw mineral materials often of man-made origin, at least 30 cm deep, and show no horizon development attributable to pedogenic processes. With time they evolve into lithomorphic soils as a distinct humose upper layer develops. Two subdivisions can be made in each group. (1) Soils with only one layer that is not humified topsoil. (2) Soils with at least two layers including a humified topsoil.
   (a) Brick rubble. Man-made raw soil evolving into pararendzina. Brick rubble often mixed with a little old soil and subsoil. Building demolition areas.
   (b) Furnace ash, slag, cinder. Man-made raw soil evolving into humic ranker. A coarsely textured non-calcareous soil associated with railway land, particularly old sidings, heavy industry, etc.
   (c) Chemical wastes. Man-made raw soil evolving into a range of lithomorphic soils. The raw material may have an extreme pH or contain highly toxic substances. Of limited occurrence, associated with the chemical industry, certain other manufacturing industries and electricity generation (pulverized fly ash).
   (d) Disturbed subsoil. Man-made raw soils evolving into profiles that have affinities with ranker-like alluvial soils in which the

humose layer passes down into an unconsolidated C horizon which may show stratification. These soils form from a stratum of artificially rearranged or transported material >40 cm thick consisting predominantly of mingled subsoil horizons (B, B/C or C). Found in some recently landscaped sites, some landfill sites.

(e) *In situ* subsoil. Raw skeletal soil evolving into a variety of lithomorphic soils. Unconsolidated or weakly consolidated mineral horizons, sometimes partly stratified, which have weathered *in situ*. Found on steep banks associated with transport corridors or sites benched for major buildings. Also widespread in non-urban areas.

(f) Hard surfaces. Tarmacadamed, paved and concrete surfaces. Roads, pavements, floors of demolished factories, angle between pavement and wall, etc.

4.3 CHARACTERISTICS OF URBAN SOILS

Despite their variability urban soils share sufficient characteristics for them to be defined as 'a soil material having a non-agricultural man-made surface layer more than 50 cm thick that has been produced by mixing, filling or contamination of land surface in urban and suburban areas', (Blockheim, 1974). This definition highlights disturbance as an important characteristic, and, owing to the ever shortening length of building-development cycles, the rate and scale of disturbance is increasing, so many soils in urban areas remain perpetually immature.

Variability

Vertical and horizontal variability at all scales is common. This variation bears little relationship to the underlying geology or to topography so detailed maps are required to reveal it. Most often the variation reflects the constructional history of the soil which may be elucidated from the number and position of lithological discontinuities or interfaces in a profile (Craul, 1985). This variation is well known to archaeologists who, by studying discontinuities, are often able to recognize multiple occupation layers. Towns, particularly in the lowlands, can be looked on as receiving sites for material which may accumulate to such a depth that the site becomes raised. In the Middle East the sites of ancient long-abandoned cities are marked by mounds known as tels which are composed entirely of urban debris. Biologically it is variation in the effective rooting zone, to 90 cm, which is the most important. Below this, impervious layers may influence the drainage class of the upper horizons, and, more rarely, the lowest layers are responsible for the production of gas which may seep to the surface and cause the death of vegetation. At the site of the Liverpool Garden Festival an extensive refuse tip producing large amounts of methane was dealt with by capping, using a half metre layer of clay; the gas was then extracted using boreholes (Cass, 1983). Methane generation from a landfill site in Washington D.C. has been sufficient to heat the glasshouses of the White House for ten years.

Structure

As they mature, soils usually develop a crumb structure in which particles aggregate together thus increasing bulk volume by the presence of large pore spaces between the aggregates. This type of structure has a favourable effect on aeration, water permeability and root penetration. Many conditions in the urban environment work against the development of a good crumb structure and bulk density is often increased through compaction. Particularly damaging are disturbance and handling of soils when they are wet, and surface traffic by foot or machine over saturated soils. This includes events such as pop festivals, circuses, temporary car parks, and fetes; recovery of soil structure following these may take many years. Structure is so delicate that it has been reported how even vibration resulting from heavy vehicles on roads can compact silty streetside soils at levels below the influence of surface compression (Craul and Klein, 1980).

Bulk density is the weight of a given volume of soil as it lies *in situ* or after careful excavation; it can be regarded as a measure of compaction in the substrate. It is expressed in g cm^{-3} with well-aggregated soils rich in organic matter having values less than 1.0 and highly compacted soil values exceeding 2. Many arable soils have values up to 1.6. Urban soils that have been thoroughly cultivated such as those of allotments and flower beds have bulk densities within the range 1.0–1.6, but mostly they are higher. Patterson (1976) found average values in Washington DC ranging from 1.74 to 2.18 cm^{-3}. Root penetration is highly restricted above 1.7. Craul and Klein (1980) reported a range from 1.54 to 1.90 in New York with most values around 1.82. Observable effects of the restricted aeration and drainage that results from compaction are not dramatic. It can reduce plant vigour, it may favour species tolerant of imperfect drainage such as creeping bent (*Agrostis stolonifera*) and creeping buttercup (*Ranunculus repens*) and the fixation of nitrogen by legumes is inhibited by poor aeration. Exceptionally hydrophytic plants such as rushes (*Juncus effusus, J. articulatus*) or marsh foxtail (*Alopecurus geniculatus*) occur on severely compacted soils where a surface water gley has formed. This phenomenon has been noticed on wasteland sites in towns in the west of Britain where in most months there is an excess of precipitation over evaporation. Many plants seem able to cope with compaction by exploiting vertical cracks and channels in the soil created by the presence of rotting anthropogenic material, old root runs or local pockets of coarser material.

Soil reaction

Urban soils tend to be more alkaline than those of the surrounding area. This is sometimes a result of disturbance bringing unleached material to the surface but more often pH is elevated by the release of calcium from mortar, cement, plaster and other components of building rubble. This effect brings calcicoles such as old man's beard (*Clematis vitalba*), Jacob's ladder (*Polemonium caerulea*), centaury (*Centaurium erythraea*) and occasionally carnation sedge (*Carex flacca*) on to urban demolition sites. Locally roadside and pathside soils have pHs around 9.0 as a result of sodium and calcium chloride applied as de-icing salt, whilst in hot climates irrigation with calcium-rich water can raise soil pH. The opposite effect is associated with furnace ash, cinders and some subsoils which may support calcifuge vegetation including broom, heather, wavey hair-grass (*Deschampsia flexuosa), Festuca tenuifolia* and common bent (*Agrostis capillaris*); it is not uncommon to see the more acid patches on an area of urban wasteland picked out by sheep's sorrel (*Rumex acetosella*) or stands of *Hieracium* Sect. *Sabauda*.

The wet and dry deposition of acid material as a component of air pollution can cause significant surface leaching. Undisturbed soil beneath shrubberies in northern towns supports a 'turf' of the acidophilous moss *Dicranella heteromalla* which develops in response to a surface pH of about 4.0 (range 3.2–5.1). This bryophyte also occurs at the foot of trees in towns where rain tracks wash down acid material into local receiving basins (Gilbert, 1968b). In the 1960s and 1970s the stimulation of grass growth along the lime lines used to mark out urban sports fields was popularly believed to be due to the correction of air pollution-induced surface leaching. This was studied at Liverpool (Bradshaw, 1980) where it was shown that old lime lines crossing acid grassland supported a richer flora than adjacent more acid turf. Additionally earthworms only occurred in the limed areas which also showed greatly enhanced levels of soil respiration and nitrogen mineralization. To summarize, in terms of soil reaction, examples of shifts in both directions are quite frequent in urban areas, but due to disturbance and soil amendments the tendency is for soils to be on the alkaline side.

Contaminants

Purves (1966, 1972) and Purves and Mackenzie (1969) were among the first workers to recognize that soils in urban and industrial areas contain elevated levels of potentially toxic trace elements such as copper, lead, zinc and boron. Working in the Edinburgh area they found that the average urban soil contained more than twice as much available boron, five times as much copper, 17 times as much lead and 18 times as much zinc as those collected from adjacent rural areas (Table 4.1). The source of the contamination in garden soils is often material deliberately added such as ashes, municipal compost and sewage sludge. In other habitats vehicle exhaust, airborne dust from breaker's yards and industrial air pollution are responsible. Paint can be a major source of lead in the older suburbs; instances are known where the former site of painted wooden huts is marked in the soil by elevated lead levels. Uncultivated green space in towns such as parks and golf courses are also characteristically contaminated with boron, copper, lead and zinc. Here air pollution is probably the primary source with local 'hot spots' being associated with pockets of litter or refuse.

Table 4.1. Mean levels of available trace elements in urban and rural soils

Origin	*No. of samples*	*Water-extractable B (ppm)*	*EDTA-extractable Cu (ppm)*	*Acetic acid-extractable*	
				Pb (ppm)	*Zn (ppm)*
Rural arable	100	0.70	2.8	0.65	2.9
Urban gardens	46	1.81	15.8	11.2	52.4

From Purves (1972).

Plants are able to protect themselves from much of this contamination by restricting their uptake of toxic ions so it is extremely rare to find plant distribution in towns influenced by this factor (but see road verges p. 148), which is more of a health hazard to humans than a factor affecting the ecological balance. However, micro-organism activity can be reduced by up to 50% for several decades as the toxic elements (except for boron) are leached only very slowly. This can lead to interrupted nutrient cycling and a build-up of litter on the soil surface. An example of the effect this has on ground beetles is given on page 118.

Other factors which have been suggested as having an important influence on urban pedology (Craul, 1985) are modified temperature regimes associated with the urban heat island, nutrient availability particularly in relation to nitrogen, low organic matter content, the development of a water-repellent surface crust and the presence of toxins ranging from herbicide residues to the results of anaerobis, e.g. methane and alcohol production. Some of these, such as fertility, are extremely important but so variable that they can only be discussed in a specific context (see later), while others rarely appear to be biologically significant.

4.4 BRICK RUBBLE

The 'clean sweep' philosophy of urban renewal has produced large tracts of land covered with brick rubble. It is often a number of years before these sites get redeveloped, and, during this time, soil-forming processes start to work on the raw bricks and mortar to produce lithomorphic soils (these are A/C soils with no B horizon) which would be classified by the soil survey as pararendzinas. These differ from rendzinas in having a parent material that is non-carbonatic, i.e. little altered unconsolidated material containing less than 40% calcium carbonate into which category brick rubble falls. Pararendzinas are shallow and free draining. The physical and chemical properties of raw brick rubble have been investigated by Dutton and Bradshaw (1982) so its attributes as a growing medium for plants are known.

Physical properties

Thirteen samples of brick rubble from demolition sites in inner Liverpool and Belfast were, after stone picking, analysed for soil texture and assigned to textural classes (Table 4.2). The results suggest that texturally the rubble substrates are as good as any topsoil. However, a large proportion of the sand-sized fraction (0.02–2.0 mm diameter) is made up of mortar particles, and, whereas sand is inert in the soil, powdered mortar rubble can aggregate with clay to form a concrete-like substance. Also textural analysis omits the coarser fraction of substrates – gravel, cobbles and stones – and rubble soils are dominated by these; half-bricks, brick fragments, broken paving slabs, lumps of concrete,

Table 4.2. Textural analysis of brick rubble compared with topsoil, after stone picking

Site material	*Texture*	*No. of sites*
Brick waste/subsoil	Loamy sands	3
	Sandy loams	2
	Sandy clay loams	6
	Clay loams	1
	Silty clay loams	1
Topsoil	Sandy clay loams	13

From Dutton and Bradshaw (1982).

kerbstones and the like made up the bulk of any profile. Bearing the above two factors in mind coupled with on-site variability, textural analysis of these soils is not particularly informative. The large amounts of granular material make the substrate difficult to compact even by heavy machinery so bulk density is usually low. To summarize, drainage and aeration are normally excellent promoting rooting and favouring a high biological activity.

Organic matter is usually measured by loss on ignition but this technique is not suitable for rubble soils as on heating there is also a loss of carbonate. For this reason few reliable estimates of organic matter content have been made. Examining profiles quickly reveals that the soils are initially very low in organic material which if present is derived from pre-existing garden soil, domestic rubbish, pieces of wood or fly tipping. However, it quickly builds up as plants colonize the site, increasing water retention, binding soil particles and forming a reservoir of nutrients, particularly nitrogen. Rates of turnover are rapid, dead plant material quickly being converted by molluscs, insects and earthworms into droppings which accumulate to form pockets of fine dark granular mull humus.

Chemical properties

Nutritionally, rubble soils are unbalanced with an excess of certain elements and a deficiency of others; this makes them ecologically interesting. Available nitrogen is commonly severely deficient (Table 4.3). In the absence of organic matter it originates from air pollution,

Table 4.3. Nutrient levels in rubble soils from typical urban clearance sites compared with garden soil and topsoil

			Available ppm			
Site	*pH*	*N (total %)*	*P*	*K*	*Ca*	*Mg*
Demolition site	7.2	0.05	36	220	3830	–
Demolition site	6.7	0.08	65	450	1070	–
Demolition site	7.0	0.05	19	290	8830	–
Demolition site	–	–	12	81	–	72
Demolition site	–	–	22	173	–	145
Demolition site	–	–	5	77	–	323
Demolition site	–	–	65	94	–	393
Garden soil	6.3	0.14	61	140	1030	–
Topsoil	–	–	30	83	–	66

From Bradshaw and Chadwick (1980) and Dutton and Bradshaw (1982).

cyanobacteria and thunderstorms. As vegetation and organic matter build up, soluble nitrogen is released slowly from the humus, is added by free-living nitrogen-fixing bacteria and is contributed by nitrogen-fixing plants such as legumes. Clays and organic matter which are colloidal and have many binding sites help to reduce leaching. Despite these inputs, the build up of nitrogen is normally slow and its availability largely controls the rate at which vegetation develops on urban clearance sites (Fig. 4.1). It is possible that during the last few decades the rate of colonization in cities that experience very acid rain has been speeded up as nitrates and nitrites are major components of this pollution.

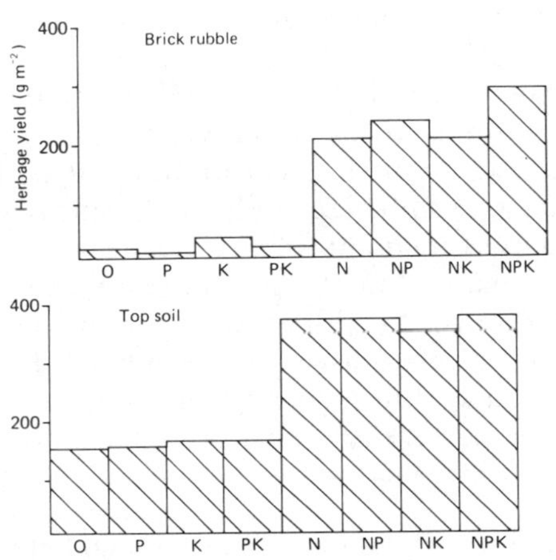

Fig. 4.1. The improvement of growth of urban grasslands in Liverpool established on brick rubble and on topsoil following the application of N, P and K (50 kg ha^{-1} N, P_2O_5, K_2O). O is the control. There is a serious deficiency of nitrogen even on the topsoiled site (Bradshaw, 1983).

Brick rubble contains large amounts of phosphorus which is present in the clay from which bricks are made. The origin of much of this phosphorus is the skeletons of organisms which lived long ago. As the brick material weathers, phosphorus is released in amounts quite adequate for plant growth even when the pH is high; see Table 4.3. These results have been confirmed by a further series of analyses which showed total P in the range 6000–10 000 ppm just under 1% of which was available. Potassium and magnesium are abundant components of the clay minerals from which bricks are made so are also freely available. Levels of the latter may be very high indeed.

Calcium, important as a nutrient and controller of pH, is readily available from the mortar rubble and maintains soil pH in the range 6.5–8. This is a level at which the nitrogen cycle proceeds rapidly and most nutrients remain available. In the urban environment more than adequate amounts of sulphur originate from the sulphur dioxide component of industrial air pollution; it is also found in gypsum, the major constituent of plaster. Most trace elements are present in the bricks and in air pollution so they are rarely limiting.

Soil fauna of brick rubble sites

The coarse structure of the parent material, which includes a certain amount of organic rubbish such as wood, polythene, cloth and cardboard, ensures a rapid invasion of the young soils by those larger invertebrates that are important in the early stages of organic matter breakdown. They increase in abundance as the vegetation develops till their droppings have almost entirely buried the rubble at which stage the macro-fauna is dominated by earthworms and ants. Hand-sorting through 1 m^2 quadrats of brick waste of different ages provided the lists in Table 4.4 which give some idea of how the faunal groups replace each other with time and of their abundance. There are few datasets to compare these results against but the number and variety of invertebrates appears large; they must do a lot to speed up nutrient cycling. Initially their droppings accumulate as a dark granular deposit, pockets and layers of which form within the rather inhospitable looking rubble. As soil formation proceeds earthworms become dominant and a humose

Table 4.4. The numbers of larger invertebrates found in or on the soil at brick rubble sites of different ages in Sheffield. Data collected in the field by hand-sorting through a single 1 m^2 quadrant at each site. For further details see Table 6.3

Site age (years) . . .	*12–15*				*4–6*				*0–1*			
Site number	*1*	*2*	*3*	*4*	*5*	*6*	*7*	*8*	*9*	*10*	*11*	*12*
Earthworms	C	C	F	A	R	R	–	–	–	R	–	–
Ants	A	O	O	A	F	–	R	–	–	–	–	–
Wood lice	C	A	A	O	F	A	–	–	–	–	–	–
Snails	F	F	–	F	–	–	–	–	–	–	–	–
Slugs	A	C	O	O	C	O	–	–	–	–	–	–
Centipedes	O	R	O	F	A	F	A	–	–	–	R	–
Millipedes	–	R	O	F	–	A	O	–	–	–	–	–
Ground beetles	R	O	–	R	R	–	O	F	R	R	–	–
Beetle larvae	R	R	R	–	–	–	R	O	–	–	–	–
Spiders	F	A	F	F	F	O	C	A	A	C	F	F

A, abundant; C, common; F, frequent; O, occasional; R, rare.

upper layer develops. After 10–15 years of vigorous biological activity a layer of topsoil 5–15 cm deep has developed. The A horizon is represented by a uniform, almost black when moist, organic-rich mineral material with a strong crumb structure. Within it fragments of brick, mortar and ash up to 10 mm across are frequent. The whole is thoroughly permeated by fibrous roots and worked over by the soil fauna. At 5–15 cm depth it passes into coarse very stony freely drained brick-mortar rubble with which there is a rather abrupt undulating junction. The whole profile is base-saturated. This calcareous lithomorphic soil is a typical pararendzina.

Plant–soil interrelationships

The previous sections have demonstrated that physically and chemically brick rubble is a fairly inocuous substrate, so urban clearance sites quickly become colonized by a considerable range of plants. However, the stoniness of the soil coupled with its initial nitrogen deficiency restrict plant growth so sites remain ecologically 'open' for many years. One result of this is that annuals and biennials can maintain viable populations and legumes such as clovers, melilots, lupins, medicagos and trefoils may become abundant. In Sheffield, Yorkshire-fog (*Holcus lanatus*) is the leading pioneer grass species on these sites where its performance is frequently related to the presence of legumes, particularly low-growing circular colonies of creeping white clover (*Trifolium repens*). Transects laid across such patches have shown that when growing with white clover the percentage cover of the grass increases greatly as does the density of flowering tillers (Fig. 4.2). Other parameters such as height of sward are also increased. This phenomenon, regularly mentioned in the literature, is rarely seen so clearly displayed except perhaps occasionally on sand dunes. Further plant–soil interrelationships on brick rubble sites are mentioned in Chapter 6.

4.5 TOPSOILED SITES

The soil of landscaped areas is created to specified standards. For example under future grassland 150 mm depth of topsoil is spread while under shrub beds and ground cover 300–350 mm is recommended. This attempt to provide a series of uniform soils suitable for a range of predetermined vegetation types fails due to poor quality control. Bloomfield *et al.* (1981) assessed 44 topsoils being used in various parts of Merseyside, Greater Manchester and Cheshire and discovered that none approached the quality of good garden soil. A bioassay, using rye grass (*Lolium perenne*) showed that half were no better than brick waste, 15 no better than raw colliery shale and one prevented plant growth

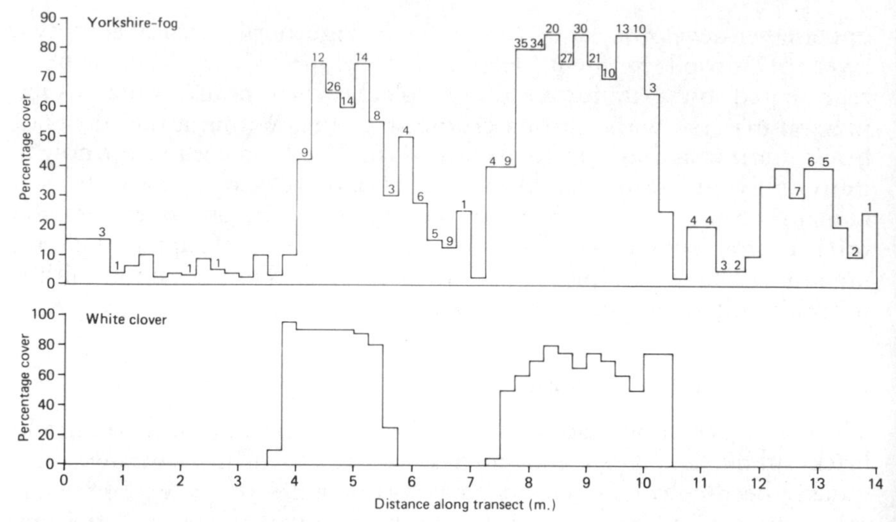

Fig. 4.2. Results from a 13 m long transect showing how the performance of a Yorkshire-fog (*Holcus lanatus*) sward developed on brick rubble is related to the presence of creeping white clover (*Trifolium repens*). Figures in the upper histogram indicate the number of flowering tillers in each 25 × 25 cm section of the belt transect.

altogether. Chemical analysis suggested their poor performance could mainly be related to lack of available nitrogen, a large number having levels of less than 10 ppm. Others had a very low pH, low calcium or low potassium. The important role of nitrogen was confirmed by adding ammonium nitrate, at the rate of 50 kg of N ha^{-1} to the soils and repeating the bioassay (Fig. 4.1). Bradshaw has pointed out how in the absence of formal bioassay analysis the severe nitrogen deficiency of many recently landscaped sites can be deduced by observing the bright green patches produced by dog urine which has a high nitrogen content.

With time these man-made soils mature but it is a slow process. A pit dug within 5 years of the soil being 'created' still reveals two highly distinct layers (Fig. 4.3). At a 4-year-old public open space site in Sheffield which had been graded, topsoiled and sown down to *Lolium* grassland, the A horizon was a uniform slightly stoney yellowish-brown clay loam showing horizontal lamination produced by the regular passage of mowing machinery. The structure was weak, porosity low and the soil dense. At the surface a 3–8 mm thick accumulation of litter showed that incorporation of organic matter was not particularly rapid and few earth worms were seen. At 12 cm depth there was an abrupt transition to an extremely compact heterogeneous layer composed of stones, bricks, clay, shale, old topsoil and fragments of china. Root penetration into this horizon was minimal and biological activity appeared to be absent. Compaction and pan formation by smearing was conspicuous at the top of this horizon; bulk density fell slightly at depth.

After several decades such soils have developed considerably, largely as a result of the accumulation of organic matter and earthworm activity both of which improve texture and indirectly nutrient status. A pit opened in a 27-year-old amenity grassland lying over cleared urban land in Sheffield (Fig. 4.3) revealed a uniformly dark brown A horizon with a moderate crumb structure. Only small amounts of fresh litter at the surface suggested that incorporation by the abundant earthworm population was rapid, their activity being sufficient to greatly reduce surface compaction from foot traffic and mowing machinery. At 15 cm the old topsoil/subsoil junction was marked by a sharp increase in stoniness but the surface of the bricks had softened due to weathering and there were invaginations, to 10 cm deep of the A horizon into pockets of coarser textured material. Lumps of raw clay and shale, however, remained little altered. Root penetration into the subsoil was chiefly at the interface between stones and finer material where fissures form during dry weather, root matting also occurs on the surface of bricks which as they rot become moisture retentive and release nutrients. Though pan formation at the lithological discontinuity is disappearing after 27 years it will take centuries for the man-made origin of the soil to become obscure.

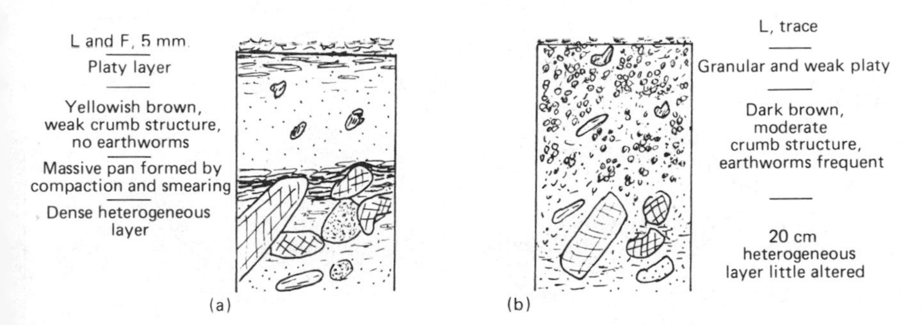

Fig. 4.3. Profiles showing the progress of soil development on sites topsoiled 4 years (a) and 27 years (b) ago.

Plant–soil interrelationships on topsoiled sites

Many of the 'problems' of topsoiled sites are only problems if one is trying to maintain hard-wearing rye-grass swards or establish woody ground cover at maximum growth rates. Specialists in reclamation acknowledge that the level to which the soil is ameliorated needs to be determined by the future use of the site. For example amenity tree belts are best established on low productivity soils as a large proportion of their nutrients are obtained via mycorrhiza, nutrient stress suppresses weed growth and a sparse vegetation supports only cool rapid fires which the trees can survive. Ecologically the range of conditions produced by a topsoil through to subsoil nutrient gradient, and clay through to sand drainage series, results in a variety of swards developing what ever the sown mix (Fig. 4.4). Another direction of variation is associated with compaction which favours the cosmopolitan species *Lolium perenne, Poa annua, Juncus tenuis, Plantago major* and *Trifolium repens*. The current expansion of *Carex hirta* in town grassland can probably be correlated with soil compaction.

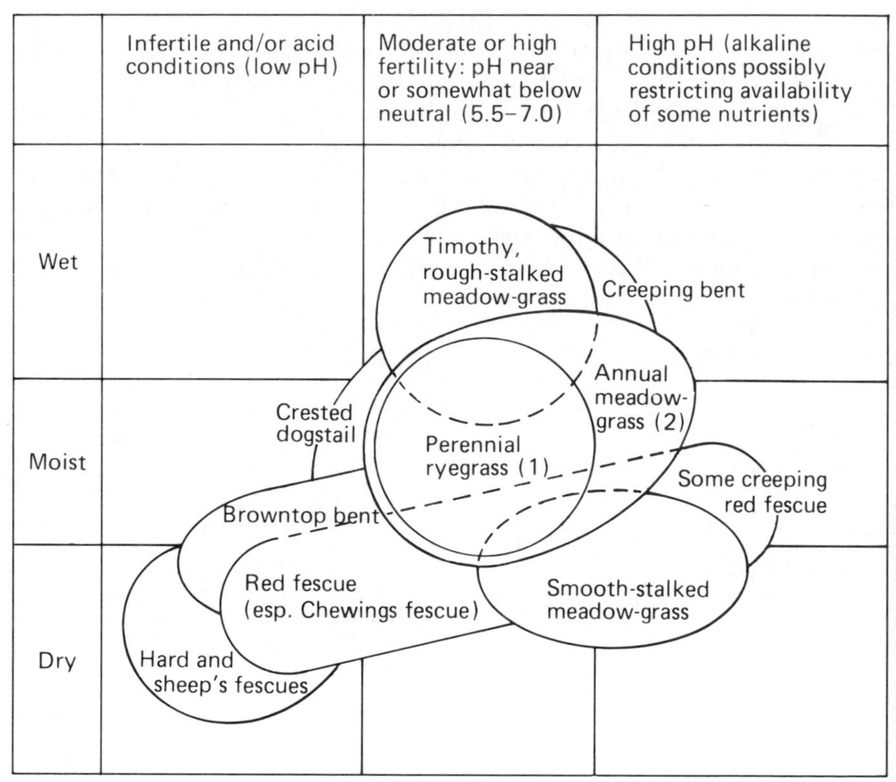

Fig. 4.4. Soil preferences of some common turf grasses. Note (1), especially with high K. Note (2), especially with high P (Shildrick, 1984).

Chapter 5

VEGETATION DYNAMICS

The landscape of urban areas is in a dynamic state because it is subjected to cycles of man-induced change associated with redevelopment. In the past these were relatively long compared to a human life span but recently the rate and scale of redevelopment have speeded up to such an extent that buildings are constructed to last no longer than 50–60 years. Ecologists are used to studying ecosystems that change with time, particularly when catastrophic events are involved, so the concepts of plant succession can be extended to cover this phenomenon.

The terminal vegetation towards which much of the urban succession is proceeding appears to be compositions that include large trees, fully grown shrubberies, areas of dense low woody ground cover and a range of grasslands, all growing on soils in which depth, organic matter content and horizon differentiation are increasing. If Odum's (1969) tabular model of trends in ecological succession are applied to well-developed suburban vegetation it is seen to possess the properties of a fairly advanced successional stage. Along the way many plagioclimax communities arise and it is most realistic to regard the whole as a biotic subclimax. In the countryside we are accustomed to equating biotic influences with grazing, trampling and burrowing animals but in towns the course of succession and its end point are largely controlled by gardeners planting, and using a wide range of maintenance equipment including mowing machines, sprays, hoes, secateurs and spades.

Clements (1916) recognized the following subprocesses of succession: (1) initiation (nudation); (2) immigration of new species; (3) establishment; (4) competition; (5) site modification (reaction); (6) stabilization. Successions are initiated by nudations which in towns are the work of bulldozers rather than physiographic processes. In the designed landscape immigration and establishment are heavily influenced by the horticultural industry which, working to specifications provided by landscape architects, seed, turf and plant up with woody species the recently bared areas. Within this framework there are varying opportunities for further diversification depending on the level of management. This exerts a controlling influence on competition and reaction which is most important since the new ecosystems contain many exotics and cultivars which would not survive for long without the energies of gardeners favouring them. This is particularly the case with pendulous varieties, dwarf cultivars, plants with variegated or coloured foliage and those growing beyond their natural climatic limit.

The factors which shape the terminal vegetation are not primarily competition which is carefully controlled but culture, fashion and the work of plant breeders coupled with the satisfaction of innate needs. Apart from providing delight and prestige it has been suggested that the most successful urban landscapes mimic the forest margin – open woodland ecotone in which it is believed man evolved and where we find rich opportunities for both prospect and refuge (Appleton, 1975).

Another feature of succession well illustrated in towns is Watt's (1947) concept of cyclical change. He illustrated this by reference to beechwoods, heather moorland and dry grasslands where non-directional cycles within the community are controlled by the life span of the dominant species. Woodland ecologists are used to studying the 'gap phase' where seed availability, light and soil toxins govern the return to high canopy forest. In towns the 'gap phase' is related to the life span of buildings. Extensive redevelopment inititiates a new succession but on sites where a single dwelling is built and the growing practice of saving topsoil and conserving major biological assets like trees and shrubs is followed, the course of change fits models of cyclical succession quite well.

The urban development cycle is not a closed system; greenfield sites are continually being built on and there is a small but significant flow of land in the other direction. One of the few studies to have quantified the flux of land uses in an urban area is by Handley and Bulmer (1988) who studied changes on Merseyside using ground survey information collected between 1960 and 1976. Their results show that during the time interval most of the land fed into the urban system was agricultural (30 km^2), followed a long way behind by rough grassland and scrub (6 km^2), and amenity grassland (5 km^2). Developers prefer to build on greenfield sites as they are less likely to encounter unforeseen problems such as disused services, old cellars, culverted streams or variable landfill which can substantially add to the cost of a scheme. For reasons such as these, previously developed sites may leave the system going over to other 'uses' particularly derelict land (4 km^2 in the Merseyside study). This contributes to inner city decline.

5.1 ECOLOGICAL CHANGE AS GREENFIELD SITES BECOME DEVELOPED

Many of the modifications that occur to farmland as it enters the development cycle are illustrated by the following case study which involved the construction of a fire station on greenbelt land at the edge of Sheffield during 1971–72. The site, a 1.4 ha field (Fig. 5.1), was originally enclosed by stone walls with a holly hedge studded with trees

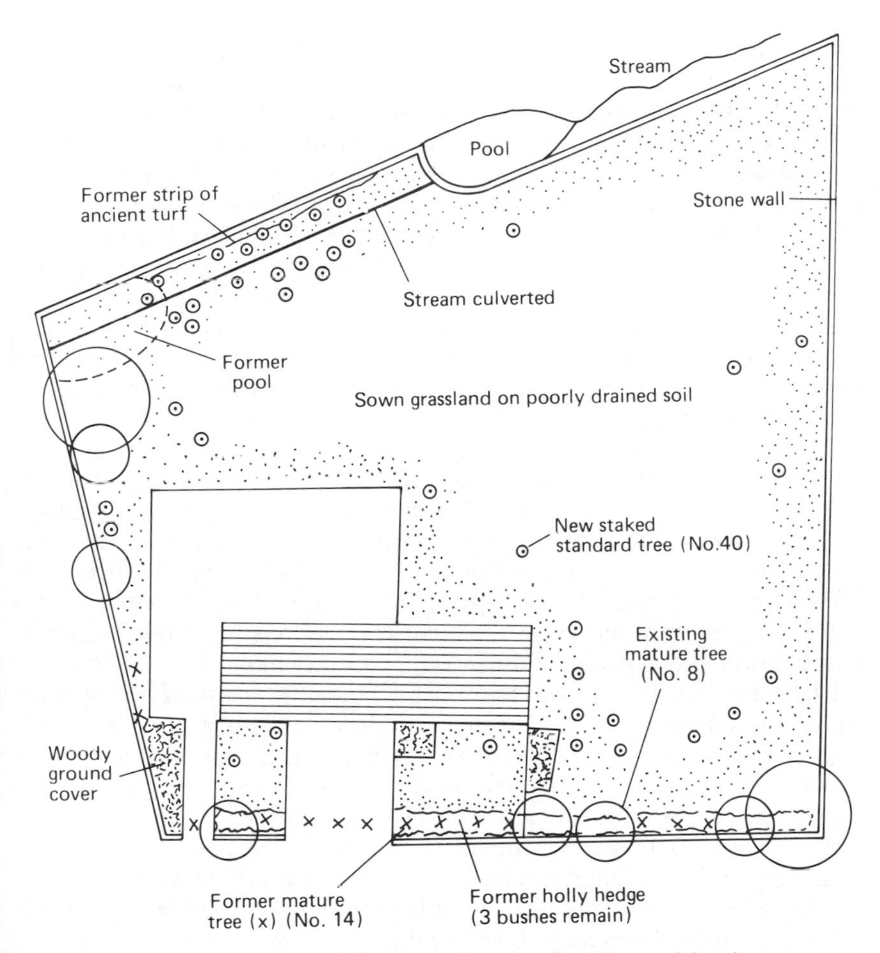

Fig. 5.1. Plan of Ringinglow fire station showing the effects of development on a greenfield site.

down one side; it had been used for grazing cattle. During the predevelopment phase there was a gap of nearly a year between purchase of the field and the builder moving in, so the opportunity was taken to let the site to a riding school. This land use, known as horsiculture, is widespread in the urban fringe but has a disadvantage: the animals are general feeders and at certain times of the year strip trees of their bark. The result of 12 months temporary horse grazing was to reduce the holly hedge to a row of stumps and the removal of bark from the inward facing side of the tree boles. This did not kill them but provided a possible entry point for disease and harmed them aesthetically. Horses are hygienic creatures; they established a latrine in a corner of their field which could still be picked out ten years later as a patch of darker green grass infested with docks. A period of interim land use like this, or no land use, often precedes development and during this time much damage can occur to assets through burning, chopping, stealing, turf stripping, rubbish tipping and other forms of vandalism.

Contract work on the fire station site commenced with the culverting of a small stream which crossed a remote part of the field. It was fringed with water plants and at one point formed a pond inhabited by moorhens and mallard. The City Engineers were unable to explain why the culverting had been carried out excusing themselves with the platitude that water in the landscape is always dangerous. As a general rule surface water features get eliminated during development or at the very least permanently polluted through being used as a sink for storm water.

Prior to development the site contained 22 mature trees, 17 of which stood over the holly hedge. After felling four to provide access to the fire station the remaining 18 were still in position when the contractors left a year and a half later. By 1974, bark was peeling off the trunks of two sycamores, and four beech supported only a thin canopy of small yellowish-green leaves. Two years later, all six had died and others were becoming stag-headed due to the death of the top most branches. Today (1988) only eight of the original 22 mature trees remain, almost all those which came under pressure during the contract have died. To enhance the landscape 40 new trees were put in as staked standards (2 m high) but many have barely doubled their height due to poor soil conditions. The majority are exotics such as horse chestnut, grey alder (*Alnus incana*), sycamore and ornamental cherry. The principle here is that there is a loss of mature features, the landscape being put back to an earlier successional stage. After urban development has encroached on to agricultural land, sites usually end up with a much greater density of smaller trees (48 as against 22), a high proportion of which are exotics.

Shrubs tend to be more vulnerable than trees, only surviving if special care is taken. In this instance three bushes are still alive from what 15 years ago was a holly hedge 100 m long. Normally new landscape planting more than makes up the loss but on this site only a few small easily maintained beds of *Cotoneaster horizontalis, Lonicera pileata* and *Hypericum calycinum* were specified.

Most vulnerable of all is the field layer; very little of the pasture turf escaped destruction by machinery or the storage of materials. A narrow strip of ancient turf, however, did persist beside a remote wall illustrating that habitat conservation is most easily achieved at the margin of development sites. This is so constant it is almost a rule. Unfortunately a maintenance gang later sprayed the boundaries eliminating this relic community of bedstraws, violets and lady's mantle.

The chapter on soil pinpoints the loss of structure that occurs to both subsoil and topsoil during development. It is rare for this to be corrected by cultivation, so soil ends up poorly drained. Today the grassland, sown down to a rye-grass mix, is composed largely of species that can tolerate wet ground, e.g. rough-stalked meadow-grass, Yorkshire-fog, wood bitter-cress (*Cardamine flexuosa*), creeping buttercup, self-heal, creeping white clover and the moss *Brachythecium rutabulum*. Species accidentally introduced during or immediately after the construction phase may become prominent; on this site *Veronica filiformis* and *Petasites fragrans* are now flourishing but neither were picked up in the predevelopment surveys.

The changes that occurred as this site entered the development cycle have been confirmed many times elsewhere. Few of the phenomena are harmful to wildlife in the long term. Though mallard and moorhen have been lost, swallows now nest in the building and the mown grounds are frequently visited by wagtail, rooks, jackdaw and in winter by fieldfare and redwing which provides interest for the firemen who also regularly put out food for a local fox. The young condition of the trees limits hole-nesting birds and the abnormal scarcity of shrubs means there is a lack of cover; however, diversity in the grassland has almost recovered and the 28-day gap between mowings enables most species to flower profusely. An unexpected small bonus has been the development of unusual epiphytic lichens in 'wound tracks' associated with the horse-damaged trees.

5.2 ECOLOGICAL CHANGE AS URBAN LAND IS RECYCLED

Once land has been urbanized it only occasionally reverts to other uses, so every 50–60 years is subjected to a catastrophic (ecologically) rebuilding programme. The effects of this are closely similar to those in the greenfield development just described but residential areas contain a greater complexity of landscape/ecological assets and the tightness of the sites leads to greater contractor pressure. Since 1970 I have, with the help of students, followed redevelopment on 15 residential sites, ranging from Victorian mansions to interwar council estates.

Although operations on a construction site appear haphazard, contractors follow a well-defined programme of work which, with minor variations, is the same everywhere. The five major divisions of their programme, which follow on from an important predevelopment phase, are siteworks, substructures, superstructures, finishes and external works. Ecologically the first and last are the most significant. The relationship of these phases to Clements (1916) stages of succession are illustrated in Table 5.1.

Table 5.1. The relationship of a building contractor's programme of work (upper line), to Clements (1916) stages of plant succession (lower line)

Pre-development phase	*Site works*	*Sub-structures*	*Super-structures*	*Finishes*	*External works*	*Maintenance*
	→			→		→
Mature vegetation			Nudation		Immigration Establishment	Competition Reaction

Predevelopment phase

The mature vegetation is open to the types of vandalism already mentioned but in addition, especially if the grounds of large Victorian residences are involved, the quality of site assets may depend on a continuation of existing management regimes. Features such as rockeries, clipped hedges and lawns are the product of many decades of constant management and can quickly deteriorate once it ceases. If this happens they may not be perceived as site assets by the contractor. In one instance a developer removed turf, walling stone, tufa, a yew hedge, large numbers of rhododendrons and topsoil to complete a previous development while using the new site to dump spoil, concrete and rubble. Demolition of existing buildings on a site is not often damaging though bonfires of a dry timber burn better than expected and may locally scorch trunks and branches. The worst contractors carry out convenience felling of trees due for retention which they fear could be awkward to work round.

Siteworks

This stage involves all preparatory operations leading up to excavation of the foundations. One of the first jobs should be the protection of features on the site which have been chosen for recycling, identified on a plan and excluded from the contract area by specifying fence lines. These may include groups of trees, ivy covered walls, water bodies, rockeries, old lawns, shrubberies, large isolated trees, intact soil profiles and complex vegetation such as ecotones and large units of trees, shrubs and ground flora. Their selection will be a compromise between what is best for the site from an aesthetic, functional and ecological stand point and the need of the developer to put up a sufficient density of buildings to make a profit.

The very first operation, the siting of cabins, often conflicts with conservation interests. They need to be placed where buildings are not going and this is often just the type of marginal location where it is most appropriate to retain mature vegetation. I have seen contractors carve places for their huts out of slow growing belts of holly, box and yew. The next job, the fencing of site assets using chestnut paling, is usually neglected for more urgent business such as site clearance and grading. This is a particularly fraught period with subcontractors taking down trees, bulldozers roaming the site stripping vegetation, lorries moving topsoil and bonfire smoke drifting across the area. This stage overlaps with grading to new contours using huge machinery attended by even bigger lorries. During this period of high activity, on a site not yet well known, with subcontractors involved, and no fencing up, many things can go wrong. Most damaging is the disposal or temporary storage of surplus fill (subsoil) which may be tipped down banks, into tree belts, spread over ground cover, pushed in hollows, heaped against walls, etc. It causes untold damage. This stage equates with nudation; on tight sites it may be total but on larger ones up to 20% survives to be recycled and form a permanent framework of mature vegetation. For this to succeed stout fencing is essential but was used on less than half the sites monitored.

Spontaneous short-lived plant successions start up immediately with what can loosely be called a primary succession on the subsoil piles (coltsfoot, annual meadow grass, knotgrass,) while a secondary succession quickly covers the topsoil store with species such as charlock, redshank (*Polygonum persicaria*), spurges and often opium poppies (*Papaver somniferum*). Many species are common to both successions but that on topsoil has the benefit of a seed bank. The siteworks phase takes only a week or two but is catastrophic.

Substructures

This involves excavating foundations, concreting footings and laying brickwork to damp-proof course. Trenches to carry storm and foul water pipes also go in at this stage. The extremely raw subsoil brought to the surface by these excavations lies about, may be used to correct grading errors and generally forms a heavy puddled layer which will later restrict drainage. Machinery starts to form a pattern of internal roads and lorries arrive regularly with bricks, pipes, sand, cement and aggregate. The destruction of site assets is now a little by little process in contrast to the trauma of the previous phase. Storage and traffic start compacting the soil over root platforms, static machinery and generators are operated under branches, heavy materials are leant against tree trunks, and fencing that collapses is not re-erected.

During this phase the most damaging operation is probably the excavation of service lines. These often have to run in predefined directions and cannot be bent round valuable features. Technically it is possible to hand-dig trenches past tree roots but in practice it is always done by machinery capable of ripping through the stoutest root systems. The main contractor has no control over work performed by statutory undertakers (water, gas, electricity, telephone) who are specifically excluded from the penalty clauses attached to Tree Preservation Orders. In America, canopies are thinned in proportion to the root spread destroyed to help reduce water stress but I have not heard of this happening in Britain. The role of infection entering through damaged roots is unknown. At this stage the site becomes fully exploited as the nudation activities spread out towards the margins.

Superstructures

This phase occupies the major part of the contract. Throughout, pressure on the environment is at its maximum. Large areas of the site are given over to the storage of building materials, up to a fifth will be taken up by internal circulation, while cement mixing points, huts and secure compounds are also competing for the shrinking area of land beyond the scaffold lines. Under this kind of pressure two far-reaching types of ecological damage occur.

The first is attrition to vegetation that it was hoped would survive to provide maturity and biomass. Under the pressure of running a contract, the conservation of ecological and visual assets rate a very low priority, even though, in the sea of mud, they have a clearly increasing value. Protective fencing gets breached which leads to immediate destruction of the field layer, substantial damage to shrubs and considerable but delayed injury to tree root systems. Tree trunks are easily harmed as quite gentle blows can lead to extensive death of the vascular cambium, an event that occurs beneath the bark so the effects may not become visible till after the contractor has left the site. Then sheets of bark peel off to reveal a vertical scar that may be several metres high and extend halfway round the trunk. The inability of callus growth to quickly seal off damage of this nature has not been fully appreciated by arboriculturists; one dumper truck colliding with a trunk can cause a huge scar, two will ring bark a tree.

The second major type of damage during this period is to the subsoil. This normally possesses a polygonal structure of vertical cracks which promote drainage, aeration and root penetration. Under the continuous impact of site traffic, storage and excavation, the structure of the subsoil is completely lost. It is deformed and smeared, churned up and compacted until the top 30–50 cm is completely amorphous. Added to this is the incorporation of broken bricks and pipes, aggregate, old mortar, wood, polythene, metal bands, the lot. There is no way the original structure can quickly be restored to these areas of subsoil. During open cast mining operations topsoil and subsoil are stripped separately, then carefully respread, cultivated and sown with special crops to restore structure and fertility, but during building construction even the topsoil is not always stripped and stored, and subsoil is never treated as an asset.

Finishes

During this period plumbers, glaziers, joiners, painters, electricians, heating engineers, telephone men and gas board workers are on site. Their presence is indicated by an increase of vehicles and sometimes of huts. There is little damage left to do, they are unlikely to be as disruptive as the main contractor.

External works

Just as sitework at the start of the contract was a crucial stage so is external works. It involves the construction of paths, steps, retaining walls, tarmacing, paving, lighting, topsoiling and planting. By this time important people have usually moved on to another site, the landscape work being trusted to less experienced staff or subcontractors. There will be almost no money left due to overspending at earlier stages, the 'external works budget' having been used as a contingency fund. By now any protective fencing will have been removed and further serious damage can occur from lighting bonfires under trees, raising or lowering the ground level beneath their spread, excavating for walls and paths through root platforms and laying lighting cables. Respreading the topsoil is often carried out in a very haphazard manner. If handled when wet, between October and early May, it quickly loses its crumb structure and earthworms are all but eliminated. The standard thickness for areas going down to grass is an even 15 cm of topsoil but this is rarely achieved as it is used to correct grading errors, being spread directly on to the compacted puddled uneven subsoil. Worse, for the sake of tidiness, topsoil may be spread over areas previously protected, destroying beds of ivy, woodland ground cover or marginal ecotones. This further simplifies the landscape structure.

At this stage additional loss of biomass occurs due to the realization that some trees have been retained too close to buildings. This is particularly the case with sensitive developments that may only have earned planning approval because of an apparently highly responsible attitude to tree preservation. It is a salutary exercise to compare the number of trees on the approved plans with those remaining at the end of a development. The worst case I have observed was 28 reduced to 7 and I have never followed a development where the intentions regarding tree conservation were a total success.

5.3 POSTDEVELOPMENT

Assets that survive from the previous development cycle may be eliminated during the early years of the next one. This is particularly true for mature trees many of which carry conspicuous scars on their trunks and are felled because they appear unsafe to people who are unaware that this type of damage is universal on trees in certain situations such as by rivers. There it is caused by floating logs which, during times of flood, bump against the upstream side of trunks causing death of the vascular cambium in exactly the same way as described for dumper trucks, yet such trees may live for centuries. Other trees may be removed because they interfere with traffic sight lines, steal light from rooms or shade drying areas. Another frequent cause of loss is wind damage as the trees are exposed to gusts from new directions.

By far the greatest subsequent loss, however, is slow death from 'contractor pressure' – a mixture of soil compaction, water table changes, root severance, alterations in ground level, hard surfacing, physical damage to aerial parts, oil leakage and a new microclimate. Under such stresses trees gradually display the following syndrome. First there is a reduction in size of individual leaves. They may also show a slight chlorosis so the canopy appears abnormally thin and yellowish. In autumn the foliage has a tendency to wither early and *in situ* because abscission processes are disrupted. This is particularly a feature of the upper branches and the following year a proportion of these fail to break bud. This phenomenon extends more deeply into the crown next year, the trees starting to appear stag-headed. From now on there is a tendency to produce epicormic shoots up the trunk and along the major branches; presumably as a consequence of apical dominance being lost due to death of the leaders. The general physiognomy is now one of senescence; it may halt and the tree gradually recover, or progress until it dies. Large old trees of beech, ash, sycamore, oak and birch are more easily killed by 'contractor pressure' than lime, sweet chestnut, alder, all shrubs, and trees up to early middle age which show considerable resistance to the syndrome. Use of an incremental borer to investigate the performance of trees that have been influenced by a development cycle usually reveals a signature in the trunk where for many years following the construction period unusually narrow growth rings have been laid down compared with trees on adjacent control sites (Fig. 5.2). Root suffocation (anaerobis) rather than root severance is thought to be the main cause of the syndrome.

Immediately following a construction period the vegetation is in what can be called 'the postdevelopment condition'. This is a sward of newly sown rye-grass out of which rise trees with scarred trunks daubed with black arboriculturists tar. Due to a building up of the ground with surplus fill the trunks do not have a natural buttressed base but rise straight out of the soil. There is little else to this simple ecosystem which is visually boring and unattractive to wildlife. New planting comprises staked standards with long clean stems and a high centre of gravity which are destined either to put on very little growth during the next 10 years or be vandalized. The above scenario is an extreme case, a more

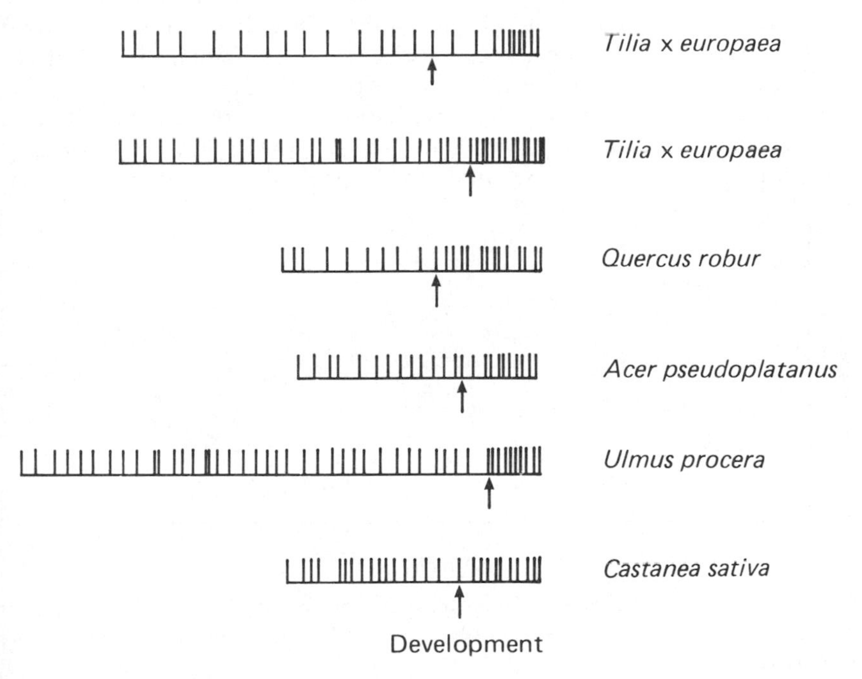

Fig. 5.2. Growth ring widths in mature trees that have been influenced by 'contractor pressure' during a development cycle (Wathern, 1976).

usual trend especially on private land, is for the postdevelopment landscape to return gradually to complex multilayered vegetation as newly planted areas mature.

5.4 DISCUSSION

Urban areas are characterized by high levels of disturbance and environmental modification, so a knowledge of vegetation dynamics is important to an understanding of their ecology. A study of the 50–60 year development cycle which pervades built-up areas has provided a convenient framework within which to study this rather variable phenomenon. The consequences for ecosystems of passing through this cycle can be summarized as follows. There is loss of the entire herbaceous community and its replacement initially by bare earth then a uniform grass sward. A major reduction in the biomass of trees and shrubs also occurs, sites normally being left with a fringe of overmature trees in uncertain condition. A zone of total destruction can be predicted by projecting scaffolding lines round groups of buildings on a plan, then joining up their corners; this delimits the core of the development where 'contractor pressure' is at its maximum. Figure 5.3 shows hypothetical changes in biomass associated with a series of development cycles on a site where the conservation of biological assets is successively taken more seriously. After increasing for 50 years the biomass in suburban areas is usually stabilized by thinning, lopping and pruning.

Topsoil is stored during construction then respread at the end converting a slow primary succession on subsoil to a faster secondary one, but even so the new man-made soil shows many signs of

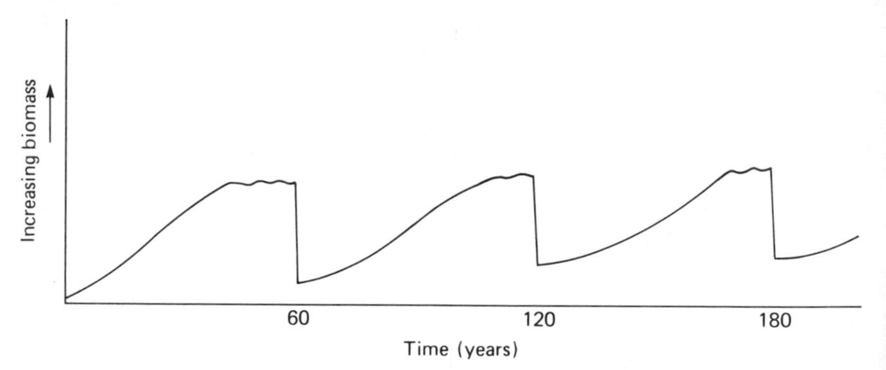

Fig. 5.3. Hypothetical changes in biomass associated with a series of development cycles on a site where the conservation of biological assets is successively taken more seriously.

immaturity. For example horizon development, biological activity and drainage are all poorly developed. At a site near Edinburgh a comparison was made between the performance of forestry-size transplants (50 cm) put into ground which the contractor had worked over and others planted at the same time into undisturbed ground. After 6 years the latter were 3 m high and flourishing while those in the 'made ground' had not reached 1 m and there were many gaps. At the beginning of a development cycle soils generally afford a poor growing medium which gradually improves.

The loss of habitat diversity is illustrated by the tendency for surface water to be eliminated, the repointing of old walls, and a deficiency of organic matter both as surface litter and dead wood. Simplification is not only at this level. Compared to their previous condition sites show severe structural modification coupled with fragmentation; ecotones and internal edges are particularly prone to destruction as they are favourite sites for the storage of materials. By the end, vegetation is no longer related to environmental conditions especially with regard to the light climate.

The effect of the development cycle on wildlife is dramatic. Diversity in all groups falls, initially almost to zero, then increases quite quickly once contractors have left the site. This relatively rapid recovery is because many urban species are highly mobile opportunists and possess a wide ecological amplitude; they do not require corridors for dispersal, advancing on a broad front. Planted ornamentals provide nectar and pollen for a wide range of visitors and nearly every plant species is host to several phytophagous insects. There are many routes for introductions. Lichens, byrophytes, spiders, birds, most insects and many higher plants are windborne. Others come in with planted container-grown material, mud on shoes, fruits stuck to clothing, or are carried by birds. As planted shrub beds mature and trees grow up the increased cover encourages a wider range of birds, and fruits provide winter food. A pond may be built, a compost heap formed, toadstools appear as organic matter builds up and within 20 years the partly contrived, partly spontaneous succession has built up an ecosystem which for diversity and interest exceeds those present in many countryside areas. By the time the site is ready for redevelopment again less mobile species and those dependent on dead wood will have arrived, soils will locally be surface leached and supporting new plants, biomass will have stabilized and the distribution of organisms will reflect habitat factors such as light, disturbance and particularly management. The ecology of this terminal vegetation, the result of directional change imposed by cultural and natural forces on land left bare by contractors is described in the chapter on gardens.

Chapter 6

URBAN COMMONS

Ecologically one of the most significant recent events in towns has been the expansion of wasteland. These newly created areas principally occupy sites in the inner cities where substandard high-density housing has been demolished, and lately, with the decline of heavy industry, vast factory complexes have also been flattened. As the housing served the industry, the two types of site run together tending to occupy flat ground adjacent to railways, canals and main roads, though having said this there is also a general scatter of wasteland in all but the most affluent areas.

Plans exist that assign almost all such sites to new housing, light industry, retail–leisure complexes, tended open space or some other general amenity use but these are frequently overambitious, so after demolition and grading out, a high proportion remain as wasteland with no positive use. Such sites green over quite quickly through colonization by spontaneous vegetation; the public wear paths across them, children play on them, bonfires are lit, fly tipping occurs, gypsy encampments may spring up and so they gradually become unofficial urban commons. The alternative method of putting urban land into store is to give it a minimum topsoil treatment and then sow the area down to amenity grassland. Subsequently this is often neglected because recreation departments have not been provided with the funds to manage it or they will not accept them, finding the grass cover unacceptably low. When this happens the natural and the initially treated sites tend to converge, though in fact plant succession on each follows its own course.

Today urban commons ranging from 0.1 to 10 ha in size are a major feature of cities, particularly the inner areas of industrial towns. Officially they are regarded as undesirable, unsightly and depressing, and the aim of most local authorities is to convert them to a conventional landscape of mown grass and trees once money becomes available. As this standard interim treatment costs around £7.50 m^{-1} and needs to be followed by on-going maintenance many remain untreated for at least a decade and exceptionally much longer. In many situations, however, they are landscape assets providing pockets of complexity and unpredictability and in certain towns they are now regarded as an acceptable form of non-intervention or minimal intervention landscape. Ecologically, urban commons are rich in types of wildlife that do not occur in the countryside; they support true urban communities. For this reason they are valuable sites where anthropogenic and synanthropic species thrive under seminatural conditions. The following account describes these urban communities, highlights the relationships to be found within them, and shows how they can be incorporated into the urban scene as a logical extension to the range of soft landscapes.

6.1 PLANTS

Succession

The traditional assumption that an area of landscape is finished after the establishment phase has never been true for urban commons, the dynamic aspects of which are one of their greatest attractions. Plant succession takes the vegetation through a series of stages culminating in scrub woodland; with practice it is possible to predict how long an area has been vacant from observing the plants. After demolition, sites are typically graded out as a slightly domed expanse of brick rubble lying in a matrix of fine material containing a lot of mortar. There may be considerable variation in the stoniness of the substrate across a site but it is usually alkaline, freely draining, well aerated and low in organic matter. Fertility is reasonably high except for nitrogen. Details of soil development over brick rubble are given in Chapter 4. An important ancillary operation is the construction of a 2 m high bank of rubble round the perimeter to discourage fly tipping and Romanys. The typical succession, and a few variations, will be described for South Yorkshire, then this will be compared with the pattern on a national scale.

The Oxford ragwort stage

The initial annual short-lived perennial plant stage which colonizes the freshly bared surface tends to be dominated by Oxford ragwort and other species with windborne seeds. A 0.8 ha site in the centre of

Sheffield, totally cleared in January 1984, had by mid-August 1984 accumulated an open vegetation composed of 41 species. Most abundant were American willowherb, annual meadow-grass, buddleia, coltsfoot, goat willow, groundsel, knotgrass, Oxford ragwort, rosebay willowherb and rye-grass, all of which had established from seed. Seven of these 'top ten' produce wind-dispersed seeds; those of late-flowering species, e.g. buddleia and rosebay willowherb, must have been blowing around the city for some time to have colonized the site so quickly. Just under half the species to arrive in the first 8 months had seeds adapted for wind dispersal, whereas potato arrived in rubbish, wheat was probably brought in by birds, and the rest had no special means of dispersal. Many were probably also carried by the wind gusting down streets, others perhaps by the starlings, pigeons and house sparrows which visited the site and still others such as knotgrass on the tyres of earthmoving equipment. Places used as temporary car parks accumulated species very fast; in Wolverhampton all recent records of black twitch grass (*Alopecurus myosuroides*) are from such sites (Ian Trueman, personal communication), while in Sheffield sand-spurrey (*Spergularia rubra*) seems to be carried around on tyres.

By August 1985 a further 13 species had arrived and only wheat disappeared; populations of great hairy willowherb, sowthistle and hedge mustard (*Sisymbrium officinale*) had increased greatly. By now the distribution of many species was strongly centred around initial sites of introduction, e.g. wall barley, *Bromus mollis*, shepherd's-purse, autumnal hawkbit. There was some evidence of a higher species diversity along the side of the site bounded by a main road, especially of species not too common in the early stages of wasteland succession, e.g. yarrow, ribwort plantain. After 20 months the vegetation was still very open with annuals and biennials contributing 40% of the species. Physiognomically, in the second season the site changed from being dominated by Oxford ragwort to rosebay willowherb, but then it was redeveloped for warehousing so succession could not be followed further.

Floristically this was a fairly typical early succession but the proportions of the pioneer species vary considerably depending on local seed sources and the fineness of the substrata. Elsewhere in Sheffield *Atriplex hastata, Chenopodium album, Agrostis stolonifera,* dandelion, mugwort, wormwood (*Artemisia absinthum*), tall melilot, white clover, weld, Yorkshire-fog and barren brome are commonly very conspicuous in the first year or two.

Across the road from the 0.8 ha site just described, an area cleared in May 1980 had a finer textured substrate which encouraged dandelion, Michaelmas daisy, tansy, white clover and *Agrostis stolonifera* in its early stages but almost no rosebay willowherb. Eight 1 m^2 chart quadrats observed there from October 1980 until the site was landscaped in June 1982 revealed that over the first 2 years the diversity of angiosperms rose steadily to a mean of 16 spp. m^{-1} at which point the cover of higher plants was 35–80% and of bryophytes 10–40%. Although all plants arrived initially by seed, a number spread into the quadrats by vegetative extension. These were particularly the rhizomatous/stoloniferous grasses *Agrostis stolonifera, Elymus repens, Poa pratensis, P. trivialis* and *Holcus lanatus,* also creeping white clover and in the first year knotgrass. Other species built up locally dense populations round the decaying remains of a pioneer individual, e.g. shepherd's purse, *Atriplex hastata*, pineapple weed, scentless mayweed and *Poa annua,* while common wind-dispersed species like dandelion, Oxford ragwort, autumnal hawkbit and *Epilobium ciliatum* were widely and densely distributed from the start.

Over the first 2 years variation between the quadrats was marked. In some, dandelions established large rosettes with up to 28 inflorescences, creeping white clover spread to form circular colonies 90 cm in diameter and plants of scentless mayweed had up to 20 capitula; other plants built up stout rootstocks and seedlings were abundant. In other quadrats, growing conditions were less favourable, plants remaining small and either not flowering or producing few inflorescences; only *Agrostis stolonifera* did uniformly well. These stressed quadrats had fewer seedlings and those present tended to be eroded out or overturned by frost heave. These variations led to a patchy appearance of the plant cover. Fertilizer experiments, using a general NPK ameliorant, showed that the stress was mostly caused by nutrient deficiency, especially with regard to the performance of grasses. If it had been possible to follow these quadrats for many years it is suspected that the nutrient-poor areas would eventually have become the most interesting floristically, owing to their open nature.

This stage is typified by the abundance of Oxford ragwort which possibly has the longest flowering period of any of our wildflowers. In Sheffield it is well in bloom from late April till mid-October or November depending on the weather. Clapham *et al.* (1962) suggest the even longer flowering period of May to December. Figure 6.1 shows how overlapping cycles of flowering occur through the season. The primary terminal corymbs flower first, then while they are fruiting secondary axillary ones develop with a peak in July. These are followed by tertiary corymbs which extend flowering into the autumn. As the plants differ in age and persistence, they may behave as annuals, biennials, short-lived perennials or (in a stressed habitat) long-lived perennials. For this reason, flowering is not entirely synchronized so at any one time there is plenty of Oxford ragwort in full bloom.

Many of the early colonizers belong to genera that were widespread

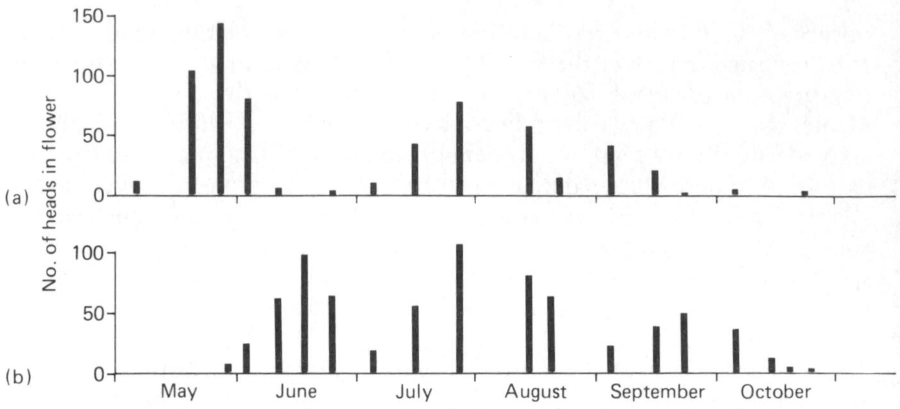

Fig. 6.1. Production of flowering heads by single plants of Oxford ragwort (*Senecio squalidus*) growing (a) on a wall and (b) on a brick rubble site. Three peaks of flowering can be discerned.

during the late-glacial period. They include *Artemisia, Epilobium, Polemonium, Polygonum, Potentilla* and *Rumex*. The frequency of these genera suggests that conditions on the wasteground which include intermittent disturbance, low grazing pressure, low competition and the presence of unleached base-rich soils must have similarities to those existing just after the end of the ice age; a number of the species may never have had it so good since.

Tall-herb stage

After 3–6 years the vegetation becomes dominated by tall perennial herbs with leafy stems, and the proportion of annual and rosette plants falls. Rosebay willowherb is one of the leading species; it spreads initially by seed, then vegetatively by rhizomes which may extend up to 1 m a year. The reason for its rise from rarity to abundance in the last 150 years is still conjectural. Popular 'explanations' such as the felling of woodland during World War II or the contemporaneous growth of car ownership, picnicking and cigarette smoking do not fit the facts documented in almost every county flora. Two aspects are of considerable importance in its history. First, it is not generally realized that the second British record made in 1666 is from 'Greenwich in the place where the ballast is taken up' (Merrett, 1666), and secondly it is an attractive plant which was formerly scarce in Britain and was for that reason grown in gardens under the name French willowherb. Both these circumstances together with its invasion history of the lowlands are highly suggestive of an introduced taxa, the spread of which was, like Oxford ragwort, helped by the railways. So in Britain we have at least two distinct races, an indigenous highland population and an introduced lowland race with a different ecological behaviour but a very similar morphology to the native type. For further discussion read Salisbury (1961), Myerscough (1980) and particularly Rackham (1986).

Joining rosebay willowherb in the tall-herb stage are many other plants of introduced origin with Michaelmas daisy, golden-rod, wormwood, fennel, goat's rue (*Galega officinalis*), lupin (*Lupinus polyphyllus*), feverfew, tansy, Shasta daisy (*Leucanthemum maximum*), *Artemisia verlotiorum, Melilotus* spp., comfrey species, Jacobs ladder and columbine being frequent in Sheffield's urban commons (Fig. 6.2). Mixed with them are native species such as creeping thistle, spear thistle, great hairy willowherb, red clover, goatsbeard, mugwort, common mallow, false oat grass, buttercups, cocksfoot and many others. Floristically this is a particularly rich phase as competitive exclusion is only just starting to operate which enables recruitment from a wide range of communities. For example, wood margin species such as figwort (*Scrophularia nodosa*), hedge woundwort (*Stachys sylvatica*) and bracken (*Pteridium aquilinum*), mix with species normally found on damper soils, e.g. hemp agrimony (*Eupatorium cannabinum*), Himalayan balsam (*Impatiens glandulifera*) and reed-grass (*Phalaris arundinacea*) while others are species of neutral grassland like hard-head (*Centaurea nigra*), yarrow (*Achillea millefolium*) and cat's ear (*Hypochoeris radicata*). The whole loosely knit assemblage is interspersed with patches still at an earlier successional stage carrying ruderals. Calcifuges are one of the few groups hardly represented.

Fig. 6.2. A 5-year-old urban common at the tall herb stage of succession.

On certain sites, biennials are very prominent particularly wild teasel (*Dipsacus fullonum* ssp. *fullonum*), mullein (*Verbascum thapus*), evening primrose (*Oenothera* spp.) and centaury (*Centaurium erythraea*). In Sheffield, members of the Umbelliferae rarely enter the vegetation at this stage though in other towns they can be codominants. Sowing stratified seed of cow parsley (*Anthriscus sylvestris*) on to several sites, where they quickly developed expanding populations, highlighted the importance of seed availability.

A few plants are well equipped to survive the trauma of demolition and grading out; prominent among these are the kitchen mints which were almost ubiquitous in the small gardens and backyards of the now largely demolished working class housing. In Sheffield spear mint (*Mentha spicata*) with its narrow pointed hairless leaves is the commonest while the broader-leaved peppermint (*M.* × *piperita*) including its var. *citrata* the lemon mint, round-leaved mint (*M.* × *suaveolens*) and apple mint (*M.* × *villosa*) are frequent. These vigorous often hybrid species rarely produce seed, spreading by long runners. In other towns I have seen tall mint (*M.* × *smithiana*) and bushy mint (*M.* × *gentilis*) on urban commons. It has been noted that certain stands originate from the disposal of garden refuse. Other species which are normally garden relics include Virginia creeper (*Parthenocissus quinquefolia*), Boston ivy (*P. tricuspidata*) and Russian vine (*Fallopia aubertii*), the rootstocks of which can survive a severe grading out and then smother large areas with sprawling regrowth.

Grassland stage

With time the proportion of grass in the vegetation increases as the smaller *Poa* and *Agrostis* species of the early years become replaced by *Arrhenatherum, Elymus repens, Dactylis, Festuca rubra* and *Holcus lanatus,* though *Lolium perenne,* unless sown, remains subordinate. So after 8–10 years the flowery meadows take on the appearance of grassland containing scattered clumps of tall herbs. Even rosebay willowherb usually loses its vigour due to unknown factors (cf. Canadian pondweed). Some of the more persistent tall herbs have creeping stoloniferous stocks, e.g. tansy, Michaelmas daisy, golden-rod, Japanese knotweed, creeping thistle and yarrow.

By this stage in the succession thickets of Japanese knotweed may be conspicuous. Very few species can persist under the dense shade cast by its 3 m high canopy which is reinforced by a litter of decayed stems covering the ground to a depth of 10–20 cm. Only greater bindweed (*Calystegia sepium*) regularly coexists and may occasionally attain a density sufficient to suppress its living support (Fig. 6.3). Japanese knotweed rarely sets viable seed in this country (Grime *et al.*, 1988) but

Fig. 6.3. The introduced greater bindweed (*Calystegia sepium* ssp. *silvaticum*) is one of the very few herbaceous species capable of suppressing Japanese knotweed (*Reynoutria japonica*).

is spread very efficiently by man carting soil containing vegetative fragments. Sites with only one small clump may, a year after grading out, contain a hundred. Spread by rhizome extension is initially rapid but slows down dramatically once a thick grass cover has established. This species and giant hogweed (*Heracleum mantegazzianum*) are the only two land plants that it is an offence to grow in the wild under Section 14 and Schedule 9 of the *Wildlife and Countryside Act* 1981.

Another form of clumping results from plants with no capacity for vegetative spread and only a limited one for seed dispersal. Ribwort plantain, cowslip and sheep sorrel are examples of this. Figure 6.4 demonstrates spread over 2 years from a pioneer rosette of ribwort plantain. The high proportion of seedlings in the 30–40 and 40–50 cm diameter zones is because they correspond to the length of the mature flower stalks, and dispersal of the sticky seeds mostly occurs by liberation from the capsules once the scapes have fallen. The maximum distance any new plant had established from the parent colony over the first 2 years was 97 cm when 5 and 28 inflorescences respectively were produced. In grazed or trampled grassland, dispersal on feet would probably have resulted in more efficient spread.

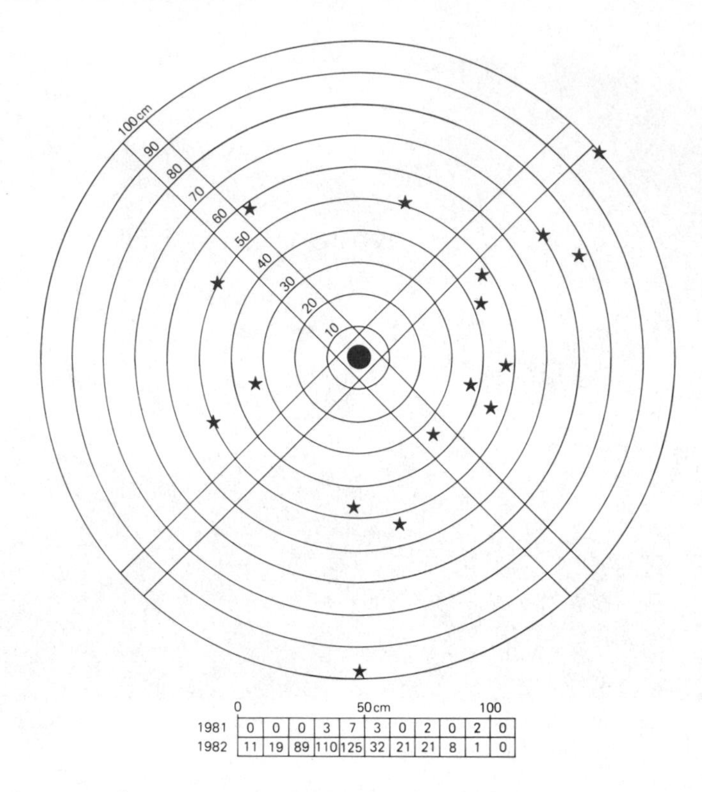

1981	0	0	0	3	7	3	0	2	0	2	0
1982	11	19	89	110	125	32	21	21	8	1	0

Fig. 6.4. Map showing two generations of offspring produced by a plant of ribwort plantain (*Plantago lanceolata*) (●). All plants originating in 1981 were plotted (★). The 1982 generation was sampled along four 10 cm-wide radial belts, and the numbers encountered were added together and are expressed in tabular form. No seedlings were observed beyond 97 cm distance suggesting a low dispersal ability on ungrazed and untrampled urban commons.

At this stage floristic variety on a site is still very high due to the 'palimpsest effect', (Fig. 6.5). A palimpsest is a manuscript in which the old writing has been partially rubbed out to make room for new but the old texts can still be read beneath the most recent one. The earlier successional stages have not quite faded away even after 10 years due to factors such as site variation and subsequent local disturbance, which enable annuals to persist particularly at the margins where Oxford ragwort, orache and shepherd's purse continue to flourish.

By now cryptogams are prominent. The bryophytes have diversified from the original carpets of pioneer, often abundantly fruiting, acrocarpous mosses comprising *Funaria hygrometrica, Barbula convoluta, Ceratodon purpureus, Bryum* spp. and *Pohlia* spp. to include *Amblystegium serpens, Calliergon cuspidatum, Brachythecium rutabulum, Eurhynchium*

Fig. 6.5. A 12-year-old urban common at the late grassland stage of succession.

praelongum, *Rhynchostegium confertum* and locally *Hypnum cupressiforme*, *Mnium hornum*, *Polytrichum juniperinum* and *Rhytidiadelphus squarrosus*. Their presence acts like a mulch helping to retain soil moisture.

Where the vegetation remains thin and untrampled luxuriant swards of terricolous lichens, particularly *Cladonia* spp., may develop as they show considerable resistance to air pollution and pollution levels within grassland are low anyway. Table 6.1 shows those likely to be encountered on the soil, pieces of wood or on mortar-encrusted bricks embedded in the ground of a 10-year-old urban common. The list contains several rather uncommon species which have not previously been found in towns. Several of these such as *Sarcosagium campestre*, *Steinia geophana* and *Vezdaea retigera* also occur on calcareous mine spoil heaps which must provide conditions broadly similar to coarse compacted brick–mortar rubble. There are also parallels with the lichen flora of intermittently disturbed soft chalk nodules found associated with rabbit warrens on downland (Gilbert, 1988).

The entry of toadstools has been followed at one site where mycorrhizal species such as *Hebeloma crustuliniforme* and *Lactarius torminosus* came in first followed by the litter breakdown species

Table 6.1. Lichens which may be encountered on a 10-year-old urban common

Substrate / species		Substrate / species	
Soil		Damp brick	
Cladonia chlorophaea	O–F	*Lecania erysibe*	R
C. coniocraea	R	*Scoliciosporum umbrinum*	O
C. fimbriata	O	*Trapelia coarctata*	C
C. furcata	R	Mortar, limestone chippings	
C. humilis	O	*Arthopyrenia monensis*	R
Collema limosum	R	*Bacidia caligans*	O
Peltigera spuria	R	*B. chloroticula*	R
P. uliginosa	O	*Lecanora dispersa*	C
Vezdaea retigera	O	*Lecidella scabra*	O
Wood		*Sarcogyne regularis*	R
Lecanora conizaeoides	F	*Sarcosagium campestre*	O
Placynthiella icmalea	F	*Trapelia obtegens*	O
Trapeliopsis flexuosa	R	*Verrucaria muralis*	C
T. granulosa	O	*V. nigrescens*	C
Clothing and cardboard		Rusty metal	
Lecanora conizaeoides	O	*Bacidia saxenii*	F
Steinia geophana	R		

C, common; F, frequent; O, occasional; R, rare.

Hygrocybe conica and *Panaeolina foenisecii*. Buried wood or other rich organic sources may bring in lawyer's wig (*Coprinus comatus*) at anytime, and *Myxomphalina maura* was collected from a bonfire site. Following the succession of mushrooms and other fungi on these sites would make a worthwhile research project.

Scrub woodland

The entry of woody plants into the succession is one of its more interesting aspects. The generally held model of succession is now that most phenomena are a consequence of differential colonizing ability, disparate growth rates and variations in size and longevity of the plants involved (Miles, 1979). This produces an appearance of communities successively replacing one another. Direct observations of urban commons through time and the examination of sites known to be of different ages bears this out.

The extremely localized occurrence of suitable seed parents is one of the factors that holds back the succession towards woodland in the centre of cities. By contrast, gap sites in the outer suburbs quickly turn into a forest of saplings where nudation has been severe. A survey of scores of urban commons in Sheffield has shown that the early woody plant colonists all possess light windborne seeds. In order of decreasing abundance they are goat willow (*Salix caprea*), buddleia, common willow (*S. cinerea*), birch (*Betula* sp.) and eared willow (*S. aurita*). Some come in during the first season, so after 3–4 years are quite conspicuous. As the herbaceous vegetation thickens, it becomes more and more difficult for these small seeded woody plants to establish successfully, so their colonization soon ceases, except where there has been subsequent disturbance or a fire. The maturing vegetation enters a stage of competitive dominance by the herbaceous layer; even sites which appear to be open are unavailable to small-seeded tree species. This 'window of opportunity' for the establishment of these species lasts only a few years so 10-year-old sites appear to be occupied by almost even-aged stands of trees and shrubs. This can be confirmed by counting girdle scars or growth rings. If for any reason these species do not come in early, the sites may remain more or less treeless. As an example of this, a female bush of goat willow shed millions of viable seeds on to a site but after the first few years they only continued to establish successfully on a bonfire patch and occasionally on the gypsy bank.

There is, however, another group of trees with larger seeds that are dispersed less efficiently. These have the capacity to enter closed vegetation so are continually recruited if they can reach the sites. They include ash, sycamore, broom, laburnum, rowan, hawthorn, elder, domesticated apple (*Malus sylvestris* ssp. *mitis*) and Swedish whitebeam (*Sorbus intermedia*), and also occasionally shrubs such as gooseberry, currant (*Ribes* spp.) and Duke of Argyll's tea-plant (*Lycium barbarum*). All but the first four have fleshy fruits and are bird dispersed; they form mixed-age populations. Once established only grazing by field voles (*Microtus agrestis*) and rabbits or digging up by gardeners threatens the young trees.

The growth rate of the woody plants is high (Fig. 6.6) exceeding that of trees planted by local authorities into topsoil. This apparently high growth rate can, however, be improved. A handful of slow-release nitrogenous fertilizer applied to 4-year-old sycamores more than doubled their growth rate over the next 2 years when compared with controls, and the colour of the spring foliage changed from yellowish-green to a dark-green. Laburnum which fixes nitrogen grows particularly well, but broom does not, possibly because the pH is too high. I have never seen a thriving native oak on a rubble site. Magpies regularly bury acorns on certain plots but these do not persist for more than 4 years; once the reserves in the acorn have been used up they slowly die. To what extent this is competition, an unsuitable soil pH or lack of the correct mycorrhizal fungus is not known.

Brambles are prominent on many sites after 5–10 years. A perusal of county floras showed that batologists have rarely studied urban

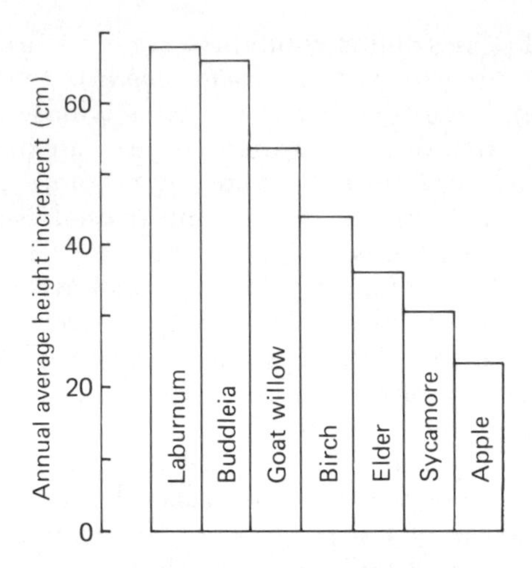

Fig. 6.6. Histogram showing annual average height growth of woody species on crushed brick rubble sites. $n = 25$.

brambles so some time was spent examining them. Most are bird sown rather than garden relics and, as might be expected, local seed sources determine the species present. Two of the commonest are *Rubus procerus* which is the 'Himalayan Giant' of nurserymen and the cultivar known as 'Bedford Giant'. The former can be recognized by its very large roundish to 'square' white-backed leaves, big panicles and stems with stout widely spaced prickles. Bedford Giant has coarse stems bearing an extremely dense armature of prickles and pricklets. Both are very vigorous and bear high-quality fruit. Another easily recognized bramble which has all the qualities of an urban specialist, being unknown in a wild state, and yet breeding true from seed is *R. laciniatus* the fern-leaved or cut-leaved bramble. It is widespread in the London area (Burton, 1983). In Sheffield, the commonest native bramble to be found on urban commons is *R. dasyphyllus* which has flowers of a deep pink and low-arching hairy stems which root at the ends; it is the most abundant prickly bramble in the UK. *R. ulmifolius*, a lime-tolerant species, is also quite common in towns.

Alan Newton has pointed out to me that it is generally members of the section *Corylifolii* originating by hybridization between *R. caesius* (dewberry) and other *Fruticosi* which have the vigour and growth habit to take maximum advantage of ruderal sites. They are tolerant of basic soils. By contrast very few of the section *Suberecti*, those closest to the raspberry, are found in towns, members tending to be confined to the least disturbed and often highly acid niches.

The deciduous woodland that eventually forms on these sites has an unusual composition. The best examples encountered are in the centres of Newcastle upon Tyne and Birmingham where closed woodland over 40 years old occurs. These copses comprise unlikely mixtures of native ash, hawthorn, willows, broom and guelder rose (*Viburnum opulus*), growing alongside laburnum, domestic apple, Swedish whitebeam, cotoneasters and garden privet. Such woods are quite unlike any other self-sown examples in the country; their composition is influenced strongly by available seed sources. They accumulate woodland plants very slowly, the ground flora remaining dominated by shade-tolerant weeds apart from ivy and ferns. Almost nothing is known of their fauna.

The role of substrate in the succession

The gradual amelioration of the rubble soil as succession proceeds is covered in Chapter 4. This section is concerned with the condition in which sites are left after grading out, as it is suspected that several parallel successions occur, each based on a different state of the substrate. On one 7-year-old site three well-defined communities could be recognized. Two-thirds of the area was underlain by the usual crushed brick–mortar rubble and there the vegetation had succeeded to an intermediate sward density containing approximately equal amounts of tall herbs and grasses among which Yorkshire-fog (*Holcus lanatus*) was prominent. This had checked the growth of many ruderals but was ideal for Michaelmas daisy, dandelion and clovers. On a slight slope where heavy machinery had not penetrated whole and half-bricks lay to a depth of several metres. This area was a stand of tall herbs, especially abundant being rosebay willowherb, common mallow, wormwood, mugwort, melilots, and nipplewort (*Lapsana communis*). Many species were 1–1.5 m tall which created deep shade at ground level where the soil was densely covered by the litter of last year's growth. Over another part of the site, grading out had exposed a granular layer of ash and steel slag, houses having been built on top of this material which now formed an intimate mixture with the rubble. Here a very open community comprising 20% higher plants, 60% bryophytes/lichens and 20% bare mineral soil had developed. There was a reasonable species diversity including annuals, but many of the species were dwarf and not flowering. Among the higher plants only autumnal hawkbit was growing with normal vigour.

It appeared that on this site three distinct successions were taking place at different rates, with different species involved, separate physignomies, distinct biomass values and leading eventually to contrasting woody plant stages. The productivity of the communities was gauged very roughly by a natural bioassay technique which involved calculating the mean height of rosebay willowherb and wormwood which were present in all three areas. The results (Fig. 6.7) appear to fit well with Grime's (1979) model of the paths of vegetation succession under conditions of high, moderate and low potential productivity. On the productive coarse rubble, ruderals are succeeded by a high biomass of competitive ruderals and competitors before passing over to stress-tolerant competitors. Rubble mixed with inert materials only supports a small biomass of ruderals followed by stress-tolerant ruderals and eventually stress tolerators. Standard brick–mortar rubble is suitable for C–S–R strategists adapted to habitats in which the level of competition is restricted by moderate intensities of both stress and disturbance.

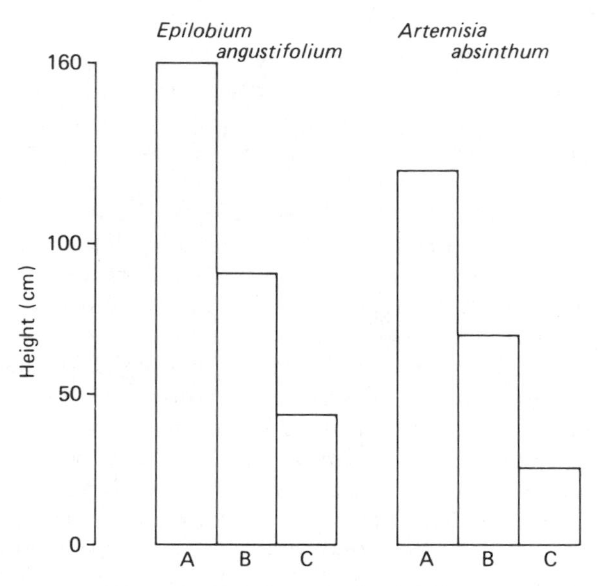

Fig. 6.7. Histograms showing the mean height of mature wormwood (*Artemisia absinthum*) and rosebay willowherb (*Epilobium angustifolium*) on three substrates covering adjacent areas of a 7-year-old urban common in Sheffield. A, coarse brick rubble; B, crushed brick–mortar rubble; C, rubble mixed with steel slag and ash.

The role of chance in the succession

A feature of the succession on urban wasteland is the part played by chance in determining variations both within a site and between sites. A vast (20 ha) area in the Ellesmere district of Sheffield was demolished between mid-1976 and mid-1977, gypsy banks constructed and then the site left to nature. Six years later it was still divided up into blocks by the old streets. A floristic survey showed that each section was distinct from its neighbours. For example there was a biennial block which supported large populations of teasel, mullein (*Verbascum thapus*) and evening primrose (*Oenothera* spp.); a hare's-foot clover block which in July was coloured pinky-grey by abundant fruiting heads of the urban rarity *Trifolium arvense,* which outside this section was represented by a single plant; another area had been well colonized by woody plants including 16 even-aged birch, again not present elsewhere in the 20 ha. Two others had an arrested succession still dominated by three species of melilot. Another had been disturbed subsequently and carried a large population of Oxford ragwort together with the American casual *Solanum cornutum.*

The cause of this variability is partly small differences in substratum or subsequent history but must largely be due to chance. In ecology, as in life, occupation is nine-tenths of the law and species which arrive first will tend to remain prominent. This was demonstrated experimentally by thinly scattering seed of the common urban species Michaelmas daisy, tansy, weld, goat's-rue, buddleia, Shasti daisy (*Leucanthemum maximum*) and feverfew on a recently cleared site. They all established readily and determined the subsequent visual character of the 5 × 10 m plots into which each had been introduced. I have witnessed one plant of centaury increase over 4 years until it dominated a site with its pink flowers and the single plant of hare's-foot clover mentioned earlier had a year later produced a patch several metres across.

The role of chance introduction by long-distance dispersal is subordinate to less spectacular recruitment from the vegetation surrounding a site, and the vegetation of the backyards, small gardens and streets immediately preceding clearance. A site of mixed housing and industry surveyed less than a year after clearance and grading out showed a far higher proportion of garden plants on the site of the former housing, e.g. *Alyssum saxatile,* honesty (*Lunaria annua*), snapdragon, lupin, Jacob's ladder, a double flowered form of feverfew and spear-mint.

The role of subsequent history in determining the succession

Not infrequently the entire course of the succession is altered by a half-hearted landscape treatment in which the site is regraded, stone picked, given a minimum topsoil treatment, then sown down to a rye-grass mix and left. In the first year or two spectacular stands of species present in the topsoil appear. If the soil has come from an agricultural source the site may be dominated by a single species of crucifer, e.g. oil-seed rape (*Brassica napus*), turnip (*B. rapa*), radish (*Raphanus raphanistrum*) or charlock (*Sinapis arvensis*). Soil of garden origin can, in the early years, be responsible for spectacular stands of species such as lupin, caper spurge (*Euphorbia lathyrus*) and cotton thistle (*Onopordum acanthium*).

After 2–3 years the *Lolium* swards collapse, due to a lack of nitrogen, and unsown species invade in a big way. These include large numbers of legumes (Table 6.2), up to 14 species per ha being recorded. How their heavy seeds arrive so rapidly and in such variety is not known. They do not appear to be contaminants in the sown mix. Many of the species spread quickly by rhizome growth so that after a few years these legume meadows are among the most eye-catching sites in Sheffield. Particularly successful are tufted vetch (*Vicia cracca*), goat's-rue (*Galega officinalis*) and locally lupin (*L. polyphyllus*), while in Birmingham and London the everlasting pea (*Lathyrus latifolius*) is frequent. Elsewhere, melilots, lucerne, meadow pea, red clover, white clover or a variety of vetches may be dominant. A plant to watch out for is the *Vicia cracca* look-alike *V. villosa*. This vetch from central Europe has slightly larger and brighter flowers and spreads vigorously to produce almost pure stands. The success of these legumes is partly a result of their nitrogen-fixing properties and also due to the lack of grazing and mowing. They constitute a distinctive urban wildflower community forming an alternative to the cut meadow in which few would persist. All the best examples arise from failed attempts at landscaping an urban common.

In contrast to major interventions which redirect the entire succession small-scale events produce local change. One of the commonest is the dumping of garden refuse containing live plants; this phenomenon is widespread but most often occurs on wasteground near public houses. Some of the species which establish particularly successfully in this way are *Bergenia crassifolia, Lamium maculatum, Galeobdolon argenttaum,*

Table 6.2. Characteristic species of legume meadows originating from failed attempts to landscape urban commons in Sheffield

Substantially present	
Galega officinalis	*T. hybridum*
Lathyrus pratensis	*T. pratense*
Lotus corniculatus	*T. repens*
Lupinus polyphyllus	*Vicia cracca*
Medicago lupulina	*V. hirsuta*
M. sativa	*V. sativa*
Melilotus altissima	*V. sepium*
M. officinalis	*V. tetraspermum*
Trifolium campestre	*V. villosa*
T. dubium	
Locally present	
Lathyrus aphaca	*Sarothamnus scoparius*
L. latifolius	*T. medium*

G. luteum subsp. *montanum* 'Variegatum' *Lysimachia nummularia, L. punctata,* montbretia (*Tritonia* × *crocosmiflora*) and mints. They may form considerable patches by vegetative growth but do not appear to spread by seed. Occasionally small natural looking rockeries develop on piles of broken tarmac, paving stones and rubble with a selection from *Cerastium tomentosum, Sedum acre, S. album, S. reflexum, S. spurium* and perhaps *Saxifraga* × *urbium* present.

The dumping of organic-rich refuse often determines the distribution on a site of high N and P demanding species such as nettle, bishop's weed (*Aegopodium podagraria*), white dead-nettle (*Lamium album*) and bittersweet (*Solanum dulcamara*). The presence of elder may also be related to high soil organic matter. All the above species are normally limited to the margins of sites where dumping is concentrated. The site of bonfires turn within a few months into sheets of the moss *Funaria hygrometrica* which act as a perfect seed bed for *Epilobium* spp. and goat willow. Excavation into the soil under isolated stands of willows often reveals a seam of charcoal.

Regional variation

Urban commons have been surveyed in 20 towns throughout England and Wales, and pronounced floristic variation detected. Although it is not quite correct to say that every city has its own characteristic community dominants, it should be possible to place a town in its regional setting from an examination of the wasteland flora. In very few towns is the habitat absent. I failed to find an example in the over-neat city of Salisbury and in London their access is difficult, being mostly boarded up. Elsewhere built-up areas are punctuated with smaller or larger examples particularly adjacent to railway land or industry. Regional floristic variation is imposed by the classic ecological factors climate, soil and the actions of man (historical), with the latter exerting the most influence as one might expect in an urban area.

The north–south summer temperature gradient appears to control the distribution of buddleia which is abundant in southern towns, moderately frequent throughout most of the midlands, but was seen only once on Tyneside. Its abundance in towns like Bristol is amazing; it forms thickets everywhere, colonizing ledges on buildings as well as wasteground; a visit to the city makes it easy to visualize the description of an early visitor to China who reported that buddleia thickets on shingle beside the Satani River provided 'famous harbourage for leopards'. Though introduced into gardens around 1890 impetus for naturalization on a large scale seems to have been prompted by the availability of bomb damaged sites following World War II. The Mediterranean bladder senna (*Colutea arborescens*) is another shrub which has become naturalized on wasteland mostly in the south. There must be a vacant niche in the UK for shrubs able to colonize dry calcareous rubble.

The east–west climatic gradient of increasing cloudiness, rainfall and humidity imposes a strong pattern. Sites on the dry east coast in towns such as Norwich, Hull and Newcastle-upon-Tyne are dominated by an *Arrhenatherum* grassland containing large amounts of hogweed, creeping thistle and often the biennials carrot and parsnip. In central England (Sheffield, Leeds) these grade into a *Holcus* grassland containing few Umbelliferae, while in the west (Manchester, Swansea) the moister soils enable an appreciable proportion of reed grass (*Phalaris arundinacea*) to flourish on the urban commons. The wetter soils also facilitate colonization by marshland plants such as the rushes *Juncus articulatus, J. bufonius, J. effusus,* and *J. inflexus*, together with Himalayan balsam, hemp agrimony (*Eupatorium cannabinum*) and dense stands of great hairy willowherb (*Epilobium hirsutum*).

The greatest distinction of urban commons in the west is the abundance of Japanese knotweed (*Reynoutria japonica*) which may totally dominate sites in a way never seen further east. It also grows so high that it is possible for a 6 ft-tall person to walk upright in the tunnels which children create through it. In places it is joined by the even taller giant knotweed (*R. sachalinensis*). Climatic control must operate through its action on vegetative growth as neither species has been known to produce effective seed in Britain. Though the flowers of *R. japonica* are dioecious the anthers mostly produce empty or aborted pollen in this country. A dwarf variety, v. *compacta*, less than 1 m tall, is sometimes male fertile and has occasionally set viable seed here. A history of the spread of the chief *Reynoutria* species in Britain is provided by Conolly (1977). She suggests that an intolerance of severe and unseasonal frosts may be important in restricting the spread of *R. japonica* in the east of the country.

The soil of the surrounding district determines to some extent the availability of seed to colonize an urban common. The Carboniferous limestone cliffs that penetrate Bristol hold large populations of traveller's joy (*Clematis vitalba*), the wind-blown seeds of which establish readily on brick rubble sites so the city centre is a good place to see this species. A distinctive set of towns occupy the southern end of the Jurassic limestone belt, e.g. Swindon, Cirencester and Dorchester; these have a particularly rich wasteland flora as physically and chemically the local soils to some extent match brick rubble. Here common St John's-wort (*Hypericum perforatum*) is a leading species accompanied by a wide range of calcicoles.

Historical factors are helpful in accounting for phenomena that cannot readily be explained by climate or soil. In the absence of precise historical records explanations offered under this heading have the added piquancy of being speculative. For example the abundance of alexanders (*Smyrnium olusatrum*) in Norwich is believed to be a result of its former cultivation as a pot-herb by monks who lived in the city. Whether this is correct will probably never be known, but currently it is the best interpretation of the observed fact of its abundance.

The frequency of Michaelmas daisy and goat's-rue in Sheffield, soapwort (*Saponaria officinalis*) in Rotherham or Canadian golden-rod in Birmingham suggests they have been naturalized for longer, or introduced on more occasions into these cities than elsewhere. These instances will be examined to demonstrate how, in the absence of direct confirmation, hypotheses can be built up using the ecological and historical facts that are available. Often it is better to acknowledge that 'we don't know', but occasionally it is possible to do better than this.

A Sheffield miner related to me that he remembered his father recounting how in the early part of this century horticultural traders used to work the poorer city suburbs selling garden plants which only just merited that description. They were aggressive species like tansy, Michaelmas daisy, feverfew and goat's-rue all of which have naturalized widely in the city. He recalled his father purchasing Japanese knotweed and how friends were invited round to marvel at the spotted stem and attractive foliage and how later the plant was divided up for exchange. In this way appreciable populations of potential garden escapes built up in small gardens and backyards so the inoculum pressure of their seed on the surroundings would be considerable.

The abundance of soapwort in Rotherham can probably be related to the fact that it has been naturalized there for a very long time, Jonathan Salt recording it as common by the River Don nearly 200 years ago. The profusion of Canadian golden-rod in Birmingham is possibly a consequence of it being very much an allotment plant and Birmingham has a long history of 'guinea gardens' and allotments. Populations originally grown for ornament would have naturalized as sites became available.

The following brief description of urban commons in a cross-section of towns gives an idea of how their vegetation varies across the country.

Bristol

Buddleia dominates the city centre forming thickets everywhere; it even grows on rooftops. A slogan on a wall reads 'Buddleia rules OK!' At a site by the Royal Hotel, which has remained undeveloped for 40 years, it is starting to be overtopped by sycamore which is probably its natural successor in towns. There is much traveller's joy (*Clematis vitalba*) in the city and local character is provided by large naturalized fig trees on the banks of the river. Other typical species are *Lactuca virosa*, red valerian (*Centranthus ruber*) and *Diplotaxis tenuifolia*. There is less rosebay willowherb than in most other cities and almost no feverfew, wormwood or Michaelmas daisy.

Swansea

Much of Swansea is dominated by Japanese knotweed which is locally called rhubarb or cemetery weed. It is often joined on wasteground by buddleia, so the combination of these two species together with abundant hemp agrimony (*Eupatorium cannabinum*) and pale toadflax (*Linaria repens*) are characteristic. Also well represented compared with other towns are red bartsia (*Odontites verna*), silverweed (*Potentilla anserina*), seaside plants (sea campion, buck's-horn plantain) and wetland species. Rosebay willowherb, Michaelmas daisy (one plant seen) and *Artemisia* are rather scarce and Oxford ragwort is still mostly by the railway. Walls support large populations of polypody (*Polypodium vulgare*) and ivy-leaved toadflax.

Sheffield

The urban commons in Sheffield contain a particularly large number of colourful garden escapes which provide a succession of blooms from mid-June with feverfew and goat's rue, through summer with tansy and soapwort, until mid-October when the Michaelmas daisies finally die down after their 2-month flowering period. Hillsides covered with pink-, purple- and white-flowered goat's rue are among the most spectacular urban natural history sights in the country. Michaelmas daisy reaches its maximum abundance later in the succession. Other plants of unusual abundance are eastern rocket (*Sisymbrium orientale*), wormwood and Yorkshire-fog, while the leading woody colonizer is goat willow (*Salix caprea*). Umbelliferae are rare.

Liverpool

Owing to the activities of the Inner City Partnership which spent large amounts of money landscaping urban commons, few mature examples are present. Perhaps for this reason species typical of early successional stages seem particularly common, e.g. yellow crucifers, lesser-hop trefoil, black medick, wall barley, melilots. By contrast woody plant colonizers are seldom seen and there are few well-developed stands of tall herbs. Species of unusual abundance are hedge woundwort, cut-leaved cranesbill (*Geranium dissectum*) and evening primrose (*Oenothera* spp.); enormous populations of the latter occur on sand dunes just outside the city.

Manchester

Urban commons in Manchester are rendered particularly distinctive by the abundance of Japanese knotweed, major clumps and thickets of which occur on nearly every site. Often the even taller giant knotweed (*Reynoutria sachalinense*) is also present. Though highly invasive initially, and very persistent, the stands almost stop spreading as succession proceeds. A second distinction is the abundance of wetland species. The reed grass (*Phalaris arundinacea*) is everywhere with some particularly fine stands near the university. At one site it is growing with amphibious bistort (*Polygonum amphibium*). At other quite ordinary looking sites up to four species of *Juncus* occur together with semi-aquatic grasses (*Glyceria fluitans, Alopecurus geniculatus*) and creeping yellow cress (*Rorippa islandica*). In Bellvue a small population of lesser broomrape (*Orobanche minor*) was found on a site less than 10 years old.

Swindon

Here a rather species-rich neutral-calcareous grassland flora develops on wasteland sites. The most distinctive features are extensive stands of the common St. John's-wort (*Hypericum perforatum*) which is supported by species like wild carrot, wild parsnip, welted thistle (*Carduus crispus*), great burnet (*Sanguisorba officinalis*), crow garlic (*Allium vineale*) and ploughmans spikenard (*Inula conyza*). Toadflax (*Linaria vulgaris*) is substantially present as might be expected in a railway town.

Hull

The older sites carry a dense *Arrhenatherum elatius–Heracleum sphondylium* community containing a good variety of other tall herbs such as creeping thistle, spear thistle, rosebay willowherb, docks, goosegrass (*Galium aparine*) and ox-eye daisy.

Birmingham

Here the abundance of golden-rod (*Solidago canadensis*) and the scarcity of buddleia are noteworthy. The local Urban Wildlife Group has adopted the larger bindweed (*Calystegia sepium*) as a motif to decorate some of their publications, as it is particularly conspicuous scrambling up chain-link fences, over walls and smothering bramble bushes.

6.2 ANIMALS

Colonization by animals is normally a continuous process with pioneer species coming in early while others only become established once suitable conditions of cover, microclimate and food have developed. The first species to arrive on wasteland belong to mobile groups such as birds, lepidoptera, hemiptera, winged beetles and spiders. These colonize during the first year establishing local populations in what appears to be an extremely hostile environment. Aggregations of species are often centred round an individual plant, for example a specimen of knot grass (*Polygonum aviculare*) in a waste of raw brick rubble supported a caterpillar of the shuttle-shaped dart moth (*Agrotis puta*), the weevil *Otiorhynchus sulcatus* and a population of plant bugs. Later less-mobile species (snails) and those requiring more mesic conditions (woodlice) or special food plants (many leaf beetles) install themselves. It is not really known how worms, woodlice, snails, centipedes and millipedes are dispersed; it is possible that, on occasion, a few individuals survive demolition and grading out and spread is from these nuclei. The impact of animals on the early succession, particularly herbivores on plants and decomposers on soil development, is complex and probably dramatic (Brown, 1985), but the details have not been worked out for wasteland ecosystems. The following descriptive account has been compiled from four sources; first, general observations accumulated over a period of nearly 10 years; secondly careful collecting from 1 m^2 plots located on sites of different ages; thirdly information gleaned from a sparse literature but especially useful regarding carabids and hemiptera; and fourthly details supplied by local naturalists and national experts.

Mammals

Few mammals are seen on urban commons apart from feral cats and dogs, though tracks and signs of other species are not uncommon. The most abundant small mammal is probably the field vole (*Microtus agrestis*) which can be detected by searching for its characteristic runs, food stores and olive-green droppings once the grassland stage has been reached. Older sites, particularly those with cover such as bramble patches, provide a habitat for wood mice (*Apodemus sylvaticus*) and bank vole (*Clethrionomys glareolus*). Stoddart (1980) reported that railway banks in Manchester held considerable populations of wood mice so these may be their primary habitat in urban areas from which colonization spreads out. The house mouse (*Mus musculus*) is rarely trapped in seminatural vegetation, but Clinging and Whiteley (1985) have reported that discarded milk bottles on demolition sites in Sheffield often contain their remains. *Rattus norvegicus* can often be found where there has been fly tipping of domestic rubbish on a large scale. There is some evidence that when factories and warehouses are demolished the homeless rat populations move into the urban commons.

The presence of rabbits (*Oryctolagus cuniculus*) can be inferred from droppings and scrapings or direct sightings in the evening. They have not adapted well to urbanization so are only present on sites adjacent to largish areas of open ground which hold the primary colony. Foxes are frequently seen crossing wasteland or occasionally searching for food there, but mostly these areas are too open; their chief breeding sites are gardens, allotments and railway banks.

Birds

Only a few birds are regularly recorded as breeding on urban commons; this is due to a lack of cover rather than high levels of disturbance. Larger sites in Sheffield support breeding pairs of ground nesting species such as meadow pipit, skylark and red-legged partridge, all of which have spread rapidly into the new industrial wasteland habitats. As local cover builds up linnet, dunnock and wren start to nest.

Throughout the year large numbers of house sparrow, starling and feral pigeon forage on the urban commons, then, as the end of summer approaches, postbreeding flocks of seed eaters, particularly goldfinch, redpoll, linnet and reed bunting, move in. Hunting kestrel tend to be ubiquitous, joined in some towns by owls and sparrow hawks.

Molluscs

Terrestrial molluscs, in general, favour damp habitats on basic soils. As most species burrow to some extent, loose or stoney substrates are preferred to compacted ones. Many species of snail cannot stand undue disturbance. Food requirements vary greatly but the majority of species feed on decaying organic matter. Given these demands it will be appreciated that urban commons are rather too dry, too disturbed and at least in their early stages have not accumulated sufficient organic matter to be a very favourable habitat for molluscs. Time spent searching through brick rubble, pieces of wood, old carpets and rotting cardboard will, however, reveal reasonably high populations of a few species. The following account refers to Sheffield.

The commonest snail is the distinctive *Discus rotundatus* with its flat coarsely ribbed shell bearing regularly spaced reddish-brown stripes. The garlic snail (*Oxychilus alliarius*) and the introduced Draparnaud's snail (*D. draparnaudi*) may also be present. These are medium-sized snails; larger ones are less common though *Cepaea nemoralis* turns up in small numbers while the hairy snail (*Trichia hispida*), strawberry snail

(*T. striolata*) and common snail (*Helix aspersa*) sometimes establish on sites from stray specimens in tipped garden refuse. The introduced species *Monacha cantiana* turns up from time to time and is one of the very few snails that is commoner on wasteland and roadsides than elsewhere. In general, snails are not rapid colonizers becoming conspicuous only on the oldest urban commons. The average garden is far richer. It would be worth investigating whether towns in limestone areas have locally been invaded by a specialized molluscan fauna of drought-resistant species from the surrounding countryside.

Many of Britain's slugs are thought to have been introduced, particularly in the northern part of their range where they behave as culture-favoured or synanthropic species. The easiest method of finding slugs on an urban common is to look under rubbish which acts as a mulch keeping the ground moist, but a careful search, especially at depth, will show they are ubiquitous even after only 3 years. The netted slug (*Deroceras reticulatum*) is well suited to urban sites as it produces many generations a year; three out of four slugs are likely to be this one. Also widespread are the garden slugs (*Arion distinctus, A. hortensis*), dusky slug (*A. subfuscus*), Budapest's slug (*Milax budapestensis*), *Deroceras caruanae* and the introduced white worm-like *Boettgerilla pallens* which is spreading rapidly throughout north-west Europe. The most urban of all our molluscs *Limax flavus* (introduced 1936) prefers cellars and outhouses where it feeds on mould and rotting matter, but extensive colonies may build up from individuals that survive the demolition and grading out operations. Though many of the species mentioned are synanthropic none are particularly typical of urban commons, being equally common in habitats such as gardens, allotments, by water-courses, in parks and especially secondary woodland.

Insects: General

As the vegetation changes with succession so does the phytophagous insect fauna. Some of the broad principles of these changes have been worked out though rarely as a result of observing wasteland. Colonizing species display a fast growth-rate expressed as the number of generations a year, a high proportion overwinter as adults, and the proportion of winged species is greatest in early succession. These features combine to enable rapid invasion and a fast build up of populations (Brown, 1985). A further important feature is that niche breadth declines as succession proceeds; in other words, generalist (polyphagous) feeders are replaced by specialists restricted to a single genus (but to more than one species) of plant. Zoological survey work on urban commons has only occasionally been sufficiently intensive to reveal successional relationships so the following account is mainly descriptive.

The broad pattern of arthropod invasion following clearance and grading out has been established by examining a series of dated sites in Sheffield (Table 6.3). Among the first arrivals are members of the aerial plankton, particularly spiders which drift in on silken kites. These aeronauts are chiefly money spiders belonging to the Linyphiidae which find even smaller animals to prey on or resort to cannibalism. Within a few months they can be present at a density of several dozen per square metre. At this stage occasional winged beetles with a wide ecological

Table 6.3. Species present in or on the soil surface of brick rubble sites of different ages. At each locality invertebrates were collected and counted from a single 1 m^2 quadrat. Figures in parentheses refer to the number of localities in each age class where the organism was recorded

Age of site (years) . . . *Number of localities . . .*	*12–15* *4*	*4–6* *4*	*0–1* *4*
Ground beetles (Carabidae)	8	13	2
Amara aenea	6(1)	1(1)	–
A. similata	–	3(1)	–
Bembidion lampros	–	2(2)	–
Harpalus affinis	1(1)	1(1)	–
Notiophilus biguttatus	1(1)	4(1)	1(1)
Additional spp.	–	2	1
Rove beetles (Staphyalynidae)	1	1	1
Harvestman (Phalangidae)	–	8(1)	3(3)
Spiders (Araneae)	54	48	45
Money spiders (Linyphiidae)	25	32	44
Centromerita bicolor	5(2)	11(1)	1(1)
Erigone atra	–	2(2)	5(3)
Lepthyphantes tenuis	–	6(2)	6(3)
Diplostyla concolor	8(1)	1(1)	–
Oedothorax apicatus	–	7(1)	18(4)
Stemonyphantes lineatus	12(1)	–	–
Additional Linyphiidae	–	5	14
Tegenaria agrestis	6(3)	6(4)	–
Xysticus cristatus	5(4)	5(4)	–
Salticus scenicus	4(1)	–	–
Clubiona diversa	1(1)	1(1)	–
C. terrestris	3(2)	–	–
Amaurobius ferox	4(2)	–	–
Trochosa ruricola	2(2)	–	–
Additional spiders	4	4	1

Table 6.3. continued

Age of site (years) . . .	*12–15*	*4–6*	*0–1*
Number of localities . . .	*4*	*4*	*4*
Woodlice (Oniscoidea)	552	110	–
Oniscus asellus	289(3)	25(1)	–
Porcellio scaber	26(3)	53(2)	–
Androniscus dentiger	105(2)	2(1)	–
Trichoniscus pusillus	132(4)	30(2)	–
Snails (Mollusca)	37	–	–
Discus rotundatus	7(1)	–	–
Oxychilus alliarius	7(3)	–	–
O. draparnaudi	23(3)	–	–
Slugs (Mollusca)	48	16	–
Arion distinctus	9(3)	–	–
A. subfuscus	2(1)	2(1)	–
Deroceras caruanae	3(2)	–	–
D. reticulatum	31(4)	13(1)	–
Limax flavus	–	1(1)	–
Milax budapestensis	3(2)	–	–
Centipedes (Chilopoda)	15	56	1
Lithobius variegatus	6(3)	26(4)	1(1)
L. curtipes	4(1)	–	–
L. mictos	3(1)	–	–
L. melanops	–	6(2)	–
Additional Chilopoda	2	24	–
Millipedes (Diplopoda)	11	24	–
Polymicrodon polydesmoides	8(3)	5(2)	–
Additional Diplopoda	3	19	–
Lepidoptera larvae	6(2)	11(2)	5(3)
Lepidoptera pupae	2(1)	2(1)	1(1)
Ants*	C	O	–
Formica lemani	R	–	–
Myrmica ruginodis	C	O	–
M. scabrinodis	C	O	–
Earthworms*	C	R	VR

* C, common; O, occasional; R, rare; VR, very rare.

amplitude, e.g. *Notiophilus biguttatus* and *Bembidion femoratum,* are present together with lepidoptera larvae which feed on the sparse vegetation and hide under stones by day. Very local populations of other species may be present, particularly close to the margins. After three years, as pockets of soil and humus accumulate, the sites become more mesic and up to four species of woodlouse may be frequent together with four or five carabids, two or three slugs, several millipedes and centipedes, possibly harvestmen, and by now pupae of lepidoptera are regularly encountered in the soil.

As plant succession proceeds the substrate becomes more compact as root mats fill up spaces between the rubble, so after 10 years the density of groups such as woodlice and carabids which like a soil full of cracks and crannies has started to fall. Now the macro-animal community is dominated by earthworms and ants. Spiders remain abundant with many families represented. At this stage, it is harder to find and capture animals by hand-sorting through the surface litter and digging into the consolidated soil, so numbers given for the older sites in Table 6.3 are only very approximate.

Lepidoptera

The commonest breeding species of butterfly on urban commons in Sheffield are meadow brown, wall brown and small skipper; all are present in large numbers. Their larvae feed on ubiquitous grasses such as cocksfoot, Yorkshire-fog and couch. Also regularly present are common blue (legumes), large skipper (grasses) and small copper (sorrel and docks). During good years for the painted lady, migrants lay eggs on thistles which produce butterflies from July onwards. Large, small and green-veined whites are also very common but though their larvae are general feeders on Cruciferae and weld, in urban areas adults tend to lay their eggs on garden plants such as nasturtium and brassicas only using the urban commons for feeding. The same is true for small tortoiseshell, peacock and red admiral as large stands of their food plant, nettle, rarely occur on recently cleared sites. Accounts of the butterflies breeding on urban wasteland in other parts of the country are hard to come by but my own observations suggest that the Sheffield list is fairly typical with small heath a notable absentee and more skippers further south.

Many other species visit urban commons to feed on the nectar of bramble, Michaelmas daisy, thistles, knapweed and buddleia in particular. At the Rally, a small site in the centre of Leicester, the City Wildlife Group have noted 23 species of butterfly feeding including fritillaries. Sixteen species regularly visited the William Curtis Ecological Park in Central London. Flowering of the butterfly bush (*Buddleia davidii*) normally extends over July and August but experimental management on wild bushes has shown that it can be prolonged into early October either by pruning hard in the spring or better by cutting off dead heads in August which stimulates axillary growth.

Day flying moths may be quite conspicuous especially six-spot and narrow-bordered five-spot burnets, the adults of which show a predilection for feeding on the flowers of tufted vetch. Cinnabar moths are also frequently seen both as imagos and even more often as populations of black and orange striped caterpillars stripping the leaves of common ragwort, Oxford ragwort and groundsel. Silver Y moths may also be seen during the day feeding at various flower heads, while a walk through the vegetation disturbs a fair number of smaller species such as yellow shell (*Camptogramma bilineata*) and the beautiful five-plumed 'gnat' (*Alucita pentadactyla*) the larvae of which feed on greater bindweed (*Calystegia sepium*).

A considerable number of lepidoptera larvae are specialist feeders, their distribution being controlled by that of their food plant. Garland (1985) has studied those associated with wasteland floras. The wormwood shark (*Cucullia absinthii*) was originally a southern and south-eastern coastal species in Britain but during the late 1940s it began to spread inland and northward as its food plant, wormwood, became common in urban areas. Mugwort and wormwood also provide food for the bordered pug (*Eupithecia succenturiata*) and a striking member of the Tortricidae, *Epiblema foenella*. Records of the elephant hawkmoth (*Deilephila elpenor*) have increased in step with the spread of rosebay willowherb and there has been a similar though less spectacular proliferation in urban areas of the chamomille shark (*Cucullia chamomillae*) the larvae of which feed on scentless mayweed and feverfew. Other plants of urban wasteland that support distinctive lepidoptera are the Chenopodiaceae (leaf miners) and willowherbs particularly *Epilobium hirsutum* (microlepidoptera), and a walk through a patch of coltsfoot during early summer should disturb large numbers of the tortricid *Epiblema farfarae* and the distinctive triangle-marked plume moth (*Platyptilia gonodactyla*), while mints are good for Pyralidii.

There are also a number of species the caterpillars of which can eat a very wide variety of food plants; these are common throughout urban areas including wasteland, e.g. large yellow underwing (*Noctua pronuba*), lesser yellow underwing (*N. comes*), lesser broad border (*N. janthina*), ribband wave (*Idaea aversata*), buff ermine (*Spilosoma luteum*), dot (*Melanchra persicariae*), knotgrass (*Acronicta rumicis*), angle shades (*Phlogophora meticulosa*), swallow-tailed moth (*Ourapteryx sambucaria*), peppered moth (*Biston betularia*), dark arches (*Apamea monoglypha*), heart and dart (*Agrotis exclamationis*).

In the absence of survey data an indirect method of assessing the possible richness of a site for breeding lepidoptera is to consult lists which show plants in order of the number of moth and butterfly larvae known to feed upon them. Such lists prepared by Harrison (*in lit.*) show that goat willow and apple are particularly rich among woody plant colonizers, but it is not known whether this holds for scattered bushes of recent origin. Similar lists for herbaceous species show 22 plants which are each used as food by over 30 lepidoptera, many occur abundantly on wasteland, e.g. knotgrass (76 spp.), broad-leaved dock (69), ribwort plantain (45), birds-foot trefoil (45), dandelion (45), greater plantain (42), annual meadow grass (35), cocksfoot (35), white clover (33), red clover (33), yarrow (31). When using these lists it should be borne in mind that the plants are not necessarily the primary food source for all the lepidoptera recorded as feeding on them. Also the situation is dynamic. Owen and Whiteway (1980) found that 11 species of caterpillar are now known to feed on the leaves and flowers of buddleia in Britain including specialized feeders such as the holly blue butterfly (*Celastrina argiolus*) and mullein moth (*Cucullia verbasci*) previously only known from mulleins and figworts, both Scrophulariaceae. At least six of these 11 species occur in China so could possibly feed on buddleia growing in the wild.

In 1987 a special effort was made to find caterpillars on Japanese knotweed in Sheffield. Though eaten leaves were reasonably common it was some time before the larvae responsible were found. Eventually numerous caterpillars of the brick (*Agrochola circellaris*) were discovered hidden in debris at the base of the stems which they ascended at night. The larvae of this moth normally feed on wych elm and may have transferred to this new food plant following the ravages of Dutch elm disease. Alternatively the caterpillars may have spread to the Japanese knotweed from eggs laid on adjacent goat willow (*Salix caprea*). Larvae of the polyphagous Hebrew character (*Orthosia gothica*) were also found at one locality and it has been reported that larvae of the knotgrass moth will feed on Japanese knotweed in captivity. Other plant-feeding insects seen on Japanese knotweed included aphids, the meadow spittlebug (*Philaenus spumarius*) and a large weevil (*Otiorhynchus sulcatus*). Occasionally the aphids were present as building populations but more often they are accidental occurrences from adjacent vegetation. Whatever their origin they are sufficiently abundant to attract predators including hoverfly and ladybird larvae as well as honeydew feeders and parasites. So less than 30 years after becoming naturalized in Sheffield an intricate foodweb is starting to build up on Japanese knotweed.

Beetles

Several studies have been made of beetles occurring on demolition sites, probably the most comprehensive being those of Lazenby (1983, 1988) who collected from 65 such sites in Sheffield. He found that ground beetles (*Carabidae*) were quick to colonize with a small number of winged species coming in early and building up large populations within the loose rubble, e.g. *Amara convexiuscula, A. bifrons, Bembidion femoratum, Notiophilus biguttatus.* Optimum conditions are reached after 4–5 years when the sites still have plenty of loose stones and open areas but dense patches of weeds provide seeds as a food source. At this stage the seed-eating *Amara* spp. and *Harpalus* spp. with their strong blunt mandibles and short legs suitable for climbing up plants reach maximum abundance. Other species of more omnivorous habit belonging to genera such as *Bradycellus, Nebria* and *Notiophilus* are also present. They have sharper mandibles, longer legs and are usually more active. They feed on mites, springtails, slugs, dead worms and carrion. As sites pass over into the grassland phase there is a reduction in seed production, a more uniform vegetation develops and the substrate becomes more closely bound together. This causes a fall off in carabid diversity and abundance though a few species such as *Amara aenea, A. aulica, Asaphidion flavipes* and *Calanthus fuscipes* flourish. The former can sometimes be found by examining the fruiting heads of knapweed and burdock. The marginal 'gypsy banks' of what Lazenby calls 'urban scree' retain a rich carabid fauna long after this group has declined over the rest of the site.

Results from a study in Berlin (Weigmann, 1982), where pitfall traps were used to collect carabids from three ruderal sites, produced broadly similar findings (Table 6.4). The carabid fauna was composed for the most part of widespread species of dry grass, fallow land and cultivated fields; only one forest species, *Nebria brevicollis*, was abundant. Work on the immigration of carabids into newly reclaimed polders in the Netherlands has provided information on their colonizing ability. This is controlled largely by dispersal capacity with brachypterous (non-winged) forms poorly represented in the early stages. The ability to fly, however, is not as important as might be expected, as flight in most species is trivial, covering only short distances, and they do not often cross habitat boundaries. The conclusion of Meijer (1974) was that migration in carabids is mostly accidental.

The 35 species found during the Sheffield survey included beetles previously considered uncommon in the area, so the assemblage was interpreted as a special urban one not present elsewhere and certainly not an attenuated fauna. Certain members such as *Amara eurynota,* a large, oval, bronze coloured beetle, appear to have a strongly urban distribution in South Yorkshire (Fig. 6.8).

The decline in carabids as plant succession proceeds is compensated for by an increase in other beetle groups particularly rove beetles (*Staphylinidae*), ladybirds (*Coccinellidae*), weevils (*Curculionidae*) and leaf beetles (*Chrysomelidae*). Some of these are very attractive and quite specific in their food plants, being tied to one genus or even one species of plant. From their study of the successional relationship of plants and phytophagous coleoptera Brown and Hyman (1986) found that the

Table 6.4. The carabid beetle fauna of demolition sites in Sheffield and Berlin

	Sheffield	*Berlin*
Amara aenea	O	C
A. apricaria	F	–
A. aulica	F	–
A. bifrons	C	C
A. consularis	–	R
A. convexiuscula	O	–
A. communis	R	–
A. eurynota	A	–
A. familiaris	O	–
A. fulva	–	R
A. plebeja	O	–
A. similata	A	–
Asaphidion flavipes	R	–
Badister bipustulatus	–	R
Bembidion femoratum	F	–
B. lampros	A	R
B. nitidulum	R	–
B. obtusum	R	–
B. quadrimaculatum	R	–
Bradycellus harpalinus	C	–
B. verbasci	A	–
Calanthus ambiguus	–	R
C. fuscipes	R	A
C. melanocephalus	R	C
C. mollis	R	F
C. piceus	R	R
Dromius linearis	R	–
D. melanocephalus	R	–
Harpalus aeneus	–	R
H. affinis	C	–
H. rubripes	–	R
H. rufibarbis	R	–
H. rufipes	F	F
H. smaragdinus	–	R
H. tardus	–	R
H. vernalis	–	O
H. winkleri	–	R
Licinus depressus	–	R
Loricera pilicornis	C	–
Masoreus wetterhalli	–	R

Table 6.4. continued

	Sheffield	*Berlin*
Metabletus foveolatus	–	R
Nebria brevicollis	O	A
N. salina	O	–
Notiophilus bigattatus	C	–
N. substriatus	R	–
Olisthorpus rotundatus	R	–
Pterostichus madidus	F	–
Synuchus nivalis	–	R
Trechus obtusus	R	–
T. quadristriatus	R	–

A, abundant; C, common; F, frequent; O, occasional; R, rare.
Data are from Lazenby (1983, 1988) and Weigmann (1982).

species diversity of herbs and weevils was closely related (on a rural site) which led them to suggest that, in general, plant-species composition may be more important than plant structure in determining the

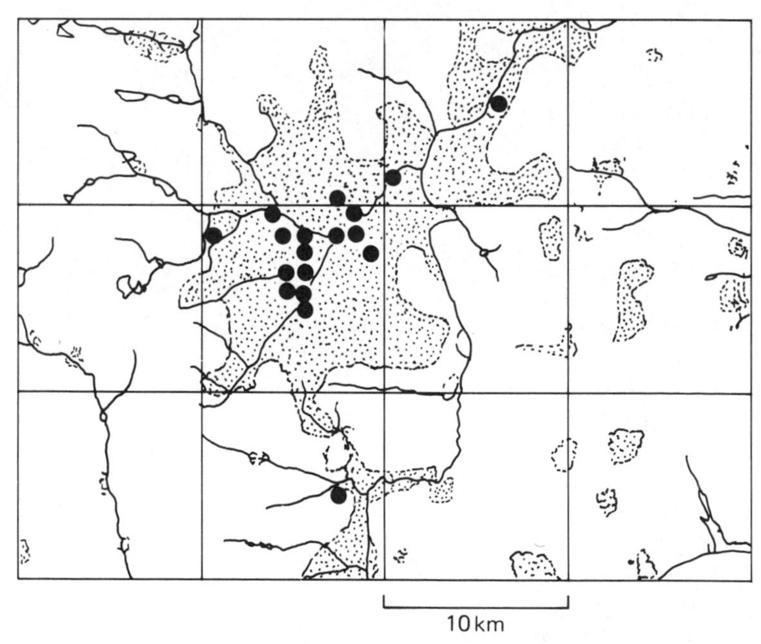

Fig. 6.8. Map showing records of the ground beetle *Amara eurynota* in the Sheffield region. The built-up area is stippled (Lazenby, 1988).

occurrence of solely phytophagous groups of insects. If this is so, urban commons, during the transitional stage of tall herbs giving way to grassland, should be particularly rich in weevils.

Hoverflies

Following the publication of a reliable identification guide (Stubbs and Falk, 1983) this group is enjoying increasing popularity but so far there have been few studies specifically directed at the Syrphidae of urban wasteland. A previously unpublished survey in Sheffield carried out by A. Brackenbury during late August 1985 recorded six species on a 1-year-old site, 16 at a site where the vegetation was at the tall herb stage and nine at a site which had progressed to the grassland stage (Table 6.5). Species diversity more or less correlated with the abundance of suitable flowers, particularly Compositae. During September, Japanese knotweed provides a valuable nectar source for late hoverflies (Fig. 6.9).

The Sheffield lists made in late August are strikingly similar to ones compiled from ruderal sites in Berlin (Weigmann, 1982) though the latter were made on ten catch dates spread over two summers (Table 6.5). The syrphid fauna in both cities consists of generally widespread species with a mixture of hoverflies the larvae of which feed on aphids or

Fig. 6.9. During September Japanese knotweed (*Reynoutria japonica*) provides a valuable food source for a wide variety of late insects including hoverflies, honey bees, wasps and butterflies (I. Smith).

Table 6.5. Hoverflies of ruderal areas at different successional stages in the cities of Sheffield and Berlin

	Sheffield			*Berlin*		
	1	*2*	*3*	*1*	*2*	*3*
Cheilosia sp.					X	
Chrysotoxum festivum						X
Dasysyrphus albostriatus		X				X
D. tricinctus						X
Epistrophe grossulariae		X				
Episyrphus balteatus		X	X	X	X	X
Eristalis arbustorum	X	X	X		X	X
E. intricarius					X	X
E. pertinax		X	X			X
E. tenax	X	X	X		X	X
Helophilus hybridus						X
H. pendulus		X		X	X	X
H. trivittatus				X	X	
Melanostoma mellinum		X			X	X
Metasyrphus corollae				X	X	
M. luniger		X	X		X	
Myiatropa florea				X		X
Pipiza festiva				X		
Platycheirus albimanus	X					
P. peltatus		X				
Scaeva pyrastri	X	X	X			X
Sphaerophoria menthastri						X
S. scripta	X	X	X			
Syritta pipiens	X	X	X			X
Syrphus ribesii		X	X	X	X	X
S. vitripennis		X		X	X	X
Volucella pellucens		X			X	
V. zonaria						X
Xylota segnis				X		
Total species	6	16	9	9	13	18

For Sheffield: 1, 1-year-old site; 2, 6-year-old site; 3, 12-year-old site.
For Berlin: 1, young site; 2, varied site; 3, very varied site.
Data were from Brackenbury (unpublished work) and Weigmann (1982).

have aquatic larvae. Only species the larvae of which require dead wood or sap runs are (almost) absent. A survey in Birmingham (Dawe and McGlashan, 1987) which investigated the occurrence of hoverflies in ornamental parks and naturally colonized wasteland concluded that abundance and species variety were remarkably similar in the two types of site. It was thought likely that adult hoverflies were sufficiently mobile for their abundance on a given site to be determined more by nectar and pollen availability than by the immediate presence of potential larval food. Hoverfly abundance and variety in Birmingham peaked during August with *Episyrphus balteatus* outnumbering all other species combined (516 flies out of a total catch of 936). In Sheffield *Sphaerophoria scripta* is the leading hoverfly of the urban commons.

Hemiptera

One of the few studies of Heteroptera on derelict land in cities has been carried out by Kirby (1984) and involved six sites close to the centre of Derby. These varied in age from less than 1 year to 6 years. Many of the bugs found were very common species of general distribution such as *Anthocoris nemorum* and *Plagiognathus arbustorum* which already had populations in the immediate area. They are no commoner on demolition sites than in other habitats and are not especially characteristic of such places. An additional group of species occurred on scarce plants, present only as garden relics, for example *Psallus falleni* and *Lygocoris contaminatus* on a single birch tree. The removal of Heteroptera in these two categories together with those found as single specimens left a set of species which may be regarded as characteristic of waste ground in Derby. All are more frequent in such habitats than elsewhere in the same area. They are *Nysius ericae, Piesma maculatum, Nabis ferus, Orius vicinus, Plagiognathus albipennis, Chlamydatus saltitans, Orthotylus flavosparsus, Trigonotylus ruficornis* and *Saldula orthochila*.

Reasons for their presence on derelict ground include association with plants common there. For example *O. flavosparsus* and *P. maculatum* are tied to members of the Chenopodiaceae while *P. albipennis* occurs on *Artemisia*. The structure of the habitat favours other species. *N. ericae, S. orthochila* and *C. saltitans* are always associated with partly bare ground and very low vegetation such as occurs in the early stages of succession. Owing to the isolation of the sites, efficiency of dispersal will be important, and those species that occur regularly must be highly mobile. Not much is known about this aspect but most of the species are fully winged and several are frequently caught in both light and suction traps. Kirby's findings are broadly similar to lists made at Tilbury dock, London (Edelsten, 1940) and from cinder-covered wasteland at Slough, Berks (Woodroffe, 1955).

The Homoptera of wasteland are less well known. The Cercopidae or true froghoppers, whose nymphs create the cuckoo spit on plants, are common on certain sites once the grassland stage has been reached. The commonest is *Philaenus spumarius* which varies in colour from drab brown to piebald. The aphids would repay study as many are specific to a single host plant. For example a collection made from a single clump of tansy (*Tanacetum vulgare*) growing on a 10-year-old urban common in Sheffield was examined by H.L.G. Stroyan. He reported four species, all of which live only on tansy. They were *Uroleucon tanaceti*, a rust-coloured aphid occurring on the underside of the more basal leaves, *Macrosiphoniella tanacetaria*, a large pale-green species with black legs which lives on the stems and inflorescences in rather large colonies, *Metopeurum fuscoviride*, a reddish species found on the stems and tended by ants, and *Coloradoa tanacetina* a very small pale-green aphid which lives scattered along the leaf margins. On the same site wormwood was supporting conspicuous colonies of the dark-reddish aphid *Macrosiphoniella absinthii* which is confined to this host, while creeping thistle held large populations of the bronzey-brown thistle aphid *Uroleucon cirsii*. This 5-minute collecting exercise off three plants on one site provides a glimpse of the rich homopteran fauna present on urban commons but unfortunately they are a very difficult group for non-specialists to work with.

Grasshoppers, crickets and cockroaches

There have been very few studies of urban grasshoppers but our two commonest species, the meadow grasshopper (*Chorthippus paralellus*) and the common field grasshopper (*C. brunneus*) both rapidly invade urban wasteland by flying in once a 50% vegetation cover has been attained. A walk across such a city centre site on a hot day in late summer will send up clouds of grasshoppers and the air will be full of their chirping. They require only small patches of grass to survive; in Sheffield there are records of *C. brunneus* from pavements, car parks, even inside city buildings. Sometimes the above two are joined by the mottled grasshopper (*Myrmeleottex maculatus*), the common green grasshopper (*Omocestus viridulus*) and the common ground hopper (*Tetrix undulata*) but they tend to prefer taller more dense vegetation and grassland–scrub mosiacs so in urban areas are more a feature of wasteland adjacent to their seminatural habitats; they do not appear to come in spontaneously like the other two. Along the south coast of England the nationally rare long-winged cone-head (*Conocephalus discolor*) has recently spread into urban wasteland in towns such as Portsmouth, Lewes and Weymouth; it is frequently accompanied there by *C. discolor* (E.C.M. Haes, personal communication).

Where sites have been extensively used for tipping domestic rubbish or were recently industrial premises, temporary populations of the house cricket (*Acheta domesticus*) may occur. There is evidence also that the oriental cockroach (*Blatta orientalis*) is now able to establish on outdoor sites in Britain (Bateson and Dripps, 1972). Certain orthoptera cannot adapt to increasing urbanization. Kuhnelt (1982) quotes the example of *Stenobothrus bicolor* which maintained strong populations in the middle of Vienna up till the 1950s but by 1980 had practically disappeared and was found only in the suburbs.

Bees and ants

Weigmann (1982) recorded the bumble-bees visiting three ruderal sites in Berlin on ten catch dates spread over 2 years. Five species were found, all of which are generally common in cities; *Bombus hortorum, B. lapidarius, B. pascuorum, B. pratorum* and *B. terrestris*. As bumble-bees seldom fly more than 2 km from their nests he assumed that all bred within the urban area which offers many suitable nests sites in the form of cracks and crannies. Less systematic recording in Sheffield and other British towns has revealed a very similar pattern, with *B. lucorum* also widespread. As with other groups, such as grasshoppers, species that are local and southern, e.g. *B. ruderarius, B. muscorum,* are sometimes found foraging on wasteland in the south of the country.

Small widely separated colonies of ants can be found on cleared sites after only 3 years, these subsequently expand in step with the food supply as sites mature till in the grassland phase every stone and piece of rubbish has a thriving ants nest underneath it. The ants spread by single mated queens flying into fresh sites and establishing themselves in suitable crevices. Once a nest is established, in the course of a few seasons further mated queens are developed and may fly off or reinforce the original nest so the species comes to occupy a large part of the site. The commonest species of black ant on the urban commons of Sheffield are *Formica lemani* and *Lasius niger*. Red ants are even more abundant with *Myrmica ruginodis, M. rubra* and *M. scabrinodis* everywhere. The ant populations must have a considerable, but unassessed, effect on the ecology of the habitat. *M. ruginodis* and *F. lemani* can often be seen tending the aphid colonies which develop on plants such as tansy, wormwood and thistles. Towns south of a line from the Wash to the Mersey are likely to possess a far richer ant fauna than Sheffield.

Woodlice, millipedes and centipedes

The alkaline freely draining warm habitat provided by partly colonized brick rubble is very suitable for slaters, sow bugs, coffin cutters, bibble bugs or whatever local name woodlice enjoy in your town. They are scavengers feeding on dead plant and animal material, and often require their food to be partly decomposed before they will eat it, so do not become abundant till pockets of organic matter have accumulated on the site. Little is known about their dispersal mechanisms though mass migration does not take place. Sutton (1980) recorded one travelling 13 m in a night.

There is some evidence (Table 6.3) that the first species to build up substantial populations are *Porcellio scaber* and *Philoscia muscorum* which will tolerate dryer conditions than most other woodlice. As sites become more mesic *Oniscus asellus* and *Trichoniscus pusillus* become commoner, migrating to considerable depths in the soil to avoid desiccation. A number of species that are native in the south of Britain are naturalized further north where they are strongly synanthropic. The rose-coloured *Androniscus dentiger* is one such species and so too is *Cylisticus convexus*; both are very much garden–wasteland species in the Sheffield area. The pill woodlouse *Armadillidium vulgare* is native in lowland calcareous grasslands and not infrequently extends its range by invading the artificial calcareous habitat presented by brick rubble. Sometimes it is joined by rarer species such as *Porcellio spinicornis*. Woodlice are often sufficiently abundant to be a valuable link in the decomposer food chain especially on sites where the soil is too shallow for earthworms to survive.

Millipedes live in the upper layers of the soil which they help to create as their diet is basically a mixture of earth, dead leaves, wood, fungi and carrion. They require calcium for their exoskeleton so are rarely found on acid soil. Knowledge of millipede distribution in Britain is poor so no comprehensive picture can be provided of those colonizing urban commons. They come in at about the same time as woodlice and may quickly build up substantial populations (Table 6.3). For example the flat-backed millipede *Polymicrodon polydesmoides* is often abundant on sites four years old as long as they are neither too dry nor too waterlogged. As succession proceeds and humus builds up two widespread snake millipedes *Tachypodoiulus niger* and *Cylindroiulus punctatus* are frequently found.

While millipedes are slow-moving scavengers, centipedes are relatively speedy predators, but as they both require a damp atmosphere and dislike the light they often lie-up together during the day under the same piece of brick, wood, rubbish or leaf litter. Many small invertebrates are taken as food such as worms, slugs, beetles, spiders and other centipedes but few millipedes are eaten, presumably because of their repellent secretions. The commonest species in brick rubble are appropriately the lithobiomorphs (stone dwellers) represented by *Lithobius forficatus, L. variegatus, L. melanops* and *Lamyctes fulvicornis*. Soil dwellers (geophilomorphs) are more difficult to locate but include the blind *Haplophilus subterraneus* and *Geophilus insculptus*. These species are unlikely all to occur on the same site but they, together with *Cryptops hortensis*, are the typical species of urban wasteland in South Yorkshire. As with woodlice, slugs and millipedes, a number are introduced so tend to be culture favoured and even some of the native species become synanthropic at the edge of their range.

Spiders

Almost nothing has been written about the spiders of urban wasteland so the 152 specimens collected during the quadrat survey of brick rubble form the basis of this account. Table 6.3 shows that after woodlice they are the most abundant group of larger invertebrates in this habitat. All but one of the 13 species collected from sites less than a year old were money spiders (*Linyphiidae*). Ballooning or aeronauting is their favoured method of dispersal. It involves letting out strands of silk which act as a kite giving them sufficient lift to carry them thousands of feet into the air and over great distances; this rapid and efficient method of dispersal must account for their regular presence on sites only recently vacated by bulldozers. Of the dozen species of money spider recorded from these young sites *Centromerita bicolor, Erigone atra, E. dentipalpis, Lepthyphantes tenuis* and *Oedothorax apicatus* were present in sufficient numbers to suggest viable populations. The latter species is rather uncommon, being found especially in shingle by rivers so the disturbed, coarse texture of rubble soil may have some similarities to that habitat. Other species such as *Lepthyphantes leprosus, L. nebulosus* and probably also *Lessertia dentichelis* are synanthropic spiders of homes, gardens, rubbish tips and sewage culverts so presumably maintain large mobile populations in towns. Visse and Van Wingerden (1982) sampled aerially dispersing spiders on a roof in the Hague over a three month period. Of the 314 specimens captured 84% belonged to the Linyphiidae of which 67% were *Erigone* spp., chiefly *E. atra* and *E. dentipalpis*.

After 4 years the assemblage of spiders has widened considerably and new groups, for example the crab spider *Xysticus cristatus*, are well represented. Members of this family (Thomisidae) have the first two pairs of legs longer and more robust than the others, these are held wide in a crab-like stance while the spider waits to ambush prey (Fig. 6.10). Under stones and penetrating quite deeply into the loose rubble are large long-legged *Tegenaria agrestis*. This spider, which spins a sheet web with tubular retreat to one side, is a typical species of waste ground and rubble as the porous substrate suits its requirements. It is also abundant along railway tracks which may form important dispersal corridors. In South Yorkshire, close to its northern limit, the species is somewhat synanthropic. Money spiders are still the dominant group. Though it is probable that most have aeronauted in, certain members, such as *Lepthyphantes nebulosus* and *L. leprosus* may have originated from the

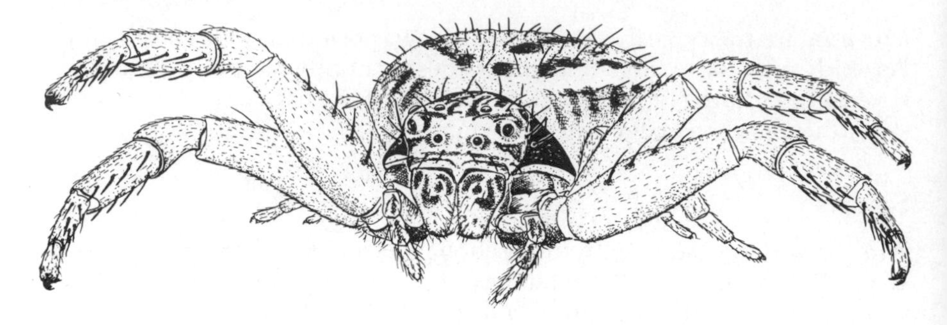

Fig. 6.10. The crab spider (*Xysticus cristatus*, female, 7 mm) is well represented on 4-year-old urban commons (Roberts and Roberts, 1985).

tipping of household or garden refuse. As with the flora, where species from widely different habitats occur together on urban wasteland, the spider fauna contains a mixture of ecological types. *Amaurobius fenestralis* a species most frequently recorded in woodland and *Helophora insignis* which is often associated with dog's mercury were both found on a 4-year-old site.

After 15 years of plant colonization the structure of the habitat has changed again and there are opportunities for a wider range of arachnids. Taller patches of vegetation are exploited by orb web spinners (*Meta mengei, M. segmentata*) while others live in dead flowerheads (*Dictyna uncinata*), but higher levels of the vegetation were not sampled systematically or many more species would have been recorded. Wolf spiders such as *Trochosa ruricola* which run after their prey at ground level and were once thought to hunt in packs are also present, as are the common *Clubiona* species which hide in plant debris during the day emerging to hunt at night. By now hunting spiders are frequent including jumping spiders such as *Salticus scenicus* (the zebra spider). This species has exceedingly acute vision provided by eight eyes, the middle pair of which have a greater range of movement than our own, elaborate focusing and binocular vision. They stalk their prey making a final leap to complete the capture. The large cribellate spider *Amaurobius ferox* is ideally suited to older rubble sites. This large species spins a web surrounding its retreat under a brick. The succession of spiders occurring on urban wasteland offers rich opportunities for further work.

6.3 RESEARCH OPPORTUNITIES

Urban commons offer a convenient habitat in which to investigate dynamic aspects of animal ecology; they are best regarded as supporting primary successions. The rate at which groups find and colonize sites is remarkably fast but there are differences so the changing composition of pioneer communities can be studied. For example, the hypothesis that rates of colonization by external chewing and sucking herbivores is higher than those shown by endophagous gall-formers and leaf miners could be tested. The balance of use, by phytophagous insects, of introduced and native plant species can be investigated with profiles of users constructed for the dominants. Many new inter-relationships are likely to be discovered among the unusual mixtures of plants and animals present. Novel methods of dispersal for invertebrates may be important on both a citywide and local scale with the tipping of garden refuse and surplus fill being particularly important for less mobile groups. The extremely rapid invasion by centipedes and millipedes suggests that they may be wind dispersed. Opportunities to examine factors such as isolation and chance, and particularly interactions between synanthropic and native species are legion. For example do culture-favoured animals decline as succession proceeds?

Chapter 7

INDUSTRIAL AREAS

It is difficult to do justice to the mosaic of habitats centred on industrial areas; this is partly due to restrictions over access which has attenuated the literature in all but the field of storage pests. Industrial areas are defined here as the manufacturing districts of towns so does not include the winning of raw materials, water supply, and only occasionally the energy industry. The industrial cycle normally starts with the acquisition of raw materials, frequently from abroad, and these are a well-known source of alien species. The methods by which some of these get dispersed *en route* to the factory are described in the chapters on railways, roads and canals, but industrial premises or rubbish dumps associated with them, are the major sites where they can be found. The buildings, processing plants and yards where manufacturing occurs provide a hard, largely abiotic and inhospitable landscape but there are a few birds, mammals, insects and plants that have adapted to life in this busy noisy disturbed environment.

Much energy is expended by heavy industry and this leads to the creation of unusual habitats. Permanently warm buildings and waste water may provide exactly the conditions required by alien plants and animals introduced with the raw materials, while the accumulation of fuel residues such as pulverized fuel ash, spent oxide and cinders create other specialized habitats. The throughput of raw materials often results in the production of further waste substrates such as alkaline materials, steel slag, cellulose fibres, woollen cleanings and sodium chromate which frequently accumulate in sites close to the works. Some are toxic. In Britain 23 million tons of general industrial waste are produced each year. Finally, storage of both raw materials and the

finished product can attract a wide range of pests if organic mate
involved. These are not peculiar to industrial premises, but m
turers object strongly to their presence and make organized ef
control them, hence their ecology is well understood. Several hundreds of species of insect, mite, spider and rodent are involved and are added to by a continual stream from abroad where standards of control are low.

The above five phenomena – raw materials, inhospitable processing complexes, waste heat, waste materials and storage – help to make industrial areas distinctive; it is impossible to generalize further. Additional habitats may be present such as lagoons, derelict buildings and even cranes and pylons provide unusual niches for nesting birds. The effects of industry may spread well beyond the factory fence especially if water courses (Chapter 15), air pollution (Chapter 3) or railways (Chapter 8) are involved.

7.1 RAW MATERIALS

One of the best known examples of species being introduced with raw materials involves the woollen industry. Refuse from the wool combing sheds, known as 'grey shoddy', is full of seeds imported embedded in fleeces which originate from areas such as Australasia, Africa, South America and the Mediterranean. Wool aliens were first noticed early this century on the banks of the river Tweed at Galashiels where woollen mills discharged untreated effluent into the river (Hayward and Druce, 1919). These discoveries stimulated a group of Yorkshire botanists to investigate the heavy woollen industry around Bradford; they discovered wool aliens at Frizenhall Sewage Farm where the urban drains discharged (Lees, 1941). Later, following the example of Dony (1953), wool aliens were hunted down all over the country and by 1960 (Lousley, 1961) had been found in the vicinity of most woollen mills. They are often concentrated around the backs of the buildings, in odd corners, below windows, anywhere cleanings and sweepings get dumped. At that time 529 spp. were known, a total which includes records from railway sidings and sites where the shoddy had been used as a manure on arable fields; the list is still growing.

As might be expected very few of the plants from warm climates, many of which are annuals, establish; they rely for their 'persistence' on repeated reintroduction. As a broad generalization 70% originate from Australia and 20% from Africa. A number of common native British plants, e.g. *Spergularia rubra, Dactylis glomerata, Holcus lanatus* and *Phleum pratense* are frequently introduced with the wool so the genetic base of these species in industrial areas may be unusually broad. Other species such as *Erodium moschatum* and *Medicago polymorpha* are native near the sea but naturalized wool aliens in towns. The pirri-pirri bur (*Acaena novaezelandica*) is one of the few Australian wool aliens to have established successfully in Britain – huge colonies occur on the banks of the Tweed, though at sites in Cornwall, Dorset and Ireland it originated as a garden cast-out.

For the manufacture of high-quality paper some firms use waste cotton rags from Africa. Dust and seeds beaten out of them in the early processing stages, when dumped, yield an array of Egyptian and North African plants. The tip of the Crown Wallpaper Company at Darwen, Lancashire produced over 100 aliens (Savidge *et al.*, 1963) in the early 1960s. Most of the species on such waste tips need competition-free conditions, the communities differing from year to year depending on the origin of the raw materials and the climatic conditions. Early warm moist periods may favour the growth of a lush native flora which swamps the alien assemblage, but a change to dry warmer conditions allows the continental and subtropical elements to develop into a sizeable community. Edaphic factors appear less important since at the same site different communities appear in different years on the same type of substrate.

The changing face of transport and industry results in changing opportunities for aliens. Last century ballast hills at ports and along the Mersey were producing a lot of unfamiliar plants (Storrie, 1886; Hogg, 1867). Material taken inland for the repair of railways or as land fill resulted in the spread of plants such as *Chaenorhinum minus* and *Senecio viscosus*. Today the oil-milling industry (oilseed rape, soya bean, castor-oil) has its own adventive flora of at least 85 foreign plants (Palmer, 1984). Many, being North American in origin, are hardier than other groups of aliens and a number have attempted to establish themselves in Kent, e.g. *Amaranthus paniculatus, Helianthus annuus, Artemisia biennis, Hordeum jubatum*. Annual lupin seed is currently being imported for crop evaluation trials as soya bean substitutes; at least two species have been recorded widely at Bristol docks where seed spills off conveyor belts, and so the changes continue in response to our shifting economy.

Most industries have contributed to the urban flora. *Cardaminopsis arenosa* was introduced to London with Swedish iron-ore; it persisted for at least 14 years. Imported timber often has seeds of balsam (*Impatiens* spp.) lodged in crevices. Grain aliens are still frequent though they were an even more important group in the days before herbicides and efficient seed cleaning. Around flour mills and breweries very large numbers were recorded, for example, 267 spp. from the Bass and Worthington Brewery at Burton on Trent (Curtis, 1931; Burges, 1946). Today breweries tend to import their grain ready malted. The bastard cabbage (*Rapistrum rugosum*) is a frequent grain alien thoroughly established on the east coast around Suffolk flour mills but elswhere in Britain behaves as a casual, while the warty-cabbage (*Bunias orientalis*) is more firmly entrenched and can stand competition from the native flora. Skins, hides and furs bring in species with hooks or spines on their seeds and fruits. Long lists have come from tanneries in Leeds (Lees, 1941) and more recently from Cornwall where the densely pubescent cupules of Valonia Oak, imported from Turkey, are crushed and mixed with oak bark for the extraction of tannins used in the production of shoe leather. Where the spent bark is dumped large crops of aliens grow up (Grenfell, 1983, 1985).

In reviewing the means by which foreign plants are brought into Britain, Lousley (1953) concluded that the very great majority came as seed accidentally included in cargoes. They then leave a trail from port or airport, along railway, road or canal to the industrial areas of Britain. Most survive for only very short periods but they faithfully reflect changes in our imports, countries of origin and advances in technology. This latter factor is responsible for a sharp fall in the abundance and variety of most industrial raw material aliens over the last 20 or 30 years. Table 7.1 shows a classification of alien plants into three categories according to their degree of persistence in the flora. Many industrial sites, including their service areas, railway depots and docks, carry more established aliens and casuals than any other habitats in urban areas, or indeed elsewhere in the British Isles. Their number is often underestimated due to foreign genotypes of well-known native species passing unrecognized.

Table 7.1. A classification of plants according to their origin and degree of persistence in the flora

Native	Species that have arrived in the studied area by natural means without intervention, even unintentional, by man, from a source where the plant is native.
Alien	Species believed to have been introduced by the intentional or unintentional agency of man.
	(1) Naturalized aliens (neophytes – new citizens). Introduced species which are naturalized in natural or seminatural habitats.
	(2) Established aliens (epoekophytes). Introduced species which are established only in man-made habitats.
	(3) Casuals (ephemerophytes). Introduced species which are uncertain in place or persistence.

Note: Species may move up or down on the scale, have a different status in different parts of a site, or deserve a multiple entry if native and introduced material are both present.

7.2 BUILDINGS, STRUCTURES, HARD LANDSCAPE

Vegetation

Open space between the factory buildings tends to be hard surfaced and in constant use for storage, circulation, loading, parking, temporary huts, skips and so on. There is little opportunity for plants to colonize but at the base of buildings and in odd corners populations of mainly short-lived common urban weeds can be found. The specialists of heavy industry tend to be yellow-flowered Cruciferae such as *Diplotaxis tenuifolia, D. muralis, Erucastrum gallicum, Sisymbrium orientale, S. altissimum, S. officinale* and *Hirschfeldia incana* together with *Melilotus alba, M. officinalis* and *Epilobium* spp. The dryness of the habitat, shallow soil and possibly enhanced temperatures encourage several thermophilous grasses, e.g. *Hordeum murinum* and *Vulpia* spp. which can become very abundant. Almost any species may turn up; I have seen patches of loose material dominated by herb-robert, broad-leaved everlasting pea (*Lathyrus latifolius*) or wall-pepper (*Sedum acre*), while goat willow and buddleia may form small thickets. Casuals tend to be concentrated in a few places such as around loading bays, conveyor belts, or at rubbish disposal sites round the back of the works, though Palmer (1984) mentions them being dispersed by wind at an oil-milling plant.

Birds

Factory areas are not particularly promising bird habitats unless they include shrubberies, rough grassland or trees. Feral pigeons, house sparrows and starlings are able to cope with the noise and disturbance finding many breeding sites in among the buildings. Small advantages may accrue, for example house sparrows have been known to feed throughout the night in flood-lit areas gathering where the workers take their tea breaks, and feral pigeons may build nests entirely of galvanized wire (Fig. 7.1). A few blackbirds and perhaps a pair of pied wagtails may also be present.

One bird has come to be associated with industrial areas in Britain; this is the black redstart (*Phoenicurus ochruros*). A survey of its numbers, distribution and breeding habits carried out in 1977 (Morgan and Glue, 1981) found that the typical black redstart chooses to nest in the industrial complexes of large urban areas. In London (33 pairs) they were most commonly reported from gas works, power stations and buildings associated with dockland. Outside London (71 pairs) docks and warehouses were important together with town centres, power stations and industrial sites in the Midlands. Williamson (1975) considers that the birds consolidation in the London area from 1940 was

Fig. 7.1. This pigeon's nest, constructed entirely of wire, was found in the rafters of a wire-making factory in Sheffield (I. Smith).

due not only to the plentiful nest sites caused by bombing but also the microclimatic effect of the city environment. By origin the black redstart is a mountain and cliff bird so uses tall structures as it would a precipice for breeding on but close enough to areas where it can forage for insects without undue disturbance from people, cats or rival insectivores like the robin. It is one of the few songbirds of heavily built-up areas. Kestrels also regularly nest high on industrial buildings where they are as at home as on the ledges of natural cliffs. The same is true of feral pigeons which are descended from cliff-nesting ancestors. These three birds have very different foraging requirements.

Surveys of towns that have been invaded by magpies (Sheffield: Roberts, 1977; Manchester: Tatner, 1982) have demonstrated that they

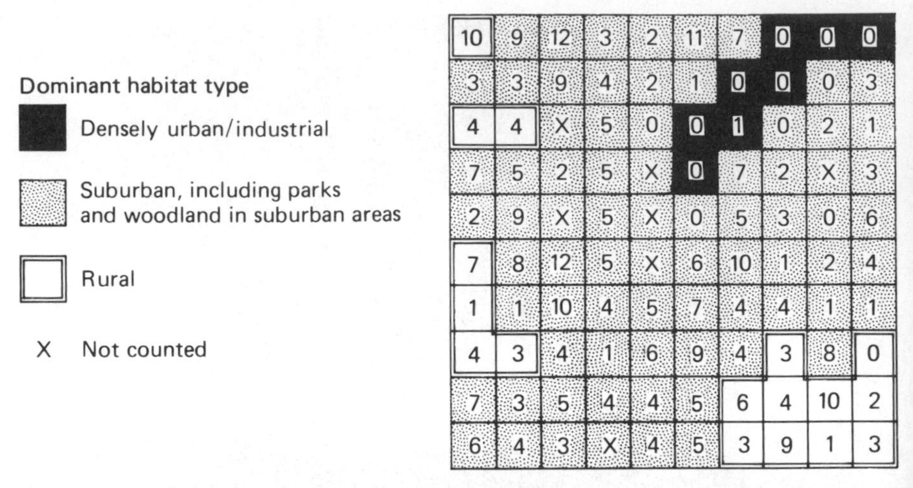

Fig. 7.2. The number of occupied magpie nests located in each 1 km^2 in Sheffield during 1976 (Roberts, 1977).

are abundant in the suburbs, their occurrence showing a good correlation with the number and variety of trees. The Sheffield survey (Fig. 7.2) revealed that the only area where they were not present, or present only at very low densities, was in the industrial zone of the city (including artisan housing) in which trees were virtually absent. Since 1976 the magpie population of Sheffield has doubled from 4.1 pairs per 1 km^2 to about 8 pairs per 1 km^2 in 1986, a phenomenon that has precipitated an invasion of the hitherto inhospitable industrial area (Clarkson and Birkhead, 1987). While in Manchester only a single nest was located in a man-made structure, a crane jib, in Sheffield the birds have invaded the industrial zone by nesting in metal pylons and watch towers (Fig. 7.3). So we are seeing a rapid and interesting example of the evolution of a bird's relationship to man. First a reduction in persecution following two world wars (Parslow, 1967) allowed the magpie to increase in numbers sufficiently to commence a range extension. An increase in numbers is likely to have been a prerequisite for this extension as the species is sedentary in Britain (Holyoak, 1971). Its success in the urban environment is thought to be due to an unspecialized diet throughout its life cycle, the maturation of postwar tree planting which increases the availability of nest sites, and importantly, the lack of persecution has reduced the birds' fear of man allowing it to feed as an opportunist on all kinds of rubbish and bird-table food. They have even started breaking open egg cartons left on doorsteps. Crows have also been recorded nesting on industrial structures so they appear to be following the magpie in becoming adapted to an urban existence following a population increase.

Fig. 7.3. Increasing population pressure due to an absence of persecution, coupled with the development of green space, has caused a behavioural change in urban magpies; they have started to colonize treeless areas of certain towns by nesting in pylons and metal watchtowers (Sheffield; T. Birkhead).

7.3 ENERGY ASPECTS

Manufacturing industries require large amounts of energy; indeed a sizeable part of industry is involved with supplying it. The environmental consequences of this are the production of heat, the release of air pollution and the accumulation of solid wastes; all have a gross and demonstrable effect on the wildlife of industrial areas.

Energy-conservation measures are reducing the amount of high-grade heat released into the environment. Its effects are dealt with in

Chapter 15 which describes how a tropical fauna and flora occurs in certain canals where water used to cool machinery is discharged. Thermal pollution of the River Don by the steel industry has enabled wild figs (*Ficus carica*) to colonize its banks, and at a more general level heated discharges from electrical power stations cause subtle changes in the balance between species. Inside buildings, heat allows pests and other visitors from warmer climates to maintain permanent populations and it is possible that thermophilous terrestrial plants such as wall barley (*Hordeum murinum*) are able to extend their northern limits in the environs of heavy industry. The modern trend is to utilize waste heat for fish farming, indoor horticulture and space heating; the days of releasing it into the environment where it enabled bizarre communities to flourish are numbered. The general effects of air pollution on wildlife have been dealt with (Chapter 3) but a further example particularly related to industry will be recounted. It involves the effect on woodland ground beetles of heavy metals lifted in thermal plumes from the Avonmouth Smelter, Bristol, then deposited on the surrounding area (Read *et al.*, 1987). Woodlands downwind of the works have, owing to reduced microbial activity, greatly increased litter depths within which communities of ground beetles hunt their prey. In this study no gross effects of pollution on the beetle populations were identified but differences relating to their ecology and life histories were. In areas of high pollution there is a decrease in the numbers of most species but an increase in the abundance of a few, e.g. *Nebria brevicollis* (Fig. 7.4); there is a parallel here with lichens where new dominants are selected under

Fig. 7.4. The ground beetle *Nebria brevicollis* shows increased abundance in woods downwind of the Avonmouth Smelter, Bristol.

pollution stress. Later dates of peak capture were found in sites with higher metal concentrations where in addition more individual Carabidae are autumn breeders. One conclusion was that due to the scarcity of prey, beetles which are active over long periods of the year, and particularly those that have plastic activity seasons, may have an advantage in areas of high pollution. Most pollution studies miss the subtle effects reported here which it is conjectured must be widespread round industrial sites.

7.4 STORES AND WAREHOUSES

Many animals that are pests inside buildings reach their greatest abundance in factories, warehouses and other industrial premises. The conditions that favour them are ample hiding places where large populations can develop unnoticed and which provide protection from attempts to exterminate them, plentiful food and a constant temperature. Pests can be grouped into 'residents' which live and breed in their food source (stored products), these include many beetles, moths and mites, and 'visitors' which live in the structure of the building such as ducts, pipe-runs and hollow walls and make journeys to visit the food store. This group comprises cockroaches, ants and crickets. One of the main limiting factors for resident pests, whether they inhabit huge silos, warehouses or domestic scale properties is the rather low moisture content of the habitat, e.g. wheat 11% humidity, cocoa beans 6–7% and malt barley 2–3%. This means that the animals best able to thrive are those adapted to rather dry environments. Many pests are typically associated with a particular type of product which must reflect their dietary preference when living under natural conditions in their country of origin. Busvine (1980) suggests the following ancestral feeding habits. Grain pests originally attacked small seeds while pulse pests attacked larger seeds. Flour and ground cereal pests lived on seeds damaged by other insects, animal hide species originally attacked dried corpses, while a miscellaneous group possibly derived from scavengers.

Many of these pests are not known in the wild and the regions they came from can only be guessed at from a knowledge of their temperature requirements. Some are found in the food caches of social Hymenoptera and rodents which must be among their natural habitats. Others infest grains or pulses in the standing crop as well as in stores or oviposit on dry seeds which their larvae then eat (Solomon, 1965). Fabric pests, the larvae of which feed on wool, fur, feathers or hides, probably originated in the following way. Before the advent of man, certain small moths and beetles found a vacant niche by acting as scavengers on the more indigestible parts of corpses – the fur and feathers which were neglected by other animals. They may also have bred in disused birds nests and animal lairs. When the opportunity arose they turned their attention to stored skins, furs, feathers, furnishings and cloths in warehouses or homes and flourished as never before. The insects concerned are clothes moths (*Tineidae*), house moths (*Oecophoridae*) and carpet and hide beetles (*Dermestidae*). All can digest keratin though washed material does not offer sufficient amino acids and vitamin B for them to thrive so only soiled goods are attacked. Their temperature preferences tend to be rather low. When infestations are bad, predators such as larvae of the window fly (*Scenopinus fenestralis*), various hymenopterous parasites and a virus disease, act to control the outbreak. Losses due to moth damage amount to many millions of pounds a year.

Hide beetles are compact, oval to nearly round, mostly rather small insects. They are common in hide, skin and carpet warehouses, bone factories and dog biscuit stores. It is the larval stage known as woolly bears that do the serious damage. Emerging adults of certain species, e.g. varied carpet beetle (*Anthrenus verbasci*) and two-spot carpet beetle (*Attagenus pellio*), fly to windows and escape out of doors where they congregate on flowers feeding on pollen and nectar, mate, then in late summer re-enter the buildings. By contrast adults of *Dermestes* spp. feed on the same substrates eaten by the larvae. Great losses are incurred by the hide and fur trade due to *Dermestes*, while carpets, clothes, leather, silk and dried meat are seriously attacked by the entire range.

The best known 'visiting' pest to food stuffs is the cockroach. Most of our ten reasonably common species originated from Africa and have spread through commerce penetrating cool regions by living inside warm buildings. An extensive survey in 1964–66 conducted by Rentokil Ltd. showed that the Oriental cockroach (*Blatta orientalis*) is about four times as common as the German cockroach (*B. germanica*) overall (Cornwell, 1968), though there is evidence that due to changes in building construction the ratio is now changing in favour of the latter species. Each has its own ecological preference and geographical distribution in Britain with, for example, *B. germanica* more common in London and the south-east. Temperature preference partly explains this with, for example, *B. germanica* choosing 25–30° C while the more northern *B. orientalis* selects 20–29° C.

Storage pests mainly keep hidden being dispersed with the infested product but a few can be observed outside. An inspection of tree trunks in the city centre will often reveal clothes moths and house moths at rest; similarly a search of flowers, especially *Spiraea* in industrial areas often discloses carpet beetles (*Attagenus, Anthrenus*).

7.5 SOLID WASTE MATERIAL

Solid wastes accumulate in industrial areas as a result of mining, energy production, and particularly as a by-product of industrial processes. Many wastes are comparatively benign materials such as cinders, concrete, steel smelter slag, pulverized fuel ash and various forms of lime, while others are highly toxic, for example chromate waste, spent iron oxide (used for cleaning flue gases) and heavy metal leftovers. Proper planning often eliminates these wastes from the landscape by ensuring that they are disposed of in holes which are later topsoiled, and increasingly, uniform wastes are being used as a raw material by other industries. Some old tips have been quarried for this purpose. For reasons such as these it is often the sites of past industry that provide the chief source of unusual substrates in urban areas.

Botanists in north-west England (Lee and Greenwood, 1976; Greenwood and Gemmell, 1978) were among the first to realize that certain types of industrial waste provide important calcareous plant habitats. Table 7.2 details nine such materials found in Greater Manchester and

Table 7.2. Types of industrial waste forming calcareous plant habitats in Greater Manchester and Cheshire

Material and origin	*Initial pH*
Leblanc process alkali waste from manufacture of sodium carbonate	Up to 12.7
Blast furnace slag from smelting of iron ores	Up to 10.6
Pulverized fuel ash from coal-burning power stations	Up to 9.5
Solvay process; recovered lime waste	Up to 8.7
Calcium carbonate slurry from chemical works	Up to 8.6
Calcareous boiler ash	Up to 8.2
Colliery washery waste	Up to 8.0
Calcareous colliery spoil	Up to 8.0
Demolition rubble	Up to 7.8

Taken from Gemmell (1982).

Cheshire. The most important is lime from the obsolete Leblanc process which weathers to produce a calcareous soil deficient in nitrogen and available phosphorus. This is colonized by a wide range of calcicoles including *Blackstonia perfoliata, Carex flacca, Carlina vulgaris, Centaurium erythraea, Erigeron acer* and *Linum catharticum* together with the common spotted orchid (*Dactylorhiza fuchsii*), marsh orchid (*D. incarnata*), northern fen orchid (*D. purpurella*) and fragrant orchid (*Gymnadenia conopsea*). The closest natural counterparts of this vegetation are calcareous dune slacks, though garden escapes such as Michaelmas daisy, golden-rod and *Sisyrinchium bermudiana* indicate the urban connection. Lime beds associated with the Solvay process support a similar flora.

Compared with Leblanc waste, pulverized fuel ash (PFA) from power stations provides a more recent and widespread habitat. Chemically the two substrates differ with Leblanc waste having a much higher lime content (40–60% $CaCO_3$) than PFA (2–3% $CaCO_3$); this accounts for the absence of strict calcicoles on the ash and for its more rapid colonization. After 16 years it can carry a dense mosaic of communities including willow scrub, marsh and grassland in which huge stands of fen orchid (*Dactylorhiza praetermissa*) and marsh helleborine (*Epipactis palustris*) occur alongside the more ruderal dominants coltsfoot, horse-tail, rosebay willowherb and Yorkshire-fog (Gemmell, 1982). Cement-like layers may form in the ash, reducing permeability and favouring species such as the reed *Phragmites australis*. Between 2 and 5% of the PFA is water soluble which gives very alkaline saline initial conditions so the early natural colonizers include halophytic species of *Atriplex* and *Chenopodium*; they also need to be tolerant of high boron.

Of the 22 sites containing large populations of marsh orchids (*Dactylorhiza* spp.) in and around Greater Manchester 20 are industrial habitats of recent origin which are clearly providing opportunities for locally rare species. At the same time the presence of hybrid swarms suggests that the new habitats are contributing to the breakdown of isolating mechanisms between species. Kelcey (1975), Davis (1976) and Johnson (1978) provide additional evidence that industrial waste tips have led to habitat diversification in heavily urbanized areas.

7.6 MODERN TRENDS

While areas of heavy industry founded prior to World War I possess many of the characteristics described in the previous pages, more recent enterprises are grouped in trading estates and have a less distinctive ecological identity. In fact all industry, as it gets tidied up, becomes less interesting for wildlife. The very features that allow a rich flora and fauna to survive – the squalor, rubbish, old buildings and machinery, derelict huts, rotting dumps, inefficient handling – are becoming unacceptable to management. The evolution of post-Victorian industrial landscapes was first towards the utilitarian trading estates of the 1920s/1930s (Slough, Trafford Park, Team Valley, Hillington) where industries were grouped into single large complexes but the advantages of handy power, labour, transport and waste disposal rarely extended to much in the way of green space except in the 'garden' factories of Welwyn and

Fig. 7.5. During the period 1920–1950 industrial sites were frequently planted up with large numbers of the hybrid black poplar (*Populus* × *canadensis* 'Serotina').

Letchworth. Any tree planting was unimaginative and usually involved large numbers of hybrid black poplar (Fig. 7.5). During the 1950–80 period, industrial estates on flat sites with single-storey buildings laid out in a grid pattern were in vogue and increasingly the units were set in a sea of mown grass interrupted by avenues, belts and clumps of trees. The most recent trend follows the American fashion for industrial parks, where there are shared facilities and the buildings are arranged in organic layouts interspersed with sports pitches, woodland, lakes and occasionally wildlife conservation areas.

Modern industrial enterprises are characterized by containerized raw materials, efficient use of heat including recycling, the bagging, skipping or incineration of waste materials and the buildings tend to be plain-clad metal frame structures offering few opportunities either inside or out for synanthropic wildlife. The green surroundings, which can occupy over 50% of the site, support a moderately rich fauna and flora but it is not particularly urban and may include countryside species such as rooks and moles. Instead of being surrounded by workers' housing, the modern industrial estate is enclosed by car parks. It is little wonder that even 15 years ago leading alien plant botanists observed that docksides, railway yards and industrial premises were hardly worth visiting. Nor have these new habitats attracted cliff-nesting birds like kestrels, black redstart or kittiwake. Fortunately for urban wildlife the ideal of perfect hygiene, efficiency and uniformity of product is rarely achieved. Bags burst, containers leak, strong winds whip seed off conveyor belts, approved methods of waste disposal are not always followed and the result is a persistent scatter of alien plants round many industries using organic raw materials. Though these do not equal the rich hauls of yesteryear, unusual plants are still found. Grenfell (1986), for example, has reported an abundance of aliens at Avonmouth docks.

Among birds, opportunists can be sensitive monitors of technological change or its breakdown. Coombs *et al.* (1981) studied the behaviour of collared doves (*Streptopelia decaocto*) over a period of 6 years at an industrial site in Manchester. At Ellesmere Port the number of doves feeding round two grain mills and a grain-storage warehouse were drastically reduced in early 1972 (Fig. 7.6) by modernization schemes which replaced most of the older facilities by a container base. This reduced the amount of handling and therefore spillage. However, later that year numbers temporarily recovered when an extractor fan accidentally blew grain and dust on to a roof, but by May 1974 grain

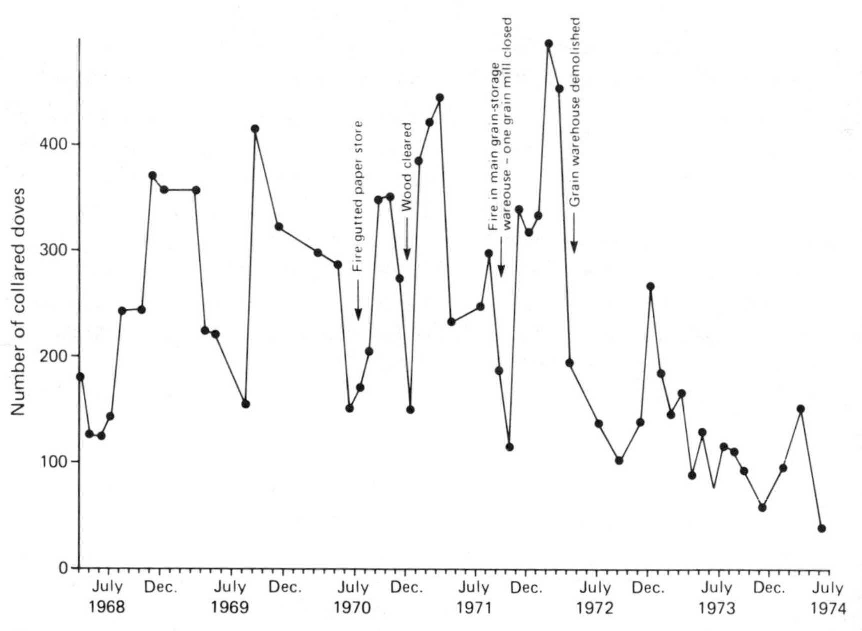

Fig. 7.6. Number of collared doves counted each month round two grain mills and a grain-storage warehouse that were modernized in 1972 (Coombs *et al.*, 1981).

spillage was very low and so were the numbers of collared doves. Feral pigeon numbers were similarly affected. This is a clear cut example of how modernization can affect wildlife. A subsidiary finding was that superimposed on food availability other factors such as level of disturbance also affected bird numbers.

The history of the collared dove in Britain illustrates the dynamic nature of urban assemblages on a longer timescale. When the *Checklist of the Birds of Great Britain and Ireland* was published in 1952 it was not recorded even as a vagrant. Soon after its arrival in 1955 it was specially protected under the *Protection of Birds Act 1954*. In 1967 this was replaced by normal protection; then by 1977 the collared dove had become so abundant that it was transferred to Schedule 2 allowing its killing or capture as it was a pest. Phenomena like this make it difficult to forecast which other birds will have joined the black redstart, kestrel, little ringed plover and collared dove as breeding species of industrial zones 40 years from now.

Chapter 8

RAILWAYS

Railway land occupies large areas in the heart of cities. In addition to mainlines and stations, extensive areas of sidings and freight depots are present, together with much masonry and brickwork; particularly good for wildlife are the awkward V-shapes and triangles created where lines converge. It has been calculated that there are about 4000 km (2480 miles) of active track in urban areas to which must be added sidings and an unknown length of disused line. Owing to access difficulties, railway property is a poorly known habitat though a walking permit can sometimes be obtained for non-electrified stretches of line. The Nature Conservancy Council sponsored a major survey of Britain's railway vegetation which took place between 1977 and 1981 (Sargent, 1984); for safety reasons it was restricted to the 14 000 km (8700 miles) of rural and semirural track. Many of the questions it asked and some of the answers have relevance to urban lines, but had these been included they would have been identified as highly distinctive and floristically the most interesting part of the railway network.

Many of the early studies in railway botany focused on adventive plants. In describing the flora of Thalkirchen Station (near Munich), Kreuzpointner (1876) provided one of the first accounts of the introduction of alien species by rail traffic; his list of 84 taxa associated with the transport of grain is dominated by annuals and biennials with Cruciferae (19 spp.) and Compositae (22 spp.) particularly well represented. Thellung (1905) showed that a large proportion of introductions into Switzerland were also associated with railways which at that time carried very large volumes of freight. Taking a different approach, Lehmann (1895) working in Latvia found that he was able to identify the origin of track ballasting materials from the alien plants he found growing on them. In Britain most County Floras cite records from railway habitats but only rarely have they been systematically studied. Two shining exceptions are Dony's (1955) account of the railway flora in Bedfordshire and that by Messenger (1968) dealing with Rutland. Both papers are analytical so, even though the counties are largely rural, important principles emerge. Dony provides a glimpse of the flora before spraying the tracks became standard procedure and was also the first person to realize the importance of railway platforms for ferns.

The railway provides a number of sharply contrasting habitats: the permanent way, sidings, embankments, cuttings, flats, masonry and brickwork. In urban areas sidings, walls, buildings, bridges and cuttings are particularly well represented. After churches and canals they are among the oldest structures to be found in towns, the majority having originated between 1840 and 1870. Until about 1960 track maintenance was carried out by small 'length' gangs of five or six men responsible for around 8 km of line; they cut back trees, bushes and brambles, hoed the ballast clean and fired the whole property; it was a matter of pride that each ganger's 'length' should be better maintained than the next. In the early sixties reorganization resulted in less attention being paid to vegetation management, and this trend has continued. So today little is carried out beyond that which is essential for safety. Currently, most management operations are performed by rail-mounted machinery such as flail mowers fitted to booms or specially designed spray trains. British Rail (BR) retains the right to use fire but rarely resorts to this traditional management tool and almost never in urban areas. Changes to this minimal maintenance programme are unlikely owing to lack of finance and a dwindling knowledge of vegetation among railway staff.

8.1 THE PERMANENT WAY

The permanent way is the name given by railway men to the ballasted road bed on which the track is laid. Its structure is illustrated in Fig. 8.1. A 30–50 cm-thick layer of cinder ballast, the cess, is laid on a slightly cambered base and strips of a much coarser ballast (the track bed) composed of limestone, granite, steel furnace slag, basalt, etc. are put down to take the sleepers and rails. Once genuine ships ballast was used but now BR have their own quarries. The key to the construction of the permanent way is free drainage; this also determines its management. Every 10 years or so the ballast is mechanically cleaned by riddling to prevent a build up of humus, the resulting fines being tipped on the verge by conveyor belt where they influence plant growth. One of the main reasons for spraying the permanent way is to prevent weeds contributing to the build up of dirt. So successful are the spray mixtures

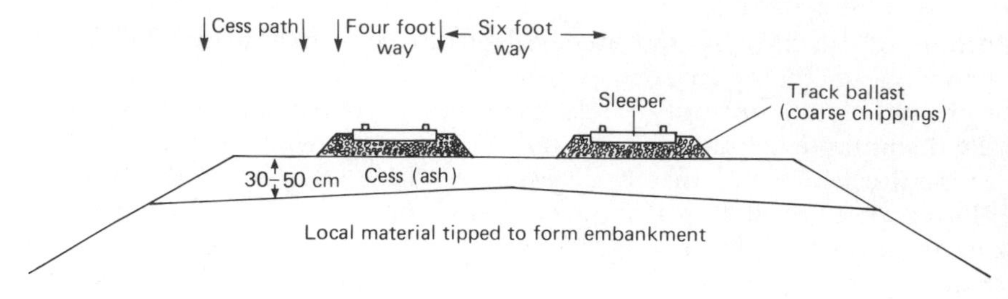

Fig. 8.1. Cross-section showing structure of the permanent way. On main lines the track ballast is deeper and extends across the six foot way.

applied in early summer each year that along mainlines very little grows on the permanent way or cess path. Altaver, produced by Chipman Ltd who spray under contract to BR, is an example of a modern railway herbicide. This persistent total weedkiller containing atrazine, 2,4-D, and sodium chlorate with a fire depressant is advertised as rapidly killing all established vegetation and persisting for over a year. For brushwood and scrub 'Garlon', which contains triclopyr, is an effective replacement for the banned 2,4,5-T. At least 14 chemical weedkillers are approved for use on BR land so a solution can be found to most problems.

Despite this formidable weaponry, a few groups of plants which show tolerance or possess avoidance mechanisms can be found on or adjacent to the permanent way, though it is rare to see anything green on the track ballast. Growing on cess at the margin of sprayed areas rhizomatous species such as bindweed, common horsetail, colts-foot, rosebay willowherb and sometimes St John's-wort (*Hypericum perforatum*) form dense pure stands as a result of their slightly greater resistance to the herbicide regime (Fig. 8.2). Horsetail is the most troublesome. Scramblers, such as bramble, ivy and creeping cinquefoil (*Potentilla reptans*), may also invade close to the track during a good growing season. Other plants I have noted which show some resistance to current sprays are yellow stone crop (*Sedum acre*), Canadian fleabane (*Conyza canadensis*), American willowherb (*Epilobium ciliatum*) and red valerian (*Centranthus ruber*). The bryophytes *Barbula convoluta, Bryum argenteum, Ceratodon purpureus, Funaria hygrometrica* and *Marchantia polymorpha* are more resistant than most higher plants but are still largely killed by the annual spray treatment.

Before the advent of herbicides on the railways (about 1932 but not general until rather later) the permanent way provided a singular very sharply drained habitat for plants. Dony (1955) and Salisbury (1961) describe the ballast community as being characterized by spring and early summer flowering annuals of dry soil. Among native species that

Fig. 8.2. The annual application of a broad-spectrum herbicide to this railway verge has favoured the development of perforate St John's-wort (*Hypericum perforatum*) which shows moderate resistance to the spray (I. Smith).

took to this habitat the most notable spread was made by the lesser toadflax (*Chaenorhinum minus*) previously an annual weed of calcareous soils, and in Ireland, by fine-leaved sandwort (*Minuartia hybrida*) a local species of dry ground. Since the community now finds its chief refuge in railway sidings, it will be fully treated in that section apart from two aspects which are most closely associated with the permanent way.

A group of track ballast species calling for special comment is an element with maritime affinities represented by sand sedge (*Carex arenaria*), dark green mouse-ear chickweed (*Cerastium atrovirens*), Danish scurvy-grass (*Cochlearia danica*) (Fig. 8.3), and strapwort (*Corrigiola litoralis*); their primary habitat is stony and sandy places near the sea. It is possible that some were introduced to the railway system in the days when ships' ballast was used in construction work. Alternatively in the days of steam trains sand was carried on the locomotives and scattered on the tracks to aid traction which could explain the origin of the seed or they may have spread by natural dispersal from their native haunts into this relatively new secondary habitat which has some of the characters of

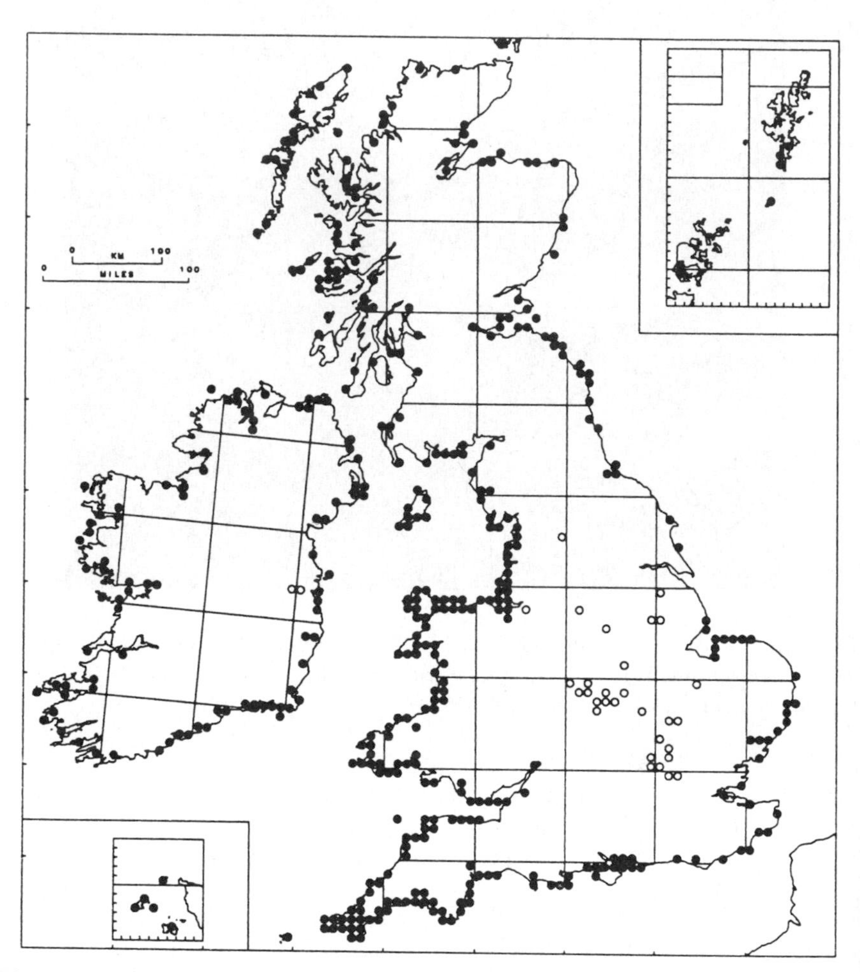

Fig. 8.3. The distribution of Danish scurvy-grass (*Cochlearia danica*) in Britain showing native records round the coast (black dots) and inland sites on railway property (open circles). Records from inland road verges omitted (map supplied by Institute of Terrestrial Ecology).

a shingle beach. Support for this latter theory is that many are particularly common on railway lines in seaside towns such as Swansea. No alien origin for the seed has been identified; they are perhaps best regarded as railway wanderers of uncertain origin.

The second aspect concerns adventives. According to continental botanists, who have concentrated on aliens, the permanent way of mainline tracks has always been relatively inferior to stations and sidings. This is mainly due to a lack of jolting and other operations which loosens seeds from the wagons. Holler (1883), working a stretch of line in Germany, noted a tendency for each side of a double track to support its own crop of aliens. In this instance they were mainly concentrated beside the track on which trains carrying grain shipments were routed. On the opposite side aliens also occurred but had a different origin. The preferred location was curves in the track. Such niceties would be difficult to observe on today's railways.

8.2 SIDINGS

Closures under the Beeching Plan reduced the length of track by about a third but had an even greater effect on freight stations which, between 1957 and 1967, were cut from 2352 to 729 and are down again due to further streamlining. While a number of unwanted urban sidings were sold to businesses such as timber, coal and scrap merchants or turned into car parks, many had the rails removed and were then left.

During goodsyard construction the Victorians usually laid the track directly on to cess so there is no 'track ballast'; consequently after asset stripping they appear as broad level cindery wastes. The main characteristic of this as a habitat relates to the substratum of firebox ash from the days of steam, which is a deep very freely draining coarse mineral soil low in humus and with a pH, when freshly disturbed, of 6.0–7.4 but capable of being leached over several decades to below 5. Most sidings are affected by low levels of disturbance associated with their use as access routes to adjacent mainlines. Operational sidings, now in the minority, are sprayed by BR personnel using knapsack equipment but not quite to the standard achieved by the mainline herbicide trains, so plants find an erratic home there especially in marginal areas used for storage or by vehicles. Whether active or abandoned, urban sidings provide an important refuge for the cess community as well as supporting successions and ecotone development up to 30 years old. Though most survey work has been concerned with vegetation, distinctive communities of animals have also been identified.

In this stressed habitat, summer drought, nutrient deficiency and moderate disturbance combine to favour small annuals and short-lived perennials. Table 8.1 (Section A) lists the more characteristic annual species found during a survey of this habitat in Sheffield while Section B gives additional species gleaned from the literature and my observations in other towns. The table is not intended to be exhaustive, only to provide the ecological flavour which is unique. The closest affinity is perhaps with 'yellow zone' sand dune vegetation but the two habitats are easily separated by the high proportion of aliens found on the cess. An advantage of studying one city in detail is that ecological and phytogeographical relationships can be more readily appreciated. The list from Sheffield is notable for containing a number of species such as Canadian fleabane (*Conyza canadensis*), blue fleabane (*Erigeron acer*), cudweed (*Filago vulgaris*) and hard poa (*Desmazeria rigidum*) which are more typical of warmer, dryer parts of the country but can penetrate marginally suitable climatic zones by exploiting this xerothermic habitat. Bermuda-grass (*Cynodon dactylon*) an adventive from warmer parts of the world, is doing the same on cess in the south of England.

Table 8.1. Annual plants characteristic of railway cess. Section A in the Sheffield area, Section B additional records from elsewhere in the country (not exhaustive)

Section A	*Section B*
Aira caryophyllea	*Apera interrupta*
Aira praecox	*Apera spica-venti*
Arabidopsis thaliana	*Cerastium pumilum*
Arenaria serpyllifolia (agg.)	*Cochlearia danica*
Cardamine hirsuta	*Dianthus armeria*
Cerastium atrovirens	*Diplotaxis muralis*
Cerastium glomeratum	*Erodium cicutarium* (agg.)
Chaenorhinum minus	*Geranium rotundifolium*
Conyza canadensis	*Legousia hybrida*
Corrigiola litoralis	*Linaria supina*
Desmazeria rigida	*Minuartia hybrida*
Erigeron acer	*Myosotis ramosissima*
Erophila verna	*Sherardia arvensis*
Filago vulgaris	*Trifolium campestre*
Myosotis discolor	*Valerianella locusta*
Sagina apetala	*Vulpia bromoides*
Senecio squalidus	
Senecio viscosus	
Sisymbrium orientale	
Trifolium arvense	
Veronica arvensis	
Vulpia myuros	

Another specialized group of plants are winter annuals such as thale cress (*Arabidopsis thaliana*), thyme-leaved sandwort (*Arenaria serpyllifolia*), hairy bitter-cress (*Cardamine hirsuta*), whitlow grass (*Erophila verna*) and the early forget-me-not (*Myosotis ramosissima*). The seeds of these species have an enforced summer dormancy, mostly germinate in the autumn, pass the winter as small rosettes, then flower early in the year so their life cycle is completed before the summer drought. The natural habitat of the group is cliffs, screes and dunes from where they must have spread to the railways; some also occur on walls in urban areas.

Biennials are also particularly characteristic of sidings where the open conditions and intermittent disturbance favour their life style. Commonly found species include carline thistle (*Carlina vulgaris*), centaury (*Centaurium erythraea*), foxglove (*Digitalis purpurea*), teasel (*Dipsacus fullonum*), viper's bugloss (*Echium vulgare*), evening primrose (*Oenothera* spp.), wild mignonette (*Reseda lutea*), weld (*R. luteola*) and several species of mullein (*Verbascum thapsus, V. nigrum, V. blattaria*). Umbelliferae apart from wild carrot and wild parsnip are scarce. Under the most unfavourable conditions the life of biennials is extended over several years, the species behaving as short-lived perennials through the production of a second rosette at the base of the flowering stem.

Lehmann (1895) distinguished between railway plants in the broad sense, which covers the total flora of a railway system, and in the narrow sense which is concerned only with plants imported over long distances. Most studies have been concerned with the latter approach. In the days when large amounts of freight were regularly handled by the railways temporary populations of exotic species could be found at most sidings. In Britain we owe our best-studied example of this phenomenon to John Dony. In 1946, when collecting records for his *Flora of Bedfordshire* (Dony, 1953), he discovered a railway siding rich in unusual aliens. On enquiry it was found that wool waste (shoddy) was unloaded at the goodsyard and delivered to local farmers for use as manure. Soon populations of wool aliens were located at other sidings in Bedfordshire, Hertfordshire and Huntingdonshire, and a visit to the Yorkshire woollen manufacturing areas around Bradford from where the shoddy was dispatched showed they were plentiful at the loading sidings too. Grey shoddy, the result of the preliminary scouring and combing process to which wool is subjected, has a high seed content particularly of *Amaranthus* spp., *Bidens* spp., *Calotis* spp., *Erodium* spp., grasses, *Medicago minima*, *Monsonia* spp., *Nicandra physalodes* and *Xanthium* spp. Most originate from much warmer countries and may not even flower here but a few become established, e.g. *Acaena anserinifolia*, and have persisted in railway yards after the handling of shoddy ceased. Certain wool aliens such as spotted medick (*Medicago arabica*), which are also native to Britain, can be distinguished as introduced populations because they are robust cultivars originating from fodder crops.

In general, German and Swiss botanists have been the most assiduous recorders of the railway adventive flora; their many works have been admirably summarized in English by Muhlenbach (1979). The studies, most of which were carried out when rail usage was at its height, agree that freight yards and switching tracks are considerably richer than main lines. On switching tracks the collision of trucks caused small seeds to fall through cracks in the wagons when, for example, bulk grain or hay was being transported, while at freight depots seed is scattered during unloading. A rough classification of aliens according to manner of immigration can be made: hay, grain, wool, citrus, bird and oil producing seeds. 'Citrus weeds' imported with straw and hay which was used as packing material round the fruits used to be particularly numerous and mainly of Mediterranean origin. Up to the mid-1930s 814 'citrus weed' species had been recorded in Germany and Switzerland though many were seen only once. This compared with 692 'hay seed' aliens for Germany alone. Factors affecting the detailed distribution of these aliens at a site were the location of the tracks used for unloading and later for cleaning out the waggons and, since around 1930, the use of weed killers. Most of the other groups of aliens, though well represented in goodsyards, were better developed at sites where the goods were processed or waste dumped.

Stratiobotany is a term coined by Thellung (1917) to cover the effects of war on vegetation. A famous example of this factor is the siege of Paris (1870–71) during the Franco–Prussian war when an abundant exotic flora of Mediterranean species developed in the vicinity of the French capital (Gaudefroy and Movillefarine, 1871). The term polemochore meaning followers of war was first used by Mannerkorpi (1944–45) to denote those alien plants introduced or stimulated to spread primarily as a result of war-related activities. As railway networks were important military targets, stratiobotany has at times been important in shaping their flora. Bonte (1930), who studied freight yards in the Rhenish-Westphalian industrial region of Germany between 1909–27, reported that flourishing foreign trade had resulted in a rich adventive flora, but when dealing ceased at the commencement of World War I the numbers of aliens quickly abated and by 1916 there were hardly any left. After the war they started to reappear. During World War II the opposite occurred; despite a great reduction in trade, the absence of spraying and increased disturbance in freight yards allowed aliens to appear once more. The role of hippochores, plants introduced during a war by the use of horses and their forage, was much less than in World War I.

The Germans seem to have enjoyed railway botany; they swept out railway wagons and planted the sweepings; removed large plants from sidings so aliens could grow better and even attributed one species to a bouquet thrown from a passenger train. Many workers elaborately classified the aliens they found into groupings with names such as ergasiophytes, ergasiophygophytes and ergasiolipophytes, having first pointed out weaknesses in the classifications of their predecessors. The picture they present is of a perpetual coming and a perpetual going of species related to changes in trade and railway practice.

An important difference exists between the American and European railway systems. In America most freight is delivered direct to industries without having to pass through a central goods depot so in St Louis for example (Muhlenbach, 1979) freight yards are simply classification yards which direct cargoes to industrial receiving points so there is much less opportunity for seeds to escape. Also there is a growing tendency to surface railway property with a coarse crushed stone which is much less favourable to the establishment of plants than the old ash surfaces. Muhlenbach, who studied the adventive flora of railways in St Louis, Missouri, USA, between 1954 and 1971, recorded 393 aliens. Most were weeds and escapes from cultivation. The results were similar to comparable European studies in that there was a preponderance of annuals (254); most species were very rare (208); the majority were ephemeral and there were great fluctuations in frequency. Eighty-nine were sufficiently well established to be considered naturalized.

For a few years after they are abandoned, or if spraying is discontinued, sidings can very quickly become colonized by opportunist annuals, in particular, sticky groundsel (*Senecio viscosus*), Oxford ragwort (*S. squalidus*) and American willowherb (*Epilobium ciliatum*), any one of which may be dominant. This is the start of a rather slow succession towards dry eventually fairly acid woodland. In practice succession proceeds unevenly, the rate being controlled by substrate characteristics and seed availability. Strongly droughted areas of near neutral pH can remain open for many years supporting a cover of less than 50% higher plants in which *Cerastium* spp., the pearlworts *Sagina apetala* and *S. procumbens*, sheep's sorrel (*Rumex acetosella*) and the annual grass *Vulpia bromoides* are prominent; many of the species listed in Table 8.1 persist in this community. If conditions are more acid, an open *Agrostis capillaris – Hypochoeris radicata* association containing extensive patches of terricolous lichens (*Cladonia chlorophaea, C. conistea, C. furcata, C. pyxidata, C. subulata, Placynthiella icmalea*) may develop, while in towns in the north of England this niche may turn over to a *Calluna vulgaris – Deschampsia flexuosa – Polytrichum* assemblage.

More mesic sites are colonized by a highly diverse taller vegetation, the composition of which is determined largely by local seed sources; the general character is as follows. Dominant grasses are *Arrhenatherum elatius, Dactylis glomerata, Festuca rubra* and *Holcus lanatus* but may also include *Bromus sterilis, Calamagrostis epigejos* and *Poa compressa*. Tall herbs include many of those one would expect on wasteground in towns but a special railway character is provided by, among other plants, the hawkweeds (*Hieracium*). Most of those on railway land in Sheffield have been determined as *H. vagum* which has no basal rosette, numerous stem leaves crowded at least below, and decreasing in size rapidly upwards. From consulting County Floras it is clear that railways are a major habitat for hawkweeds many of which favour freely drained moderately acid ungrazed habitats; the construction of the railways must have enabled many species to extend their range. Toadflaxes also distinguish railway vegetation and are usually well represented at disused sidings from where they may spread into surrounding habitats. Toadflax (*Linaria vulgaris*) is the commonest species with its primrose coloured flowers and orange palate. Another native species which appears to be spreading along the railway network but is still rare in the east of the country is the pale toadflax (*L. repens*) which has a white or pale-lilac corolla striped with violet veins. A third, the purple toadflax (*L. purpurea*), is a garden escape but competes well and persists in railway sidings; it has violet or rarely deep-pink flowers. What makes railway property interesting is that the three species hybridize to produce offspring with a wide range of corolla size, colour and veining differentiation. The Shropshire Flora mentions that both the hybrids *L. purpurea* × *repens* (*L.* × *dominii*) and *L. repens* × *vulgaris* (*L.* × *sepium*) occur on cess near Telford (Sinker *et al.*, 1985). Other railway species which may become locally abundant at this stage include sneezewort (*Achillea ptarmica*), often present as the double form grown in gardens known as bachelors buttons, lupin (*Lupinus polyphyllus*), soapwort (*Saponaria officinalis*) and tansy (*Tanacetum vulgare*). At this 'meadow stage' no two areas of siding are alike which makes them exciting places to botanize. Occasionally, locally rare species such as yellow-wort (*Blackstonia perfoliata*), lesser broomrape (*Orobanche minor*) and bee orchid (*Ophrys apifera*) turn up.

After a while abandoned sidings turn over to a patchy scrub and eventually woodland. The speed with which this happens depends almost entirely on the availability of suitable seeds, so in towns, where these are few, sidings remain 'open' for longer than in rural areas. Initial woody invaders may be any combination of birch, bramble, broom, buddleia, elder, goat willow, gorse, hawthorn, sycamore, whitebeam and wild rose. A woody species that identifies the sere as being on railway land, at least in the London area, is bladder senna (*Colutea arborescens*), an alien from south-east Europe. It has spectacular inflated pods and was in cultivation as an ornamental for a long time before naturalizing along the railways. The Duke of Argyll's tea plant (*Lycium barbarum*) which is more widespread had a similar history but is not quite so faithful to the cess. The commonest woody invaders, however, are birch on dry acid sites and goat willow on more mesic ones. These two seres can lead to rather different woodland types; on the one hand a species-poor dense birch, wavy hairgrass, hawkweed, wood sage, foxglove assemblage; on the other to a rather open goat willow scrub standing among a diverse field layer in which legumes, rosette plants and patches of tall herbs are conspicuous; there is a variant in which brambles become dominant.

Animals of the cess habitat are not well known but nectar-feeding insects such as bees, hoverflies, butterflies and moths are well represented. At the Feltham marshalling yard in West London which closed in 1968, 23 kinds of butterfly are regularly seen (Goode, 1986) while at 'The Rally' site in Leicester, which has been colonized by buddleia and is now managed as a butterfly garden, 22 butterfly species were seen in 1 year. The freely draining substratum is ideal for ants so every railway sleeper, brick, rusty piece of metal and discarded item of clothing supports an ants nest, while ant hills are common once dense vegetation has established. Species I have recorded include red ants (*Myrmica rubra M. ruginoides*), black ant (*Lasius niger*), yellow ant (*Lasius flavus*) and the negro ants (*Formica fusca, F. lemani*); all are common and widespread. Ground beetles tend to be scarce and, if found, of a banal nature, e.g. *Nebria salina* and *Notiophilus biguttatus*: the staphylinid *Staphylinus olens* (devil's coach horse) is occasional. Snails are also rare in the cess habitat though two species with undemanding requirements, *Oxychilus alliarius* and *O. draparnaudi*, occur at a low density. In contrast rabbits are often so abundant that they hold back the succession. In the south of Britain lizards and slow-worms regularly occur, but further north these reptiles are rarely encountered within the city limits.

8.3 BRICKWORK AND MASONRY

Adjacent land-uses press in on urban railway property which in consequence is frequently bounded by impressive retaining walls, buttresses, bridge abutments and the like. Most members of the urban flora have at some time been seen on railway walls but Segal (1969) in his major account of wall vegetation in Europe did not pick this particular habitat out as having any special qualities. Botanists in the east of England, however, have identified them as a particularly valuable habitat for ferns, and Dony (1955) made a special study of their occurrence on brickwork under station platforms. A list of ferns growing on the supporting trackside walls of two platforms in Bedfordshire is given in Table 8.2 from which it can be seen that several are restricted to this niche within the county. Dony suggested that steam escaping from pistons at lower wall level, by creating a perpetually moist habitat, enabled these ferns to spread into the dry east of Britain. Fifteen years later Segal (1969) described a new community, the *Sagina procumbens* with *Gymnocarpium robertianum* and/or *Cystopteris fragilis* typical of walls near water, for example ones rising up out of rivers. Floristically the Bedfordshire platform vegetation closely matches this association.

Table 8.2. Ferns found on the supporting walls of two station platforms in Bedfordshire

Leagrave Station	*Ampthill Station*
Asplenium adiantum-nigrum	*Asplenium ruta-muraria*
Asplenium ruta-muraria	*Asplenium trichomanes*
Asplenium trichomanes	*Athyrium filix-femina*
Athyrium filix-femina	⋆*Ceterach officinarum*
⋆*Ceterach officinarum*	⋆*Gymnocarpium robertianum*
⋆*Cystopteris fragilis*	*Phyllitis scolopendrium*
⋆*Gymnocarpium dryopteris*	
⋆*Gymnocarpium robertianum*	
Phyllitis scolopendrium	

⋆ Only on railway walls in Bedfordshire.
Taken from Dony (1955).

The establishment of many wall species is noticeably favoured by a decrease in the angle of inclination, and an unusually high proportion of railway walls possess a batter. Examples are wallflower (*Cheiranthus cheiri*), flattened poa (*Poa compressa*) and hawkweeds (*Hieracium* spp.); the latter include taxa with basal rosettes which have spread from cliffs and other rocky places. In Britain *Hieracium* Sect. *Amplexicaulia* contains three species, all of which are introduced and very local in their occurrence. Two of them, *H. amplexicaule* and *H. pulmonarioides*, occur on walls associated with railways, e.g. in South Lancashire (Savidge *et al.*, 1963). Pugsley (1948) commenting on their distribution remarks that 'The occurrence of three species of this group as naturalised plants in widely scattered localities in Great Britain does not admit of a ready explanation.' Their predilection for railway property suggests both a method of introduction and spread.

Banal species from arable land, general ruderals and to an extent garden escapes are all well represented on battered railway walls causing a loss of character. Conversely a number of very typical wall species such as wall-rue (*Asplenium ruta-muraria*), common spleenwort (*Asplenium trichomanes*) and yellow corydalis (*Corydalis lutea*) require completely vertical structures so are under-represented on railway walls.

If required to name a typical railway plant many people would select Oxford ragwort (*Senecio squalidus*) which is abundant on walls and waste ground along most lines. Payne's (1978) survey of the flora of walls in south-east Essex identified it as the railway wall plant *par excellence*. Kent (1955, 1956, 1960, 1964a, 1964b, 1964c) has traced the history of its spread from Oxford along the railway system till it is now present in almost every urban site in England and Wales. A native of Southern Italy, it was cultivated in the Oxford Botanical Garden for over a hundred years before it eventually escaped (1794) and was soon reported as plentiful on almost every wall in the town. About 1879 it reached the Great Western Railway system where the plumed seeds engaged in a new form of dispersal, being carried along in the vortex of air behind express trains, or even inside them. Druce (1927) described how he observed a seed drift into a railway carriage near Oxford, remain suspended in the air, then drift out again many miles down the line. Once it had reached the railways, spread continued to be erratic but definitely speeded up. Not only did they provide a means of dispersal but cess and smoke blackened walls offered a habitat not unlike the volcanic slopes of Mount Etna, one of its native haunts. By 1940 Oxford ragwort was established almost throughout the railway network and, though today the transport of soil is possibly its most important method of long-distance dispersal, if it were not for the railways we might still need to travel to Oxfordshire to see this spectacular plant.

No lichens have any special affinity with urban railway brickwork but several mosses are known to. The rare and extremely inconspicuous introduced moss *Trichostomopsis umbrosa* was for a time almost restricted to damp walls such as those supporting railway bridges where it forms pale-green patches. *Funaria hygrometrica* and *Leptobryum pyriforme* often pick out lines of mortar on railway walls; both are fast growing weedy mosses which have a special, but so far unidentified, nutritional requirement satisfied by urban conditions and recently burnt ground.

British Rail inspect their walls and bridges once every 4 years to check whether repointing or other repair work is necessary. Despite this, cavities and crannies are present, which in towns are exploited as nesting sites by starlings, house sparrows, pigeons and sometimes pied wagtails, while in London sand martins have been breeding in non-functioning weepholes for over 60 years. One of the more remarkable examples of an animal colonizing railway property is the population of common wall lizards (*Podarcis muralis*) living on two railway bridges near Hampton in South Middlesex (Stiles, 1979). These large green lizards, probably originally imported from Italy as pets, can often be seen scaling the walls during the summer months as they search for spiders, flies and grasshoppers. Introduced sometime before 1957 the colony is reported to be well established and has survived several climatic extremes in addition to large-scale renovation work on one of the bridges.

8.4 VERGES

The verges, which comprise cuttings, embankments and flats, are the most conspicuous part of a railway. Aware of this, BR recently commissioned a survey of the mainline through Winchester in order to assess the visual and ecological contribution that lineside vegetation made to the town (McNab and Price, 1985). In relation to wildlife the consultants found that chalk grassland in one cutting was of SSS1 quality and an area of high diversity associated with a junction should be conserved. In general the ecological value of the line lay in its provision of habitats for common species of plant, insect, bird, etc. which were once more widespread. From my experience the conclusions – a general value, but the periodic occurrence of sites with a much higher interest – are generally applicable to stretches of urban line.

Sargent's vegetation survey of rural and semirural lines involved the collection of data from 3502 stands. After computer analysis 31 verge noda, and one covering the cess, were identified and a conspectus of the vegetation types published (Sargent, 1984). I have encountered all but four of her noda along urban stretches of line, so her work has some relevance to the urban scene, but a zealous phytosociologist would construct many new pigeon holes in which to accommodate the more varied railway vegetation of inner cities. Sargent's classification works in the urban fringe and outer suburbs where railway verges enable rural types of grassland, tall herb communities and scrub and woodland to penetrate our towns; approaching the inner suburbs, however, increasing disturbance and an influx of aliens alters the character of the vegetation.

On flats and the sides of low cuttings large patches of competitive species (Grime, 1979) develop often in response to the dumping of fertile material such as domestic waste or builders rubble on to the verge. This contamination, disturbance and nutrient enrichment promotes dense stands of *Convolvulus arvensis*, *Epilobium augustifolium, Epilobium hirsutum, Lolium perenne, Pteridium aquilinum, Reynoutria* spp., *Rubus idaeus, Solidago* spp. and *Urtica dioica*, all of which grow with a performance rarely achieved along rural stretches of line. These species, which are characterized by a tall stature, extensive lateral spread, the rapid expansion of leaf and root surface areas, and often bulky perennating organs, introduce a large scale of pattern and coarseness to the vegetation.

A high proportion of aliens is also a characteristic of urban railway verges. Many of these are the widespread ruderals of towns which in addition to several of the species mentioned above include buddleia, comfrey (*Symphytum* × *uplandicum*), Michaelmas daisy, lupin, tansy, honesty and horseradish (*Armoracia rusticana*). A further group includes showy plants of garden origin which are not particularly common in towns as their chief method of spread in this country is by vegetative reproduction. These enter the railway vegetation through householders dumping garden refuse on BR land and are particularly common by bridges and where gardens and allotments back on to the line. Commonly observed species originating in this way are iris, daffodil, montbretia, shastri daisy (*Leucanthemum maximum*), creeping campanula (*Campanula rapunculoides*), perennial sunflower (*Helianthus rigidus*), broad-leaved everlasting pea (*Lathyrus latifolius*), geraniums including *Geranium endressii*, spotted dead-nettle (*Lamium maculatum*) and garden strawberry (*Fragaria* × *ananassa*); a comprehensive list of garden plants which have become established on railway verges in this way would be extremely long.

Engineers have difficulty draining cuttings so wet areas frequently develop; Sargent classified the vegetation of such sites in the Scirpo-Phragmitetum association. One lineside marsh near the centre of Sheffield is dominated by great reedmace (*Typha latifolia*), common sallow (*Salix cinerea*) and soft rush (*Juncus effusus*); it also contains considerable amounts of the 'nitrophiles' codlins and cream (*Epilobium hirsutum*), nettle (*Urtica dioica*) and bittersweet (*Solanum dulcamara*) which suggests a eutrophicated water supply. A further feature of urban cuttings is that rock outcrops frequently support alien species of *Arabis, Cerastium* and *Sedum*. A narrow continuous band of vigorous vegetation near the foot of slopes often marks where the disposal of ballast cleanings have ameliorated poor soil.

While cuttings are natural receiving sites the top of embankments are shedding sites for water, nutrients and fine materials. Their drought-prone soils carry an open species-rich grassland dominated by *Arrhenatherum*, *Festuca* or *Poa* spp. in which *Hieracium* Sect. *Vulgata* and dwarf ruderals are prominent. In such sites *Poa angustifolia* is common; it differs from *Poa pratensis* in being densely tufted and possessing longer narrower leaves which are brighter green. It may be a declining species, as Grime and Lloyd (1973) point out that it favours grassland subject to burning. Railway verges also support a wide range of animal life but surveys detailed enough to typify the characteristic species have yet to be carried out.

Simms (1975) comments that urban railway lines are not especially exciting for birds, particularly those regularly used by high-speed trains. In autumn, flocks of greenfinch, linnet and goldfinch can be seen hunting for seeds, and, where scrub has invaded, blackbird, robin, wren and hedgesparrow regularly occur. In the outer suburbs hedgerow species such as yellowhammer, cornbunting, whitethroat and willow warbler can be seen feeding and singing on railway property but their links with the urban scene are tenuous.

Railway triangles deserve a special mention; they are usually around 3 ha (7 acres) in size. The wooded Gunnersbury triangle in West London became the subject of a public enquiry in 1983 when BR wished to develop it for warehousing. Their proposal was rejected by the inspector as the site was highly valued by the local population as the only sizeable piece of natural habitat for miles around. It is now managed as a nature reserve so willow warblers, blackcaps, foxes and butterflies continue to breed there (Goode, 1986). While the Gunnersbury triangle is occupied by 40-year-old woodland, other sites are completely different. In Nottingham they tend to hold wetland communities while one in Sheffield carries heather moorland to within 1.8 km of the Town Hall.

The only item of railway folklore I have come across concerns the reason for caraway (*Carum carvi*) being limited to railway verges in Scotland. The explanation is as follows: caraway cake topped with fresh seeds was provided at post-funeral teas with assurances that mourners would not eat much; however, some pocketed a slice. On the way home they tried to eat it, gave up and threw it out of the window.

8.5 BOUNDARY FEATURES

British Rail delimit their property with a standard concrete post and wire fence. At first sight this does not appear to offer a very promising habitat for wildlife but close inspection with a hand lens reveals that the top of the posts are invariably colonized by an ecosystem in which the producers are small crustose lichens, while higher levels in the food chain are occupied by arthropods. The lichen assemblage of two to six species, collectively known as the Lecanorion dispersae, is the pioneer community of artificial calcareous substrata throughout much of Europe and was first described from London (Laundon, 1967). In towns this community is often permanent since levels of air pollution eliminate most of the larger species, e.g. *Xanthoria parietina*, which might overgrow it. Small brown oribatid mites (Ameronothrus maculatus) can often be observed grazing on and burrowing into the community where they live under the lichen in the space between the thallus and the concrete. Carniverous anthocorid bugs, e.g. *Temnostethus pusillus*, have been reported active among and preying on these lichenicolous mites

(Gilbert, 1976). The decomposer element is usually represented by Collembola which live among and feed off the dead organic matter and droppings which accumulate under the lichen mat. To the non-ecologist it is surprising to find such a well-stocked ecosystem on top of an apparently barren concrete post.

The other characteristic boundary feature of railways is a type of hybrid black poplar. The name hybrid black poplar covers the numerous hybrids and backcrosses that have occurred between the European *Populus nigra* and the North American *Populus deltoides* and *Populus angulata* since about 1700. They are collectively known as *P.* × *canadensis*. A number of distinctive varieties exist, one of which is *Populus* 'Regenerata' thought to have originated in a nursery near Paris in 1814 (Figure 8.4). The bole and larger limbs are covered in snags, burs and sprouts, which recalls the name 'Regenerata', the boughs ascend then arch over and fan outwards to form a broad 'vase shaped' crown; long, slender shoots hang from them. It is a female clone but largely sterile so trees drop less 'white wool' than many other poplars. Late April is the

Fig. 8.4. The railway poplar (*Populus* × *canadensis* 'Regenerata') which screens many urban lines and goods yards.

best time to set about spotting this speciality as the branching pattern is still visible despite the twigs being covered with bright pale-green catkins and pale-green to brownish green (no hint of red) emerging leaves. It was I think Alan Mitchell who first suggested that it should be named railway poplar as it is so common by mainlines into cities, in sidings and planted as a screen at the foot of embankments.

Chapter 9

ROADS

The ecology of roads is the ecology of roadside verges and most studies of these have been carried out in the countryside where their importance for wildlife is considerable. By contrast, verges in urban areas are rarely studied or mentioned in the conservation literature. Priorities here are concerned more with ensuring good visibility, providing an access way for public services and helping, through landscape work, to integrate roads and footpaths into the surrounding built-up area. Any benefits to wildlife are incidental. However, living systems even when composed of a rather limited selection of planted rigorously maintained species develop their own ecology, and urban road verges are no exception. There are as many interrelationships to be discovered in a line of town trees growing out of a grassy verge by a busy road as in better known linear habitats such as hedges or canals. Some of the features exhibited are unique to towns, other such as the effects of de-icing salt and exhaust emissions are shared with rural areas.

The proportion of land occupied by roads and verges has been calculated for Germany, where in the mid-1960s it was estimated they occupied 4% of the country with forecasted increases expected to take this to 5–6% (Rothschuh, 1968). Studies in Hamburg showed that as a general rule about one-third of the urban area was taken up by roads and pavements. Figures for Britain (Department of the Environment and Department of Transport, 1977) suggest that at the lowest building densities 15%, and at the highest 35%, of urban areas are occupied by roads, pavements and footpaths.

9.1 ROAD DESIGN

British roads are classified into trunk roads including motorways, principal roads, and other roads; they are present in the ratio 1:2:20. About 11% of principal roads, which are often dual carriageway, fall within urban areas where by far the greatest proportion are 'other roads' entirely financed by local government. This latter category covers an extremely wide range of highways from local and residential distributors which are usually bus routes with a carriageway 6.0–6.5 m wide flanked by 1.8 m pavements, through residential collectors 5.5 m wide plus 1.8 m pavements, to access streets and finally accessways and mews courts with shared surfaces.

While roads in the older suburbs and interwar housing estates were built to rather uniform dimensions and are often tree-lined, there has increasingly been a tendency to break away from the traditional approach. This culminated in the Department of the Environment and Department of Transport (1977) publishing an influential *Design Guide* which tried to strike a balance between housing, highways and public

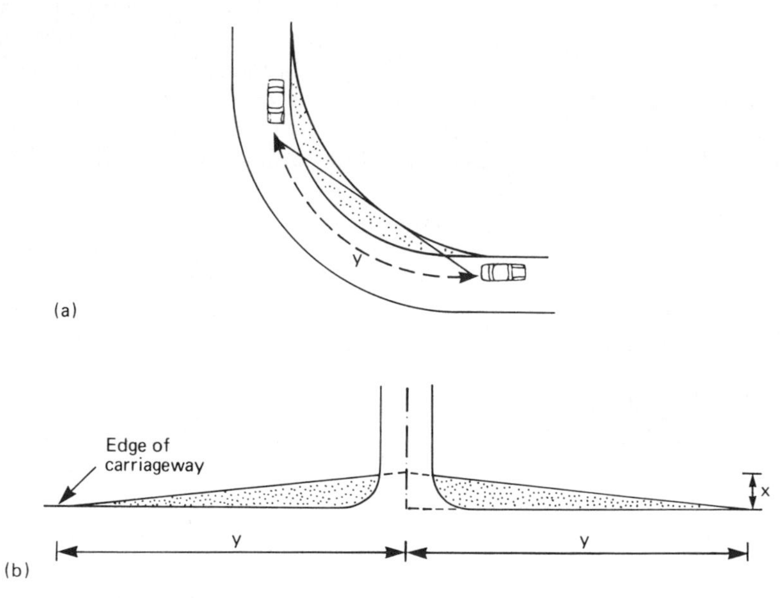

Fig. 9.1. Visibility along roads. (a) Land within visibility curves must be cleared of all vegetation over 600 mm in height so the excessive use of bends effectively sterilizes large areas of verge. (b) Visibility splays at road junctions must be kept clear of vegetation exceeding 1.05 m. Stopping distance Y is 60–65 m and dimension X is 9–10 m on roads designed to carry traffic at 30 mph (Kirkham, 1982).

utilities, though, owing to a heavy emphasis on safety, roadside vegetation suffers. For example, on roads designed for a traffic speed of 30 mph forward visibility on bends needs to be commensurate with the stopping distance of vehicles (60 m) with the result that it is recommended that land within this substantial visibility curve is cleared of all obstructions, e.g. vegetation, over 600 mm (2 ft) in height (Fig. 9.1). Sightlines at T-junctions and even the entrances to private drives are also determined by traffic speed and can 'sterilize' very large areas of verge. On 30 mph roads visibility splays of 60–65 m are required in each direction, rising to 155 m on 50 mph roads; these should be cleared of all vegetation exceeding 1.05 m in height. These recommendations take precedence over, but are in direct conflict with, the appearance of the roadscape which recent surveys have shown is among the most important determinants of resident satisfaction. The permitted ground cover of low shrubs is normally too expensive to maintain except in prestige areas so large expanses of verge are maintained as treeless mown grass or tarmac.

Adding to this monotony are restrictions relating to underground services which are normally routed under footpath or verge. Service runs may only be planted with low ground cover which in practice means sown grass. Owing to expectations for neatness and tidiness, the local authority usually mow all verges at least six times a year and the base of signs, lamp posts, trees, walls and gutters are sprayed annually with herbicide. Other factors contributing to the stresses of the roadside environment are pollution from vehicle exhaust, dust, salt and disturbance caused by car parking, road works and the activities of gangs maintaining the service runs. Despite or sometimes because of this association of adverse influences urban road verges support specialized ecosystems.

9.2 EXHAUST EMISSIONS

Roadside pollution is not restricted to urban areas but its effects are particularly pronounced in and around towns. Eight distinct products are contained in vehicle exhaust, fortunately not all are harmful to plants; carbon dioxide, carbon monoxide, water vapour and unburnt petrol fall into this category. Of the remainder the most important pollutants are oxides of nitrogen (NOx) and particulates such as finely divided carbon (smoke). These may act separately on vegetation or together, and under conditions of bright sunlight an 'atmospheric soup' formed by a range of secondary reactions can form in streets or even over an entire region. Apart from these products of the incomplete combustion of petrol, lead compounds from 'anti-knock' agents are found in exhaust. The impact of these pollutants on natural communities is difficult to assess as few studies have been undertaken. Most of the considerable research has understandably been with crop plants, often in the laboratory and usually selecting particularly sensitive strains to work with. Broadly the results suggest that along the central reservation and a narrow strip of verge on either side of the road plant life is significantly affected by lead and particulates, while the influence of NOx and its derivatives spreads further.

Lead

Numerous papers have pointed out the increased levels of heavy metals particularly lead in soil in the vicinity of roadsides. This accumulates in the surface layers where concentrations may reach several thousand ppm, so germinating seedlings are exposed to severe selection pressures. The first report of roadside plants evolving lead-tolerant races came from a city centre. Wu and Antonovics (1976) set up a transect perpendicular to the busy main street in Durham, North Carolina, extending it into an intermittently mown lawn. At 0.45, 4.0 and 80 m from the roadside they sampled the soil for lead, recording total levels of 2850, 772 and 200 ppm; a control site away from busy roads had 52 ppm. At each sampling site adults and seed of ribwort plantain (*Plantago lanceolata*) were collected and tested for lead tolerance by growing in nutrient solutions to which lead had been added. The results showed (Fig. 9.2) that roadside plantains had a higher lead tolerance than material collected further along the transect or at the control site. Further tests confirmed that lead tolerance is an inherited character by showing it was passed on to seed progeny. Bermuda grass (*Cynodon dactylon*) collected along the same transect showed generally a higher lead tolerance than plantain but there was no evidence of a greater resistance in the roadside populations when compared with populations from the control site.

These results suggest that the lead level found in the roadside soils was sufficiently high to impose selection pressure for the evolution of tolerance in a sensitive species (*P. lanceolata*) but no overt effect was seen on a species with a greater inherent tolerance (*C. dactylon*). This example of a rapid and localized evolutionary change in plants due to an urban influence is a parallel from the plant world of the well-known peppered moth–sooty bark–bird predator phenomenon.

It is clear from the above work that the effects of lead pollution are species dependent. Subsequently a lead-tolerant race of red fescue (*Festuca rubra*) has been found growing by the edge of the hard shoulder along the M6 motorway in the Midlands (Atkins *et al.*, 1982), and Briggs (1972, 1976) has demonstrated that similarly adapted races of groundsel (*Senecio vulgaris*) and the thallose liverwort *Marchantia*

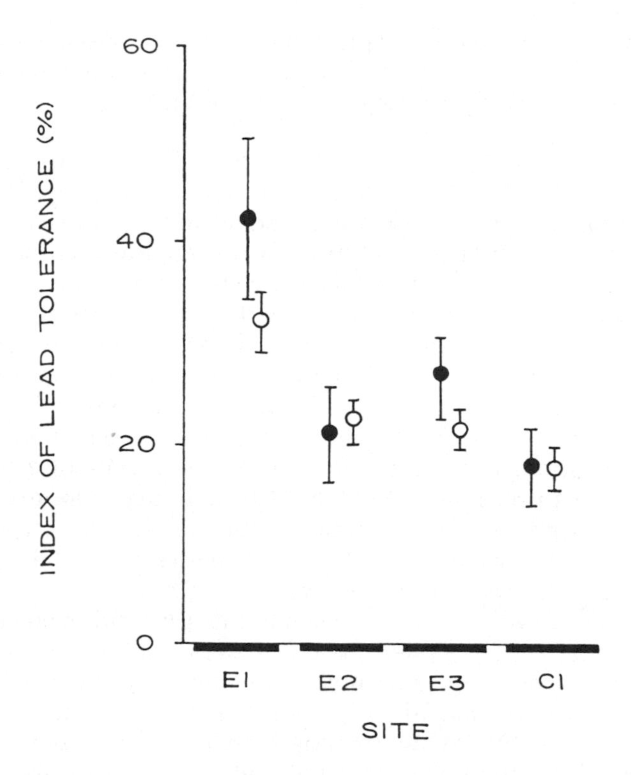

Fig. 9.2. Lead tolerance of adult (●) and seed (○) samples taken from populations of *Plantago lanceolata* at different sites in Durham, NC. Sites E1, E2 and E3 represent a transect from the roadside into a lawn and are at 0.5 m, 4 m and 80 m from the edge of Main Street, Durham. The control site, C1, was situated away from any heavily used roads. Bars on the graph represent S.E. of the means (Wu and Antonovics, 1976).

polymorpha occur on the pavements of Glasgow. Lead contamination is probably best visualized as occupying a strip, a few metres wide, on each side of a road verge; it may sometimes be skewed downwind and more complex patterns are present by city roads. Zinc derived from the attrition of tyres, and chromium originating from the erosion of yellow road markings are also present in elevated amounts but levels are not yet harmful to plants.

Lead concentrations have been determined in a wide range of animals living by highways in an attempt to determine if they are adversely affected. Though elevated levels have been reported in tadpoles inhabiting highway drainage systems (Birdsall *et al.*, 1986), starlings nesting on road verges (Grue *et al.*, 1986), barn swallows (*Hirundo rustica*) living under highway bridges (Grue *et al.*, 1984) and a range of small mammals (Clarke, 1979), no adverse effects on their breeding success or that of their predators have been found.

Photochemical pollution

Photochemical pollution containing ozone and peroxyacetyl nitrate (PAN) results largely from reactions between nitrogen oxides and hydrocarbons emitted from vehicle exhaust. Notorious in the Los Angeles basin, it is now known to be widespread in Europe. As usual most work has been done on crop cultivars known to be highly sensitive, for example, tobacco (especially *Nicotiana tabacum* 'Bell-W3'), several species of pine, bean, potato, tomato and petunia. Undoubtedly if one knew what to look for visible symptoms could be found on native roadside vegetation. In the mid-1960s Ten Houten (1966), working in Holland, learnt to recognize PAN and ozone injury symptoms on the leaves of *Chenopodium murale*, small nettle (*Urtica urens*) and annual meadow-grass (*Poa annua*); on the latter species it caused a banding injury. Affected plants were widespread.

Ozone is not really an urban problem but it affects whole regions such as S.E. England. At the moment, as effects are modified by the weather, the presence or absence of other pollutants, the age of the plant, genetic variation and the presence of pathogens it is perhaps best to sum up a complex situation by saying that in Britain at least, the impact of photochemical pollution generated by road traffic has yet to be proved of significance to horticultural plants or native vegetation.

Ethylene

This product of the incomplete combustion of petrol may have phytotoxic effects close to roads where these contain accelerating vehicles or are sheltered. The Dutch found a problem with carnations in florists shops near traffic lights – the flowers would not open as a result of ethylene pollution.

9.3 DE-ICING SALT

The best-known roadside pollutant is salt (NaCl) used for de-icing. Most people know of at least one instance where trees have been killed when a new salt-storage pile was created in a lay-by or adjacent to a garden containing trees, the roots of which penetrated under a wall into the area of the deposit. At around £30 per ton, rock salt is the cheapest chemical that will melt snow and ice; besides sodium chloride it contains a small amount of marl which imparts a brownish hue and a little sodium ferrocyanide is added to prevent caking during storage. To minimize the risk of chemical corrosion, urea is used to de-ice major bridges but is much more expensive than salt. Usage of urea runs to many tons per year; it would be an attractive project to assess its influence on soils and vegetation.

The first report of maritime species invading British roadsides was from north-east England (Matthews and Davison, 1976); since then there have been reports of similar invasions in Kent, Norfolk, Bedfordshire, Warwickshire and several other counties. The most successful genus has been *Puccinellia*, with *P. distans* now easily the most widespread maritime species by roads in Britain. The same phenomenon has been reported from the continent where Jackowiak (1982) has made a particularly detailed study of the expansion of *P. distans* in the Polish city of Poznan. All the sites are beside roads but in addition to spreading into saline sites it is reported as colonizing habitats rich in nitrogen. Jackowiak has described two new vegetation subtypes, the Lolio-Plantaginetum puccinellietosum and the Lolio-Potentilletum anserinae puccinellietosum to accommodate these *Puccinellia*-rich urban roadside communities. The species zonation imposed largely by the steep gradient in salt concentration across the verge is shown in Fig. 9.3.

To return to the British examples these have been comprehensively

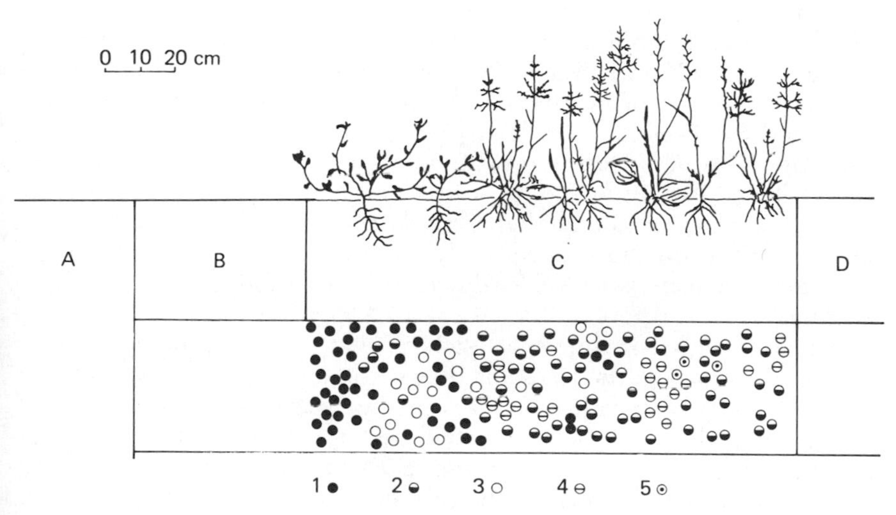

Fig. 9.3. Transect across a road verge colonized by the Lolio-Plantaginetum puccinellietosum association, Poznan, Poland. A, edge of tarmac; B, bare zone; C, vegetation with *Puccinellia distans*; D, ruderal vegetation. 1, *Polygonum aviculare*; 2, *Puccinellia distans*; 3, *Plantago major*; 4, *Elymus repens*; 5, *Lolium perenne* (Jackowiak, 1982).

Table 9.1. Maritime plants on British roadsides

Armeria maritima	*Parapholis strigosa*
Aster tripolium★	*Plantago coronopus*★
Atriplex litoralis★	*Plantago maritima*★
Bupleurum tenuissimum	*Puccinellia distans*★
Cochlearia danica	*Puccinellia fasciculata*
Cochlearia officinalis★	*Puccinellia maritima*★
Desmazeria marina	*Puccinellia rupestris*
Elymus pycnanthus	*Spergularia marina*★
Halimione portulacoides	*Spergularia media*★
Hordeum marinum★	*Suaeda maritima*★
Juncus gerardii	*Agaricus bernardii*★

★ By roadsides in north-east England.
Taken from Scott and Davison (1982) and Scott (1985).

summed up in papers by Scott and Davison (1982) and Scott (1985) which highlight their own important work in north-east England. Here, in heavily salted roadside verges around Tyneside 12 saltmarsh species are present (Table 9.1). In addition, many common roadside plants such as *Atriplex* spp., *Matricaria perforata, Polygonum* spp. and *Senecio vulgaris* are also known from saline coastal habitats and probably possess a degree of salt tolerance. While *Puccinellia distans* is now present on roads throughout much of northern, central and eastern England the other maritime species occur elsewhere principally in Kent and Norfolk. The rapid spread of *P. distans* is a result of its pre-adaption to this relatively new habitat. Its natural home seems not to be true saltmarshes, but rather the edges of saltmarshes which are disturbed, often compacted and may be poorly drained. *Spergularia marina* which shares many of the ecological characteristics of *P. distans* is also spreading comparatively rapidly.

All roads with maritime species show evidence of salt damage to the verges. On roads to the north of Newcastle-on-Tyne there is a greater incidence of damage and openness in the sward on the south-bound carriageway than on the north one because de-icing salt is usually applied in the early morning and traffic density immediately after this is higher going south because of the rush hour into Newcastle. All verges in the north-east containing maritime species were constructed after 1965; the explanation for this is probably connected with the fact that it was in the late 1960s that really heavy applications of de-icing salt began (Table 9.2). Major roads predating this period usually have couch grass (*Elymus repens*) growing thickly right to the edge and salt appears to

Table 9.2. Estimated amounts of salt used annually for winter road maintenance in Great Britain 1956–1985. Figures in thousands of tons

1956	132	1966	813	1976	775
1957	51	1967	610	1977	1675
1958	173	1968	1312	1978	1775
1959	188	1969	1626	1979	2590
1960	198	1970	1727	1980	2200
1961	203	1971	1422	1981	1075
1962	467	1972	914	1982	2400
1963	1102	1973	1016	1983	1412
1964	411	1974	975	1984	1691
1965	914	1975	775	1985	2340

have little effect on it. This species is nothing like so common on more recent roads possibly because it cannot get established when salt applications are high. Factors additional to salinity also control the distribution of the species, for example *Plantago maritima* does best on freely drained verges while *Spergularia marina* seems to occur only in wet sites. *Atriplex* spp. appear to be particularly prone to insect infestation on roadsides which may be a further example of the hawthorn–green apple aphid phenomenon described later.

A fascinating aspect of the phenomenon is the question of where the plants came from and how they got to their roadside sites. The reason for their fairly recent arrival is probably answered by Table 9.2. The favoured explanation for their origin is that primary colonization has been by seeds carried from coastal sites on vehicles and that once established local distribution is in the slipstream of cars. After due consideration deliberate introduction has been dismissed as has introduction by contamination in de-icing salt (the species do not occur by local salt piles) or in the grass-seed mixture used to sow the verges. The most likely source remains introduction from local coastal sites where cars frequently park on or drive over the beach.

The effect of salt on roadside trees is often difficult to separate from other stress factors. Experiments that have isolated its influence by applying realistic concentrations of salt either to the roots or shoots of container-grown shrubs (Thompson and Rutter, 1982) showed that sea buckthorn (*Hippophae rhamnoides*), grey willow (*Salix cinerea*) and dogwood (*Cornus sanguinea*) were particularly resistant to salt while goat willow (*Salix caprea*) and hawthorn (*Crataegus monogyna*) are rather sensitive. Other workers have reported London plane (*Platanus* × *acerifolia*), silver maple (*Acer saccharinum*) and birch as showing sensitivity while most poplars, willows and oaks are tolerant. Resistance depends not only on the plant species but varies considerably between clones and provenances. Species selection for planting along the central reservation of dual carriageways needs care but in general the death of roadside trees and shrubs from normal levels of salt application are rarely reported in Britain, nor is this fact often taken into consideration by those who create and manage tree-lined roads in towns.

There is considerable variation in salt application rates in England not only from year to year but from area to area. Within a given winter, motorway application rates in north-east England can be over ten times greater than in the south. The situation is further complicated by winter rainfall which leaches salt from the surface layers of the soil. In practical terms this means that the effect of a given salt application will be greater in the dryer south and east than in the wetter north and west. Thompson *et al.* (1979) have developed a model which, taking these variations into account, can be used to describe and predict salt contamination in soil profiles down to a depth of 50 cm. Field measurements have shown a very steep concentration gradient across the central reservation of dual carriageways and verges. Overall levels, derived mostly from salt spray, are much higher on the central reservation than on verges where, at distances greater than 2 m from the edge of the road, salt concentrations never exceed about 500 $\mu g\ g^{-1}$.

High salinities in roadside verge soils result in open swards and sometimes strips of bare ground immediately adjacent to the road. Such bare patches are often referred to as salt burn. For sometime I suspected that the magnificent early summer displays of dandelions in the salt-burnt margins of roads were the result of high salinity and this was confirmed by a letter in *The Times* on 14 June 1986 from dandelion expert A.J. Richards. He wrote that while many microspecies of dandelion are intolerant of salt, a group known as *Taraxacum* Sect *Hamata* are relatively salt tolerant and it is members of this section that predominate next to the road in the grass-free zone. This effect can be noted beside many urban roads.

9.4 TRAFFIC STRESS

Convincing evidence that traffic stress is currently affecting roadside ecosystems comes from a study of aphid populations in streets and along motorways. Braun and Fluckiger (1984a,b, 1985) investigated the cause of unusually severe infestations of hawthorn by the green apple aphid (*Aphis pomi*) which occur every year beside a busy motorway near Basle, Switzerland; similar plagues have been reported beside other roads in Switzerland and France. One of the experiments they carried out involved placing a pair of perspex chambers, each containing eight hawthorn plants, on the central reservation of the motorway. One

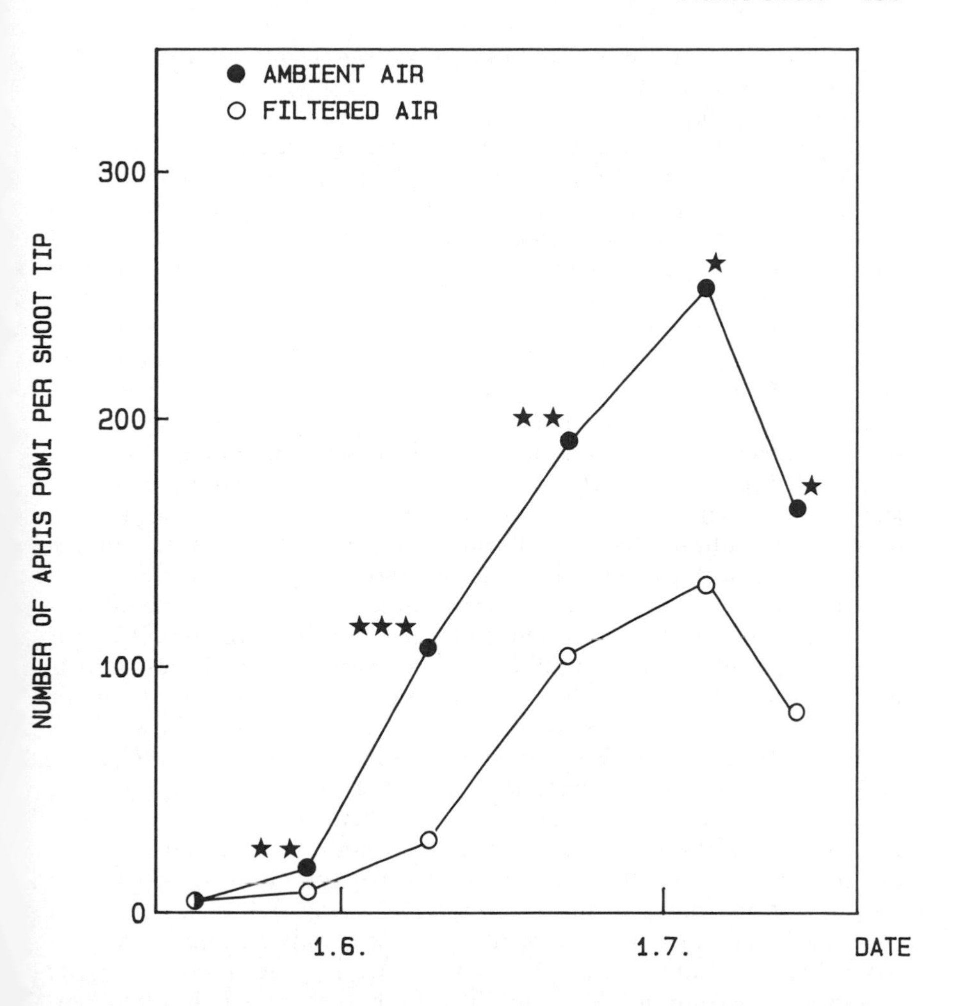

Fig. 9.4. Population development of *Aphis pomi* on potted hawthorn in ambient (●) and clean air (○) chambers. Differences significant at the 0.05 level (★), the 0.01 level (★★) and the 0.001 level (★★★) (t-test, n = 24) (Braun and Fluckiger, 1985).

received filtered air, the other ambient air. After 4 weeks aphids were introduced to the chambers and the development of the populations monitored over the next 10 weeks. Figure 9.4 shows how only 3 weeks after infestation the population in the ambient chamber was 4.4 times greater than in the clean air one. Analysis of phloem exudates revealed a significant increase in the amino acid glutamine relative to the sugar content in the plants grown in the polluted air; substantial changes in phenolic compounds were also recorded. The total nitrogen content of leaves increased in the plants receiving ambient air, presumably because they were able to utilize NO and NO_2 as nitrogen sources; this may also explain the increase in glutamine.

Several possible mechanisms were advanced to explain the rapid development of the aphid population in the polluted chambers. From agricultural research it is known that plants growing under a high nitrogen regime are in general more susceptible to aphid attack; also ethylene, a component of vehicle exhaust, accelerates senescence and senescent plants contain increased levels of soluble nitrogen which may also have played a part in the increased aphid infestation. The importance of amines (asparagine, glutamine) for insect performance has been emphasized by McNeil and Southwood (1978) and Van Emden (1972) so the amino acid differences may also have played a role. The increased zinc concentration derived from car tyres may have been important too as this metal is an essential element for aphids and has to be included in artificial nutrient media. Whatever the causative factors, and they are likely to be complex, the results of the chamber experiments indicate that air pollution along roads can alter host plant–parasite relationships to increase the susceptibility of the plant.

The investigation of complex plant–animal–pollution interactions is unusually difficult. If this was an isolated example, some doubt might be expressed over Braun and Fluckiger's experimental methods or their interpretation of the results but followers of the literature on air pollution are aware that it is increasingly being reported how plants growing at sites where they come under pollution stress (HF, SO_2, O_3, NO_x) are often more susceptible to attacks from beetles, lepidoptera, sawflies, red spider mite, greenfly and other arthropods. Subtle shifts in plant susceptibility to pests could be a major largely unstudied feature in the ecology of built-up areas. A further elegant example concerning road pollution comes from the centre of Warsaw where Czechowski (1980) observed that there were many more aphids on trees bordering roads than on similar trees in parks. This he put down to traffic pollution increasing the subsceptibility of the trees, mostly limes (*Tilia*), to phytophagous insects such as the lime aphid (*Eucallipterus tiliae*). These aphids are tended by black ants (*Lasius niger*) which are more numerous where the greenfly are most abundant. Marszalkowska Street, the main thoroughfare in Warsaw, is flanked by a double line of trees, the first close to road, the other further away. Ant nests occur under 60% of the trees in the first row but under only 42% of the second. From this and other observations Czechowski concluded that the complex of vegetation, aphids and ants is a very sensitive indicator of the degree of pollution in the city. Not all arthropods are favoured by vehicle exhaust gases; another Polish worker (Przybylski, 1979) has, through careful transect work, provided evidence that certain ladybirds (Coccinellidae), Lepidoptera larvae, click beetles (Elateridae), Hymenoptera, thrips (Thripidae), weevils (Curculionidae) and mites (Acarina) are reduced in abundance in crops and/or orchards alongside busy roads. Meanwhile greenfly (Aphididae) and certain bugs (Heteroptera) were favoured by the roadside environment.

9.5 STREET TREES

A survey of street trees in Sheffield has show them to be widely distributed along all classes of road but there is large lacuna in the middle and east of the city where high levels of air pollution in the past militated

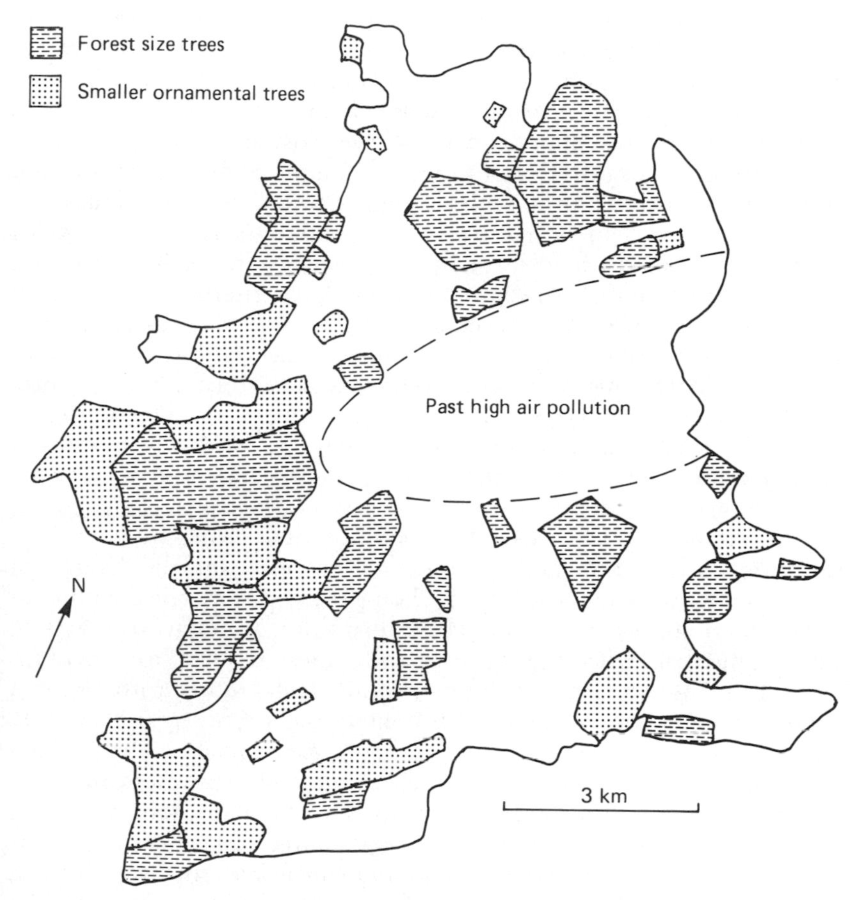

Fig. 9.5. Map showing the distribution of major plantings of 'forest size' and smaller 'ornamental' street trees in Sheffield; the latter characterize post-1950 developments. High past levels of air pollution discouraged street tree planting in the centre and east of the city.

against all tree planting (Fig. 9.5). Trees pick out several of the major routes and it is City Council policy to encourage this. Elsewhere the detail of their distribution is largely controlled by such factors as the whim of the developer, age of the development and width of verge. No attempt has been made at any time to relate species to soil type, altitude or other ecological factors. Their pattern of distribution is mostly chance, though a simple division can be made into plantings up to about 1950 which were invariably of 'forest' trees; after that date there was a change to smaller ornamental varieties at least in residential areas. Today two-thirds of Sheffield's 15 000 street trees are forest size and since most originate from the Victorian era or early part of this century they are mature specimens. Their order of abundance is lime, plane and sycamore, ash, horse chestnut, elm, beech and much more rarely common alder and poplar. Wherever native oak (*Quercus petraea, Quercus robur*) is present it originated as a field tree.

Lime, easily the most abundant, is also the most unsuitable street tree. Its popularity is an example of our culture-bound aesthetics. It may look magnificent planted as an avenue leading up to a country house but along a suburban road it has many shortcomings. In most parts of Britain it is the tallest broad-leaved tree available so it needs continuous lopping and mutilation to keep it in check. This interferes with apical dominance so the entire bole, especially at the base becomes covered in epicormic shoots which continually need to be removed. If left to develop a canopy this is usually rendered unsightly by the centre becoming filled with huge masses of sprouts which look like a disease. Other disadvantages are that it drops its leaves early with a minimum of autumn colour and the most commonly used clone has a propensity for forming huge burrs on the trunk.

On the credit side, from a wildlife point of view, limes support huge populations of the lime aphid (*Eucallipterus tiliae*); at the height of summer there may well be over a million per tree. They tap the phloem of leaves and shoots extracting large quantities of sugar which is in solution together with the amino acids they require for growth. However, as the sap is rather poor in amino acids, in order to obtain a sufficient amount the aphids have to take up far more sugar than they require, the surplus being ejected through their anuses. This ejected sugar is the familiar honeydew which during fine summers makes a sticky mess beneath lime trees as motorists know to their cost. The sugar released by the aphids is an energy source for other organisms such as the sooty moulds *Fumago* spp. and *Capnodium* spp. which form an unsightly black deposit over the leaves. Many insects feed on the honeydew including ants and, especially when nectar is scarce, hoverflies and moths. It has also been reported to stimulate fungal and bacterial populations in the soil beneath the trees (Dighton, 1978). The aphids themselves attract predators including the larvae and adults of ladybirds, the larvae of hoverflies and a range of parasites. The lime aphid is regarded as such a nuisance on urban shade trees in North America that parasites such as the hymenopteran *Trioxys curvicaudus* which is monophagus on *E. tiliae* has been introduced from Europe to contain its numbers (Olkowski *et al.*, 1982).

The commonest street lime *Tilia* × *vulgaris* is a hybrid between our native small-leaved lime (*T. cordata*) and the large-leaved lime (*T. platyphyllos*), it shows intermediate characters and hybrid vigour. The hybrid is rarely encountered in the wild as there is a difference in flowering time between the parents, so like the London plane this is a truly urban tree with roadsides its main habitat. The large-leaved lime is also frequently used in street plantings for which it is better suited being more shapely and of cleaner growth; often one only becomes aware of its presence in the autumn when the leaves with their hairy undersides and petioles fall to the pavement. The small-leaved lime which has a neat crown is less often used in streets. Increasingly recent plantings of lime have employed the Caucasian species *T.* × *euchlora* which is smaller and remains free of insect pests. It can be recognized by its green twigs and the upper surface of the leaves which are a deep rich glossy green.

In many suburban areas limes are the principal source of honey for bees and if a judicious selection is planted a supply of nectar can be provided from early June (*Tilia platyphyllos*) till late July (*T. petiolaris*). The pasturage from each tree lasts around a fortnight. In spite of their appeal a degree of toxicity is present in the blossoms, particularly those of the pendant silver lime (*T. petiolaris*) and its hybrid *T.* × *orbicularis*; it is common to see bees lying on the ground under these trees in a soporific state. Many of the bumble-bees die while the honey bees recover. The reason for this seems to be that the honey bees collect only nectar while the less specialized bumble-bees also eat pollen which is where most of the toxicity lies. Consequently the pendant silver lime and its hybrid are not recommended for extensive planting from a wildlife point of view.

In Sheffield, sycamore (*Acer pseudoplatanus*) has commonly been used as a street tree because of its resistance to industrial air pollution so there is some truth in the statement by Peace (1962) that it is the typical tree of industrial areas. Sycamore displays many of the characters of an urban specialist being originally introduced, then a garden escape with a preference for disturbed fertile soils; it is a prolific and regular seed producer. Many wildlife enthusiasts are disparaging about sycamore as they believe that, being introduced, it has only a small number of insect species associated with it. Kennedy and Southwood's (1984) authoritative paper on the subject attributes 43 phytophagous insects and mites

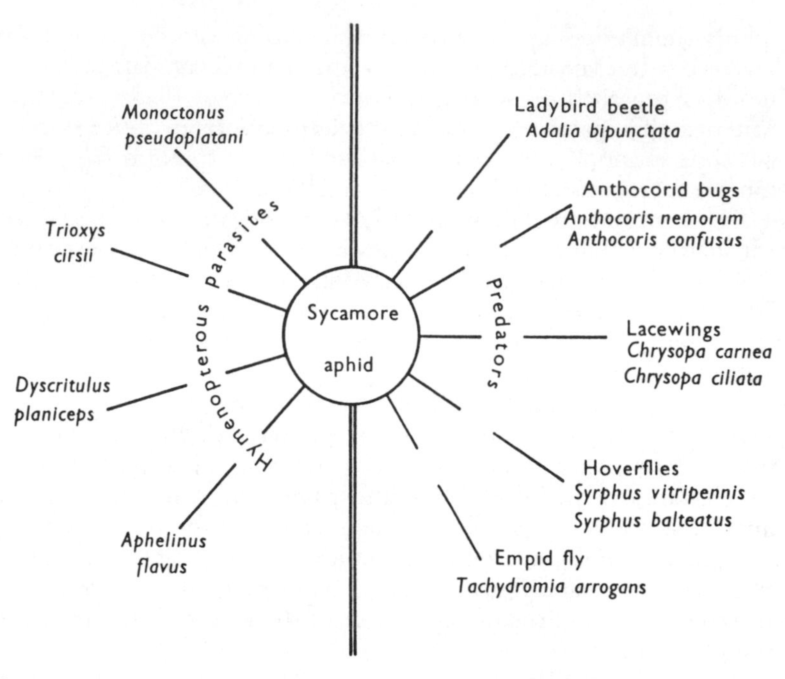

Fig. 9.6. Predators and parasites that attack the sycamore aphid (Dixon, 1973).

to sycamore, only a few less than to well-loved natives such as field maple (51), hornbeam (51), lime (57) and rowan (58). However, in terms of wildlife value, total phytophagous insect species richness is only part of the story. Sycamore supports very large populations of the sycamore aphid (*Drepanosiphum platanoides*) and also other aphids such as *Periphyllus testudinaceus*. As can be seen from Fig. 9.6 these provide food for a wide range of insect predators and parasites; some birds also eat aphids; if one examines the underside of sycamore leaves, this whole interconnecting web of life can be seen.

Highway engineers find that every forest-size tree used for street planting brings its problems. Those associated with lime have been dealt with. Ash is prone to 'elephants foot', a swelling of the trunk at ground level which lifts footpath surfaces and pushes kerbstones out of line; poplars have widespreading superficial roots which break up footpaths, and crack walls and bay windows; acacia throws up suckers in adjacent gardens; horse chestnut is too large and complaints are received around conker time; beech does not grow well after pruning and steals light; all can penetrate and fracture service runs. Pavements containing mature forest trees require maintenance every 3–5 years, those without only at 10–15 year intervals. These are some of the reasons for the decline in popularity of forest-size street trees.

The smaller ornamental trees which have mostly been in use since about 1950 characterize housing estates in the outer suburbs and environmental improvement areas closer to the centre (Fig. 9.5). Initially old favourites such as Japanese cherry (*Prunus* 'Kanzan'), Pissards plum (*P. cerasifera* 'Pissardii'), purple crab (*Malus* × *purpurea*), Japanese crab (*M. floribunda*), birch, rowan and Swedish whitebeam were used. Despite their small size these trees also attract a lot of complaints; 'Kanzan' when grafted on to *Prunus avium* stock badly damages pavements; other species catch in peoples' eyes, fruit gets trampled into houses or children use it as ammunition, regular pruning is needed to give the desired 5 m clearance for high-sided vehicles. Around 3000 complaints are received each year by the Recreation Department in connection with Sheffield's 15 000 street trees, so new species and cultivars are continually being tried in an effort to find the perfect tree for each site.

The range of small trees currently in use is considerably wider than at any previous time; they have greater aesthetic merit and many exhibit an upright habit so require less maintenance than their predecessors. The traditional genera, cherry, crabapple, rowan–whitebeam still provide most of the stock but the recommended list now contains twelve *Prunus,* ten *Malus* and eleven *Sorbus* taxa for use in footways up to 4.8 m wide, and a further 22 taxa for verges over 4.8 m. The main omission from the Sheffield list are hawthorns (*Crataegus* spp.).

An attempt has been made to discover whether the distribution of street trees shows a regional pattern. Only 'forest' trees were considered, these normally comprising around half the holding. The results indicate a clear gradient of declining diversity as one moves north through Britain. In the south a wide range of trees are employed including hickory (*Carya*), Kentucky coffee tree (*Gymnocladus dioicus*), honey locust (*Gleditsia triacanthos*), Indian bean tree (*Catalpa bignonioides*), pride of India (*Koelreuteria paniculata*) and many other slightly unusual species. Even highly urbanized boroughs such as Westminster which is obliged by the government to preserve and perpetuate its existing stock of London planes is planting *Gingko, Metasequoia* and *Betula jacquemontii*. In the Midlands, trees such as hornbeam and robinia are counted as unusual though there are surprises such as the 50 *Ostrya carpinifolia* in Leicester. Aberdeen is the poorest town with regard to the diversity of street trees, over 90% being sycamore and lime. Oaks, beech and conifers are poorly represented in this habitat throughout the country. The regional pattern is determined partly by climatic and partly by cultural factors.

9.6 GRASS VERGES

Only about one-third of urban roads are designed to have a grass verge.

The normal method of establishment is to use topsoil together with a rye-grass-based seed mixture containing some very aggressive cultivars. Even before a proper sward has established disturbance in the form of vehicle transgression starts and continues with excavations to gain access to service runs, contractors set up their encampments, materials storage, and so on; most verges are regarded as general utility areas, an urban 'hard shoulder' rather than a landscape feature. Mowing 6–10 times a year prevents long grass accumulating and promotes tidiness.

The resulting sward is very variable so just a few of the main directions of variation will be dealt with. Frequently the sown species persist for a long time as the fertility, and height and frequency of cut are ideal for perennial ryegrass; however, disturbance levels allow in a regular component of weedy species. Where pronounced soil compaction occurs, plants that thrive under conditions of slightly impeded drainage become prominent, e.g. rough-stalked meadow-grass (*Poa trivialis*), creeping buttercup (*Ranunculus repens*), creeping bent (*Agrostis stolonifera*) and greater plaintain (*Plantago major*). Rye-grass may turn yellow due to the high watertable. Trampled verges and verge margins beside pavements have much in common with heavily used lawns containing large numbers of rosette and mat-forming species. Along the road margin disturbance is often sufficient to maintain bare areas; here an international community of trackside species can be found comprising shepherd's purse, knotgrass, pineapple weed, annual meadow-grass and greater plantain. One of the few places where slightly unusual species may be seen are banks too steep for easy mowing or vehicle incursion. In such sites fox and cubs (*Hieracium aurantiacum*), bird's-foot trefoil (*Lotus corniculatus*), sweet violet (*Viola odorata*), and similar attractive introduced and native species can sometimes be found.

9.7 ANIMALS

Dogs

The most frequently observed animal in streets is the dog (*Canis familiaris*). Though the large class of 'terriers' has been bred specially to destroy rats and other vermin their most conspicuous influence is on vegetation. If the slightest nitrogen deficiency is present in verge soils the deposition of faeces or urine stimulates the grass to grow taller and turn a darker green so the sward takes on a randomly spotty tussocky appearance. This is particularly noticeable where new grassland has been sown on poor-quality topsoil.

A more subtle effect of dogs can be observed on the base of street trees against which they urinate. This area, known as the canine zone, carries a different epiphytic flora to the rest of the trunk (Fig. 9.7). In towns with moderate air pollution the zone appears dark in contrast to the

Fig. 9.7. The canine zone on a street tree; darker patches are mats of *Prasiola crispa.*

remainder of the trunk which carries a pale grey–green cover of the lichen *Lecanora conizaeoides* or a bright green one of the alga *Pleurococcus viridis*. The effect is best seen on isolated pavement trees, trees at corners, those outside blocks of flats or at the entrance to parks which are used by dogs as signal posts. When well developed the epiphytic flora of the canine zone is dominated by the green alga *Prasiola crispa* present as its juvenile filamentous *Hormidium* stage. It forms a velvety felt on the bark, almost black when wet, a glossy emerald green when dry. Additional algae such as *Hormidium flaccidum* and *Stichococcus* sp. are usually present together with a nitrophilous form of the lichen *Lecanora dispersa* and the moss *Bryum capillare*. Marginal to the strongly eutrophicated zone the lichens *Physcia tenella* and *Phaeophyscia orbicularis* may be present together with the moss *Ceratodon purpureus*.

This community, known as the Prasioletum crispae, has been studied by the German botanist Knebel (1936) and in Holland by Barkman (1958). The association is species poor, prefers bark with a high water-holding capacity, reaches its maximum abundance during autumn and winter, and above all is nitrophilous requiring large amounts of nitrate and ammonium salts. In urban areas it will develop on the base of most non-coniferous trees but away from towns shows a preference for rain tracks and nutrient streaks on elm. It also occurs on trees where guano from birds' nests and roosts gets washed down the trunk.

Dog faeces have attracted the attention of dipterists. A study in London (Erzinclioglu, 1981) reported that the commonest flies visiting dog dung in streets were the bluebottle (*Calliphora vicina*), a greenbottle (*Lucilia sericata*), fleshflies (*Sarcophaga*), the housefly (*Musca domestica*) and the lesser housefly (*Fannia canicularis*). Flies reared from 74 samples of dung were mostly *Mus domestica*. A study in the city of Bath (Disney, 1972) found an entirely different assemblage breeding in dog excrement; this disparity is probably due to the different degree of urbanization in the two localities. Two generalizations can be made. Fewer flies emerge from dry than from moist dung and all workers agree that the majority of species seen visiting the dung for feeding do not breed in it. Smith (1973) has calculated that dog faeces in Britain could produce up to 130 billion (130^{12}) flies per annum and the number visiting it will be much greater. Their role in the transmission of enteric disease is not known.

One of the few groups of birds to show an association with urban roads are scavengers. Busy routes where there are lay-bys or bus stops with litter bins are worked over in the early morning by rooks, crows, magpies, starlings, gulls and house sparrows, some of which may nest in roadside trees. Grass verges are seldom allowed to grow long enough to support sufficient small mammals to attract kestrels. Several years ago Bagnall-Oakeley (in Simms, 1975) advanced the theory that under certain circumstances passing traffic creates vibrations in the ground which brings earthworms to the surface. He provided a certain amount of evidence in support quoting in particular a roundabout near Kings Lynn.

Roads have several negative effects on the wildlife of towns. Many years ago motor cars were described as 'smelly sparrow starvers' having replaced horses on whose dung the sparrows fed. Now they cause many casualties – young blackbird, song thrush, chaffinch and sparrow suffering particularly badly. Squashed hedgehogs, amphibians, grey squirrels and rats are also frequently observed. Colin Howes (1985) has pointed out that black cats are anything but lucky, over 60% of the cat road casualties found in his survey were predominantly black. Most cat casualties occur between March and May which is when mature toms are wandering during the mating season. This is also the peak for hedgehog road deaths with squashed males outnumbering females at this time of year by two to one. Despite the number killed, roads are not thought to represent a major threat to hedgehogs as in two experiments using marked animals only 2% and 4% respectively ended up as traffic casualties (Morris, 1983). The most positive way to view animal casualties is to remember that if large numbers of a species are observed dead on the road, this must reflect a particularly thriving population in the surrounding area.

Chapter 10

CITY CENTRES

City centres provide a challenging habitat for wildlife as, in addition to severe environmental restraints, nearly every animal is regarded as a pest and most plants as weeds. The types of continuously built-up site considered here are dense urban complexes where the often large buildings do not have gardens; open spaces are mainly streets, pedestrian precincts, squares, courtyards and car parks. Soft landscape is mostly confined to the curtilage of buildings, raised planters, tubs, window boxes and street trees though occasional communal gardens may be present. The general atmosphere is enclosure by buildings and paving underfoot. The consequences for the climate have been dealt with in Chapter 3 – basically these are an increase in temperature, air pollution and dust, and a decline in humidity. Soils are highly variable with regard to depth and texture; at ground level they tend to be eutrophicated, drought prone and compact.

10.1 ANIMALS

Birds

Ecologically city centres have been likened to cliffs, a comparison that is justified with regard to birds, a number of which have forsaken their rock faces and adapted to breeding in the continuously built-up area. These include feral pigeon, kestrel, starling, common swift, house sparrow, gulls and, in America, nighthawk and chimney swift. Most at home are pigeons which were noticed to be setting up free-living colonies in London as early as 1385. Descended from rock doves, which in the wild live among cliffs nesting in crevices and caves, they thrive in this new habitat where they favour institutional buildings constructed in the Victorian or Gothic style; these provide ample ledges for communal roosting, and dilapidated upper stories give access to covered sites for nesting. The birds are either fed by the public or forage in shopping areas, markets, squares, stations, even tube trains; most of their food is scrounged from man. Goodwin (1952) distinguished seven colour varieties of feral pigeon in London; blue (10%), blue-chequer (55%), velvet (10%), black (5%), grizzle (5%), pied and white (10%), mealy-red chequer-red (5%); all have irridescent green and lilac patches on either side of the neck. It is thought that this conglomeration of domestic strains may eventually stabilize and be given taxonomic recognition; compared with wild rock doves most individuals tend to be much less muscular, to have larger wattles, thicker bills and a different shoulder shape. They also show tameness but it is not known if this has a genetic basis.

A study of the comparative ecology of pigeons in inner London has

Table 10.1. Feeding areas of wood pigeons (*Columba palumbus*) and feral pigeons (*C. livida*) in inner London

	Large open spaces in parks	*Small parks and squares*	*Squares without grass*	*Private squares with grass but without bread*★
Wood pigeons	xxxx	xxx	x	xxx
Feral pigeons	xxx	xxxx	xxxx	x

	River edge at low tide	*Busy streets*	*Quiet streets*	*Inside stations*
Wood pigeons	x	–	x	x
Feral pigeons	xxxx	x	xxxx	xxx

	Lighted streets, stations at night	*In large trees*	*In small trees, shrubs, privet hedges*
Wood pigeons	–	xxxx	xxxx
Feral pigeons	xx	x	–

xxxx, very large numbers; xxx, considerable numbers; xx, small numbers; x, individuals only. ★ Bread covers all artificial food provided by the public.
Taken from Goodwin (1960).

Table 10.2. Prey items in 105 kestrel pellets from Urban Manchester

	No. of individuals	*Estimated proportion of the diet (%)*
Passeriformes (total)	52	64
? House sparrow	45	59.2
? Goldfinch	2	1.9
? Dunnock	2	1.9
? Starling	1	1.0
? Blue-tit or redpoll	1	0.7
? Pied wagtail	1	0.5
Columbiformes (pigeon)	4	9.9
? Laridae (Gull)	1	1.2
House mouse, *Mus musculus*	5	4.6
Wood mouse, *Apodemus*	2	0.9
Rat, *Rattus norvegicus*	4	14.8
Unidentified rodentia	2	1.2
Earthworms	(in 6 pellets)	0.9

Taken from Yaldon (1980).

demonstrated how the two most abundant species, the feral pigeon (*Columba livida*) and the woodpigeon (*C. palumbus*), maintain large populations by exploiting slightly different niches (Goodwin, 1960). Table 10.1 shows how feral pigeon prefers the harder landscape of squares without grass, streets, stations and small green spaces, while the woodpigeon population is mostly found where there are trees, shrubs, hedges, larger open spaces and some grass. Here they feed off buds, flowers, young leaves, berries and other fruits while the feral pigeons depend much more on bread and weed seeds. Nest site availability regularly limits the numbers of feral birds, the increasing destruction of old buildings and their replacement by ones of simpler design means every possible nest site is usually occupied. They always choose buildings. It will be interesting to see if, in the future, they revert to the (presumed) pre-rock dove ancestral habit of tree nesting which is currently where most of the woodpigeons breed. The same habit controls choice of roost site, woodpigeons in trees and feral pigeons on buildings where they may come into competition with starlings since both prefer ledges which provide protection from wind and rain. The pigeons usually win. Goodwin observed that since starlings, even more than pigeons, arouse the wrath of bureaucracy they may also be unwittingly responsible for control measures which cause feral pigeons to lose both roosting and nesting sites.

During the mid-twentieth century the kestrel adapted to urban conditions so it is now equally at home on the parapets and cornices of tall buildings as on the ledges of natural cliffs. The transformation has been gradual; for example, in London the first sighting was in 1931 rising to five or six pairs after the war. The real expansion began in 1972 when ten nests were located, now over 100 pairs regularly breed within Greater London. Today kestrels can be seen throughout the year in most of our cities; their nest sites vary from church towers, steeples, high-rise blocks, factory roofs, gas holders, cranes and pylons, to nest boxes specially constructed for them. An adaption which facilitated their colonization was a change in diet from voles and insects to small passerines. These are not hunted by hovering, instead the birds drift along roof tops looking for their prey. A detailed study of the diet of urban kestrels has been carried out in Manchester using pellets collected from the city centre (Yaldon, 1980). The results showed that in terms of weight, birds contributed 76% to the diet, small mammals 22%, insects and earthworms 1% each. The specific identity of the prey is listed in Table 10.2 from which it can be seen that house sparrows are the major item with feral pigeon, rats and house mice also significant. Less rigorous diet studies from other cities indicate minor variations such as a greater proportion of starlings in Southampton (Glue, 1973), and Pike (1979) mentions a pair in Birmingham which in spring fed largely off frogs. Despite kestrel numbers being close to saturation level there is no evidence that they are depressing populations of passerines; if ecological theory is correct the ecosystem should benefit from the presence of a predator.

The other breeding bird that maintains large populations in town centres is the house sparrow which uses smaller nooks, holes and crannies for breeding than do pigeons. Hudson (1898) and Goode (1986) have provided touching accounts of the pleasure they give to Londoners. Of all British birds the sparrow is the most dependent on man; if humans disappeared they might well become extinct as the St Kilda house mouse (*Mus muralis*) did after the evacuation. This is the perfect test of an urban specialist. Though common enough in streets, sparrows reach their densest concentrations in city parks and gardens, railway depots, allotments and other service areas. The factor that appears to control their numbers is the supply of food that can be obtained without much trouble. The replacement of the horse by motor traffic hit their populations hard on both sides of the Atlantic. At sites where birds are fed, gangs of sparrows compete well with pigeons, gulls and mallards expeditiously darting down to seize and carry off food; owing to its smaller size and chubbier build it often gets fed preferentially.

Certain ecologists have designated inner city areas 'the swift zone' (Sukopp *et al.*, 1979) but this is not quite correct as, though these birds

Fig. 10.1. Enormous flocks of starlings roost in towns where they show a preference for institutional buildings constructed in the Gothic or Victorian style (M. Birkhead).

occasionally hunt insects over the central areas, they are primarily birds of the inner suburbs. It is difficult to know exactly which birds to include as breeding in town centres as this is determined by the structure of the urban core. Feeding opportunities provided by small patches of grassland or wasteground bring in starlings and only a small amount of cover is required to attract songbirds such as goldfinch, blackbird and thrush all of which breed at a low density in inner cities (Smith, 1985). A discussion of the black redstart is reserved for the chapter on industrial areas. Despite rather few species being involved, Dorney (1979) has compiled data from Ontario, Canada, which shows city centres supporting a higher density of breeding birds than any other urban zone.

In addition to holding breeding populations of up to a dozen species of bird, city centres are used as autumn and winter roosting sites by enormous flocks of starlings (Fig. 10.1) and to a lesser extent by pied wagtails and gulls. This relatively recent phenomenon has been carefully studied in London where the influx of starlings was first brought to citizens' attention by a letter in *The Times* on 3 November, 1894. Research by Max Nicholson showed that the town roosting habit undoubtedly arose among British resident starlings rather than continental immigrants. The birds come from no further afield than the suburbs where they gather into small groups around sundown and fly into the centre gradually forming larger and larger flocks till when they reach their roosting sites they almost darken the sky. 'Starling London' contains an outer zone where the birds stay all the time or fly away from the city to roost, an inner zone where they feed by day but fly into the centre to roost at night, and a central zone where they roost in large numbers at night but very few spend the day there. The preferred roost site varies with the season; in August, September and October deciduous trees in parks are favoured only to be abandoned after leaf-fall for buildings. Warmth and shelter seem to be the requirements sought, but Fitter (1945) has suggested a simpler explanation for the change in habit – the loss of traditional roosting sites to suburban development.

Another bird that has changed its habits to take advantage of the urban core is the pied wagtail. Outside the breeding season pied wagtails mostly roost in reedbeds, but since about 1930 an increasing number of flocks have adapted to roosting in town centres. Dublin contains the most famous example but a number are now known in London and other towns where they select trees, bushes or buildings which provide some protection from the weather. The largest roosts involve several thousands of birds.

Under section 74 of the 1961 *Public Health Act,* local authorities are empowered to cull starlings, house sparrows, house doves and pigeons which are all believed to represent a health hazard through harbouring pests and diseases. Though it is debatable whether the risks outweigh the pleasure they bring to town dwellers, it is a fact that many of the complaints received by environmental health officers are directly or indirectly associated with birds, particularly the obvious nuisances of noise and guano.

More serious are two fungal diseases of man, cryptococcosis and histoplasmosis, for which pigeons and starlings respectively are the main carriers (Rippon, 1982). Cryptococcosis is caused by the yeast *Cryptococcus neoformans*; though not pathogenic to pigeons, it is frequently present in their droppings where it may remain viable for several years. The first record for Britain came from pigeon roosts in central London (Randhawa *et al.*, 1965) where the disease had been contracted by those clearing up deposits of pigeon droppings. The fungus flourishes in the accumulated filth and debris of pigeon roosts where it is often the dominant organism, thriving in the desiccated alkaline nitrogen-rich conditions. Current estimates of the incidence of cryptococcal disease in USA range from 200 to 300 cases of cerebral meningitis a year to 15 000 subclinical respiratory infections in New York City alone. It has been called the 'awakening giant' of fungal diseases. Histoplasmosis is caused by the fungus *Histoplasma capsulatum* which grows in soil enriched with the guano of birds and bats. The highest incidence of the disease in America is also the area with the greatest concentration of starlings. Starlings were successfully introduced into New York in 1890–91 as part of a project to establish in the States all the plants and animals mentioned by Shakespeare. They flourished mightily and outbreaks of histoplasmosis are particularly

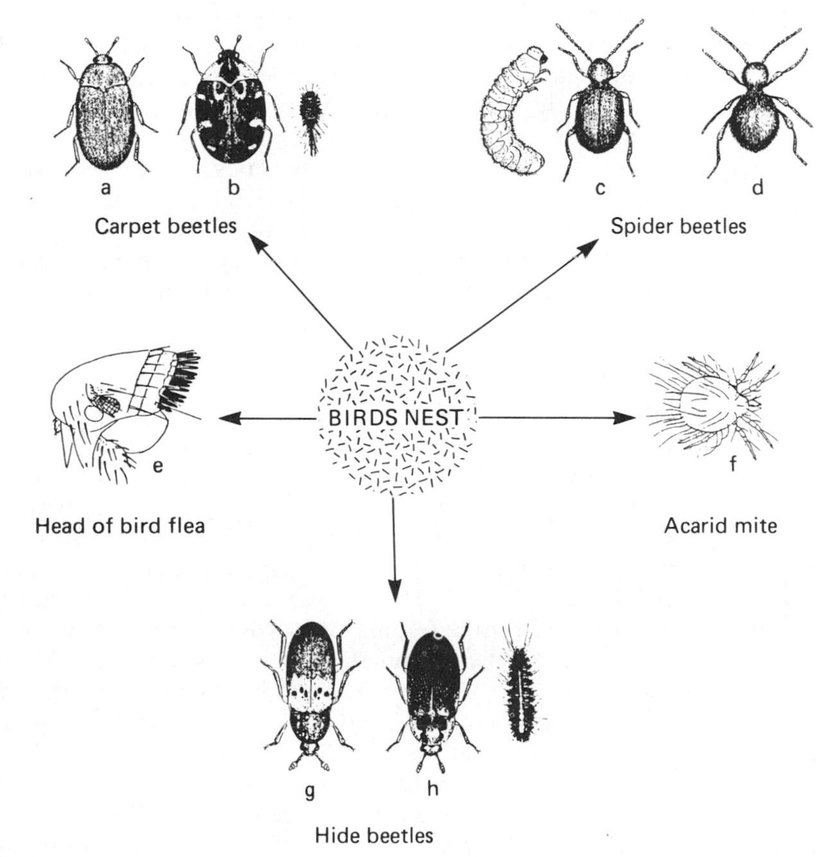

Fig. 10.2. Commensurals inhabiting birds nests which establish secondary colonies inside buildings once the birds have left. a, *Attagenus pellio*; b, *Anthrenus scrophulariae* with larva; c, *Ptinus tectus* with larva; d, *Niptus hololeucus*; e, *Ceratophyllus gallinae*; f, *Glyciphagus domesticus*; g, *Dermestes lardarius*; h, *Dermestes maculatus* with larva.

associated with their roosts. Epidemics of this disease which affects the reticuloendothelial system and enters via the lungs are frequent in the USA and include a troop of Boy Scouts that cleaned a park in Missouri, workers clearing herring gull detritus at a port in Michigan, and the inhabitants of a building where an air conditioner intake was located near a pile of starling droppings.

Of considerable interest to ecologists are the large number of commensurals associated with bird nests. Over a hundred species of beetle are known to be linked with them; they reach maximum numbers just after the young have left. Many are useful to their hosts as they scavenge the nest and may prey on the fleas and their larvae which are usually present. After the birds have left, the insect fauna tends to disperse in search of new food. Since pigeons and house sparrows nest inside city centre buildings the influx of arthropods searching for a new food source can be considerable and gives rise to many complaints. Spider beetles, hide beetles, carpet beetles, larder beetles, mites and bird fleas are the species mainly involved (Fig. 10.2). They set up secondary colonies associated with woollen carpets, furs, vegetable matter and especially dry animal protein. Many of the most troublesome are introduced species, for example the Australian spider beetle (*Ptinus tectus*) and the golden spider beetle (*Niptus hololeucus*) from Turkey.

Mammals

Mammals have shown less adaptability than birds or beetles; the only ones regularly found in town centres are brown rat, house mouse, feral cat and feral dog. In towns the brown rat frequents sewers, warehouses and docks where they form large underground colonies from which they emerge at dusk to forage. If disturbed, by, for example, reconstruction work, they may be seen above ground during the day which gives rise to complaints. The brown rat (*Rattus norvegicus*) probably reached England in ships trading with Russia around 1730 and over the next 200 years largely exterminated the highly urban black rat (*Rattus rattus*). Today the black rat is only found in any numbers in the dockside areas of large cities such as London, Liverpool and Glasgow where it is a sewer and service duct species. This is a further example of the dynamic nature of urban ecosystems. Matheson (1944) studied the domestic cat (*Felix catus*) as a factor in the ecology of Cardiff. He estimated that the cat population was at least 13% of the human population and around a quarter were feral. This agrees with proportions suggested by Hudson (1898) for London. Today the capital is thought to contain half a million stray cats which in central areas reach a density of 12 per ha (1 cat per 0.2 acres). Tabor (1983) has some surprises when it comes to describing the diet of feral cats which was investigated by examining stomach contents and direct observation. Seventy-five per cent of their diet comes from auxiliary feeding, in other words food put out by dedicated cat lovers, often elderly ladies or groups of workers in shops and factories. Nearly all the rest (25%) is obtained by foraging and scavenging garbage from dustbins and paper or plastic sacks. The average household disposes of about 2 kg of food per week which provides rich pickings. Hospitals also produce a great deal of garbage which is often accumulated in huge containers on wheels which stand out in yards. Cats can frequently be found in these searching for titbits. When a city centre hospital in Sheffield closed recently 60 cats were discovered living in its basement. Negligible amounts of food which had been caught were found in their stomachs. Around 10% of a cat's diet is vegetable matter originating from garbage and the natural grazing of leaves which contribute a roughage element.

Tabor's picture of the harmless feral cat may not completely sum up their ecological influence as Colin Howes of Doncaster Museum has provided me with the results of a survey he conducted entitled 'What the cat brought in', it involved several thousand animals. From his study it appears that in towns, cats kill a lot of prey which they do not eat. Urban cats brought back to their owners a wide range of prey items which numerically comprised 56% birds, three-quarters of which were house sparrows, 34% mammals, mostly mice and rats but also a few shrews, and 10% invertebrates. Compared to suburban cats there was a considerably higher proportion of birds and fewer mammals.

10.2 VEGETATION

The bulk of the vegetation in town centres is composed of planted ornamental species ranging from forest trees to collections of annuals in window boxes. The small widely scattered islands of green in the hard landscape offer only a limited starting point for food chains as most species, being aliens, have just a small attached fauna. Structural complexity tends to be low, as layouts have to cater for ease of maintenance, but this is to some extent compensated for by intricacy of the build form. In its openness, instability, highly discontinuous character and wide range of growth forms the vegetation has similarities to that found on cliffs, but the comparison should not be taken too far. The tiny fraction of the biomass contributed by 'weeds' has been closely studied in towns throughout Europe. These communities on walls and pavements are very suitable for ecological research as their simplicity often enables controlling influences to be identified.

Trees

The largest trees in most town centres are associated with churches and to a lesser extent civic buildings and squares as these are the only sites where there is sufficient space for bulky vegetation. Ash, elm, sycamore and London plane are the leading species. All except for the plane are equally common in the countryside, so it deserves to be considered the characteristic tree of town centres. Being entirely domesticated with no native habitat the epithet 'London' is particularly appropriate. A deciduous tree growing to an immense size, it is well-known because of its bobble-like fruits and habit of shedding large plates of smooth bark which leave the trunk conspicuously mottled with pale yellow. After the hot summer of 1976 planes shed their bark in prodigious quantities. There is considerable disagreement regarding the origin of the London plane. School children are taught that it is a hybrid between the Oriental plane (*Platanus orientalis*) and the American plane (*P. occidentalis*), and that its correct name is *Platanus* × *acerifolia* (syns. *P.* × *hispanica, P.* × *hybrida.*) The hybrid, so the explanation goes, probably arose in southern Europe around 1650, and was first planted in England about 1750. The alternative view is that the London plane is a seedling variant of *P. orientalis* that became fixed by cultivation and should be known as *P. orientalis* var *acerifolia*. Though the majority of authorities accept a hybrid origin because of its intermediate morphological characters, its great vigour, a common feature of first-generation hybrids, and the variability of its seedlings, others are equally adamant that it is not. Several cultivars exist which are readily propogated by cuttings so a keen eye can often detect clones. The London form, which is also widespread on the continent, has a characteristic crown; the branches are somewhat tortuous and the perimeter is intricately branched giving a surprisingly delicate winter silhouette. By comparison most other cultivars are nondescript, being smaller, coarser and with less interesting bark patterns; one clone does not shed its bark. It is thought that the London plane got its name after Repton used it in his design for Bedford Square; prior to that it was known as the Spanish plane (Gorer, 1985).

Apart from its noble proportions the London plane has other virtues which have pre-adapted it to the urban scene, such as a resistance to air pollution caused by smoke and sulphur dioxide, an ability to survive and regrow to a good shape after lopping, and it scarcely ever blows over. However, every tree has some disadvantages for landscape work and those of the London plane are that it can grow too large for the relatively small spaces available among buildings; it should be noted that the original trees planted at Ely and Barnes over 200 years ago are still increasing in size; in early summer it sheds clouds of downy hairs to which certain people are allergic; the tree is also suspected of being unusually sensitive to de-icing salt. Recently the fungus disease anthracnose caused by *Gnomonia platani* is believed to have been responsible for a number of tree deaths originally put down to drought. If a new virulent form of the pathogen has evolved this could be catastrophic for town spaces. The modern wave of tree planting in city centres employs medium-sized cultivars of *Alnus, Sorbus, Betula, Prunus, Robinia* and *Fraxinus* which are sited in paving or raised beds and, owing to the harsh microclimate, are little better than planes for wildlife.

In a study of urban Ohio, USA, Whitney and Adams (1980) found that the central city areas were characterized by tree of heaven (*Ailanthus altissima*), Norway maple (*Acer platanoides*), silver maple (*A. saccharinum*) and white mulberry (*Morus alba*). These 'weed' trees are noted for their ability to withstand stress. Lund (1974) has suggested that a general preadaption to disturbed habitats may favour certain species in the urban environment; *Acer saccharinum* for example is a common inhabitant of flood plains in the mid-west of the United States. This hypothesis would be worth following up as white mulberry has a long history of cultivation for the silkworm industry in China and both the tree of heaven and willow-pattern tree (*Koelreuteria paniculata*) have long been associated with highly disturbed open field conditions in the north of China.

Flower beds, tubs, window boxes

Turning now to herbaceous plants, flower beds, tubs and window boxes are planted with a conservative range of short-lived ornamental species selected for their extended flowering season and an ability to thrive in dryish soil. The native habitat of the most popular plants has been determined to discover if they form a distinctive ecological group. The ubiquitous blue lobelia (*Lobelia erinus*) is endemic to South Africa which is also the home of the genus *Pelargonium* that includes our red garden, ivy-leaved and regal geraniums. They inhabit dry rocky slopes and flats where *Cineraria, Helichrysum* and *Mesembryanthemum* also have their headquarters. Arid rocky places in Mexico have provided the popular French and African marigolds (*Tagetes patula, T. erecta*) together with *Agaretum houstonianum*. The very common white alyssum (*Lobularia maritima*) comes from strand lines in the eastern Mediterranean, while nasturtiums (*Tropaeolum majus, T. minus*) originate from scrubland and high mountain forests in central South America where they are believed to be pollinated by birds. The most popular summer bedding plants in city centres therefore are predominantly annuals derived from rocky arid habitats in South Africa and Mexico. Apart from the flowers providing a nectar and pollen source for common insects and the foliage supporting occasional larvae of widespread moths such as the large yellow underwing (*Noctua pronuba*) or lesser yellow underwing (*N. comes*), these plants play only a minor role in the city centre ecosystem.

Natural vegetation

Many people are unaware that well-defined communities of spontaneous natural vegetation penetrate into the heart of the world's major cities. One of the clearest demonstrations of this wild flora is the study of central London carried out by Hadden (1978) who recorded 157 species in the small (<4 km^2), heavily built-up tract covered by postal district W1. She found that the habitats where plants occur could be grouped roughly into three types. The first, 'cultivated sites', included gardens, windowboxes, raised planters and tended soil at the base of the trees. The second, 'uncultivated sites', were mainly building sites used as car parks – the vacant lots of Americans. The last, 'stonework', included pavements, roads, cobbled mews and walls. Table 10.3 lists the seven species that occurred frequently in all these habitats and a further nine that were reasonably widespread in two of them. The majority of these have seeds adapted for wind dispersal and it is likely that wind funnelling down streets on dry days could also be the main dispersal agent for seeds of the other species, though alternative mechanisms do exist. The remaining 141 plants were much less widespread: 77 occurred at a single site each, 31 of these were present as one individual and eight probably never flowered. Some species were totally unexpected such as knotted hedge-parsley (*Torilis nodosa*) at the footings of a church (the nearest population is in the grounds of Buckingham Palace), burnet saxifrage (*Pimpinella saxifraga*) in a lawn, and lesser spearwort (*Ranunculus flammula*) in a flower bed. Most, however, were the expected ruderals, aliens (*Digitaria sanguinalis, Datura stramonium, Ameranthus* sp.) and garden escapes [*Lobelia erinus, Lobularia maritima, Oxalis corniculata* and *Petroselinum crispum* (garden parsley)]. With the

Table 10.3. The most frequent wild flora of the London W1 postal district

Commonest species	*Slightly less frequent*
Chenopodium album	*Achillea millefolium*
Epilobium angustifolium	*Bellis perennis*
Plantago major	*Buddleia davidii*
Poa annua	*Cirsium arvense*
Sagina procumbens	*Conyza canadensis*
Senecio squalidus	*Dryopteris filix-mas*
Sonchus oleraceus	*Epilobium ciliatum*
	Pteridium aquilinum
	Rumex obtusifolius

Taken from Hadden (1978).

large number of species as single plants, or at one site only, the dynamic nature of the vegetation becomes apparent. City centre vegetation tends to be rather impermanent and variation from year to year is a major characteristic.

The most well-defined and stable community of town centres is found colonizing the joints and cracks in paved surfaces where an assemblage known to phytosociologists as the Sagino-Bryetum argentei is widespread. The severe environmental conditions, particularly trampling, make for a simple structure and low species diversity but the harshness of the habitat is somewhat tempered by the soil which, though compacted, tends to be highly fertile. The constant species in order of decreasing frequency are *Poa annua, Sagina procumbens, Plantago major, Taraxacum officinalis* (agg.), *Polygonum aviculare, Capsella bursa-pastoris, Lolium perenne* and the bryophytes *Bryum argenteum* and *Ceratodon purpureus.* All the higher plants show considerable phenotypic plasticity so can adapt to growing in declivities among ever so slightly uneven paving stones. Initially they take on the form of minute rosettes of leaves or flattened shoots which may later expand into prostrate mats. Elastic leaf and flower stalks aid survival. According to Blum (1925), on the continent, they exhibit several flowering periods controlled by the seasons, for example March–May, August–September and October–December; in Britain, however, the species flower throughout the summer.

In very heavily trampled sites an impoverished community consisting of *Poa annua, Plantago major, Sagina procumbens* and *Bryum argenteum* occurs with a reduced cover. Conversely, where trampling is relaxed such as at the base of walls and lamp posts, or in courtyards and service-yards, a wider range of species including many nitrophiles join in, for example *Stellaria media, Matricaria matricarioides, Chenopodium album, Artemisia vulgaris, Sonchus* spp., *Hordeum murinum, Galinsoga parviflora, Buddleia, Epilobium angustifolium.* Periodically these get sprayed out or weeded out by hand but sufficient survive to provide possibly the major seed reservoir for the Sagino-Bryetum argentei which along trampling gradients shows transitions to adjacent stands of taller vegetation. Several methods of seed dispersal rely on people. Clifford (1956) found that seeds were present in great numbers in mud attached to footwear and from many samples tested discovered that *Poa annua* was the most frequent species, though all members of the pavement community could be demonstrated. Dispersal over longer distances may be associated with mud caught in the tread of vehicle tyres. Darlington (1969) scrubbed the tyres of a car clean, drove 100 km then carefully washed the wheels; from the sediment thirteen species germinated, the most numerous being *Poa annua* (387 seedings), *Stellaria media* (274) and *Matricaria matricarioides* (220).

Using Ellenberg's (1974) Tables of the indicator values of vascular plants it can be deduced that ecological conditions in the cracks of typical pavements in north-west Europe are half-light, and the soil is reasonably moist, has a neutral reaction, and is rich in mineral nitrogen. The mesic soil conditions are perhaps the most surprising feature but here the abundance of *Sagina procumbens* is revealing. The seeds of this pearlwort need damp soil for germination and establishment but in the latter stages of development exhibit optimum growth on drier soils so it is suited to considerable fluctuations in soil moisture content. With increasing dryness the frequency of *Sagina procumbens* falls and that of sand-spurrey (*Spergularia rubra*) rises, as this species prefers well-drained sites.

Geographically the communities of trampled sites are believed to gradually change round the globe with species of similar growth-form and taxonomic relationship substituting for each other in a remarkable way (Segal, 1969). This is demonstrated in Table 10.4 which shows how *Plantago* spp., for example, change from *P. major* in the lowlands of western Europe, to *P. alpina* in mountainous areas, to *P. crassifolia* in the warmer regions of Europe, to *P. coronopus* in coastal sites, while in Asia *P. asiatica* takes over. In continental Europe the pavement community is structurally, ecologically, taxonomically and physiognomically strongly related to that occurring in Britain but *Poa infirma, Sagina apetala, Capsella rubella, Plantago crassifolia, Desmazeria rigida* and *Herniaria glabra* replace our eu-atlantic assemblage. Other more local directions of variation can be seen in moist sites near pipe overflows, below leaky gutters, etc., which become rich in the thallose liverworts *Lunularia cruciata* and *Marchantia polymorpha*, while a sure sign that trampling has been relaxed is the appearance of the nettles *Urtica dioica, U. urens* and annual mercury (*Mercurialis annua*).

Botanists find town centres exciting places as the diversity of seed sources and range of microhabitats make for the unexpected. For example the *Flora of Inner Dublin* (Jackson and Skeffington, 1984) records a temporary population of lentil (*Lens culinaris*) and chickpea (*Cicer arietinum*) outside a health-food shop, an event which is entirely explicable – a rare circumstance in ecology. Trafalgar Square in London is an area of hard landscape dominated by people, pigeons and a few small trees, yet it provides a promising site for birdseed aliens. Vendors sell a mixture of barley and maize to the public who entertain themselves by scattering it for the birds; some falls down tree-grills where it germinates but it is difficult to identify as the leaves get worn away by trampling feet. From 100 g of seed I purchased there recently, C.G. Hanson identified *Lolium temulentum, L. multiflorum, Panicum mileaceum* and rough corn bedstraw (*Galium tricornutum*) which has not been seen in the London area for over 30 years. Prodigious numbers of

Table 10.4. Physiognomic and taxonomic relationships among species occurring on trampled sites in a range of geographical provinces

Lowlands of W. Europe	*Poa annua*	*Sagina procumbens*	*Capsella bursa-pastoris* *Coronopus didymus*
Mountains of Europe	*Poa supina*	*Sagina saginoides*	
Warmer regions of Europe	*Poa infirma* *Eragrostis pilosa* *Sclerochloa dura*	*Sagina apetala* *Spergularia rubra*	*Capsella rubella*
Coastal regions of Europe	*Catapodium marinum*	*Sagina maritima* *Spergularia media*	*Cochlearia danica* *Coronopus squamatus*
Asia		*Sagina japonica*	
Lowlands of W. Europe	*Lepidium ruderale*	*Plantago major*	*Lolium perenne*
Mountains of Europe	*Lepidium densiflorum*	*Plantago alpina*	
Warmer regions of Europe	*Cardaria draba*	*Plantago crassifolia*	*Desmazeria rigida*
Coastal regions of Europe		*Plantago coronopus*	*Parapholis strigosa*
Asia	*Lepidium virginicum*	*Plantago asiatica*	

Taken from Segal (1969).

seeds must be introduced in this way so any town centre where pigeons and sparrows are fed will contain temporary populations of birdseed aliens. These should be looked for in flower beds and particularly on paved areas which, owing to their ability to store heat, have a microclimate sympathetic to plants originating from further south; autumn is the best time to search for them. A former seed source, which appears not to have been studied by botanists, was associated with horses' nose-bags. Introductions originating in this way are likely to have been considerable.

A very different urban habitat is provided by the sunken areas in front of houses which allow light to reach basement windows and pavement coal-hole shafts covered by cast-iron gratings. These sheltered, moist, often frost-free niches can support unusual ferns. Bracken and male fern are the commonest but several others such as lady fern and hart's tongue are reasonably frequent in the north and west. Occasionally ferns, the spores of which must have originated from indoor cultivated plants, are found, for example maidenhair fern (*Adiantum capillus-veneris*) and *Pteris* spp. in the home counties and *Cyrtomium falcatum (Polystichum falcatum)* in the centre of Cheltenham.

In a historical context, bombed sites have played an important part in the development of the urban flora. Less than 3 years after the blitz Salisbury (1943) described the plants that had colonized the ruined houses in London. Among the most frequent species were rosebay willowherb, Oxford ragwort, stickey groundsel (*Senecio viscosus*), Canadian fleabane, gallant soldier (*Galinsoga parviflora*) and buddleia. All these plants were at that time undergoing a period of steady expansion which was given renewed impetus by the sudden availability of the bombed site habitat. This acted like a catalyst. As populations built up they exerted a tremendous inoculation pressure on the urban area and consequently were able to spread into new habitats. After the war they were found to have become permanent members of the urban flora in most of our heavily bombed cities where previously they had often been rather rare. It is speculated that fires and the presence of scorched earth may in some way have aided their spread. A parallel had occurred many years earlier when, following the Great Fire of London in 1666, Ray recorded a population explosion of *Sisymbrium irio* which was subsequently named the London rocket. Soon it became established in towns over a wide area but it is now extremely rare and one of the few urban species to be included in the *Red Data Book for Vascular Plants* (Perring and Farrell, 1977). After a century's absence it reappeared in London at the end of the war where it still persists near the Tower and around Regents Park Zoo. It is to be expected that a number of today's common ruderals will have shown a similar decline 300 years from now.

Another botanist who made a close study of London's bombed sites was Lousley. He was the first to notice American willowherb (*Epilobium ciliatum*) which is now ubiquitous and he also recorded a new hybrid *Senecio (S. squalidus × S. viscosus)* subsequently named *S. × londinensis*; being of low fertility it has not spread a great deal. One of the more surprising features of the bombed sites was the rapid appearance of wild figs (*Ficus carica*). They were first noticed as young plants (Lousley, 1948) many of which have now grown into large trees. Prior to this, occasional naturalized fig trees were known on river walls below sewage works (Fig. 15.4). Today naturalized figs are being recorded in increasing numbers away from rivers. It is generally assumed that they originate from discarded pieces of fig included in picnic lunches (Burton, 1983); seeds from both dried and fresh figs bought in the high street germinate readily. Certain trees, however, occur in inaccessible places such as high walls where a bird-sown origin is more likely; these may have been derived from discarded greengrocers stock, as enquiries have shown that up to 20% of each consignment has to be thrown out. Though no convincing explanation for the origin of the wartime figs in London exists, the fruit was very scarce in Britain at that time, the period definitely encouraged them by providing plenty of brick bats, porous stone and lime rubble which is what horticultural books recommend they are planted in. *Ficus carica* is a native of West Africa and the eastern Mediterranean from Syria to Afghanistan where it grows in stoney rocky soils and clefts in cliffs so can be regarded as a further example of a cliff-inhabiting species taking to town centres.

10.3 INTERACTIONS

The wildlife of town centres is particularly labile, reflecting not only man's changing activities but also accommodating native species of animal (kestrel) and plant (rosebay willowherb) which through small adjustments become capable of exploiting the conditions. The history of spread of many of these opportunists has been carefully charted by naturalists who often record an initial slow and uncertain start to the invasion which gradually accelerates through a 'grand period of expansion' till the habitat is saturated whereupon normal limiting factors start to operate. What is less often recorded are the ripples, or interactions, which accompany these invasions. Owen (1978) relates how the population explosion of rosebay willowherb which culminated in its rapid colonization of bombed sites in London and other European capitals was accompanied by a dramatic increase in the number of elephant hawk-moths (*Deilephila elpenor*) whose caterpillars feed on its leaves. In the 1940s and 1950s there were elephant hawk-moth caterpillars all over London. At first they were relatively free from the attacks of parasitic flies and wasps which in rural areas are responsible for the deaths of a high proportion. Later the parasites arrived in numbers and they, together with redevelopment of the wasteland, probably accounted for the decline in elephant hawk-moths, though they are still reasonably common. During the period of maximum abundance many caterpillars turned from the juvenile green to the mature black colour after the fourth rather than the normal fifth moult. Experiments showed this phenomenon was related to crowded conditions which also caused the larvae to grow faster and pupate earlier, a behavioural change which could be interpreted as advantageous. Another species of moth which increased its range and abundance along with its wasteland foodplants (*Artemisia absinthum, A. vulgaris*), and then declined, is the wormwood shark (*Cucullia absinthii*). Plant–animal interactions such as these have largely gone unrecorded.

Chapter 11

CITY PARKS

Many of today's older parks originated as the private gardens of eighteenth and nineteenth century houses donated by philanthropic individuals to the town where they had made their fortune. The layout of these gardens was heavily influenced by current fashion which drew on the concepts of French landscape painters such as Claude, Poussin and Salvador Rosa who depicted harmonious scenes of man within nature. At that time nature was seen as beautiful, and the answer to the problems of the world were believed to lie in man's harmony with nature. Many town parks still contain echoes of this design philosophy – the picturesque style of landscape gardening – in which an enlarged corrected refined and idealized portrayal of nature is presented at an aesthetic level. The wildlife component of the scene was introduced by the construction of aviaries, aquaria and sometimes a small zoological garden. So, in these early days the idea of nature in the park was seen as entirely appropriate, provided it was well under control.

The social origin of the city park lay in the Victorian zeal for reform. They were seen as places for exercise to promote the health, comfort and moral well-being of the lower classes who, it was decided, would also benefit from contact with natural beauty as a relief from squalid overcrowded housing conditions. Alongside these purely altruistic reasons was the belief that they might also dampen down social unrest and coincidentally increase efficiency and productivity. Medical theory at that time postulated that diseases were the result of bad air which gave further impetus to providing the city with lungs in the form of open space. So by the end of the nineteenth century parks were widely accepted as being an essential component of all cities. At the same time the poor physical condition of recruits for the Boer War suggested the need for more active recreation. Games became part of the school curriculum and facilities for sports were added to parks in the form of playing fields, tennis courts, swimming baths, boating lakes, bowling greens and small golf courses. Civic pride was now a determinant of the character of parks and a very high standard of maintenance was achieved so that weeds, dead flower heads, even moss under benches, were all but eliminated.

Around this time the evolution of parks slowed right down; they gradually adapted to the increasing need for active recreation but the training of the superintendents was still horticulturally based. By the mid-twentieth century social conditions had changed beyond recognition with more leisure time, higher incomes, greater mobility and better education. However, the parks had not changed greatly, their role in society was now uncertain and they were becoming starved of finance as first crematoria then multipurpose sports halls became prestige areas. Today they are still being run down in terms of resource allocation which is providing an opportunity for pressure groups to introduce new types of low-cost informal management which favour wildlife. This is likely to herald an era of park redesign that will change the face of public open space in Britain as surely as the addition of active sports facilities did nearly a hundred years ago. On the continent, particularly in Holland, wildlife has been encouraged in parks for several decades, while an even earlier experiment took place in Germany during the Third Reich when exotics became enemies of the state and for a short time naturalistic planting flourished anticipating the main thrust of this style by 40 years.

When discussing parks in an ecological context it is necessary to remember that to be successful they need to fulfil a number of sometimes conflicting uses. These include active recreation, strolling, sitting, courting, dog walking, sunbathing, fishing, visual delight, display of historical features, appreciation of nature and enjoyment of ornamental horticulture; all against a background of financial restraint and park keepers who have received a conventional training. Changing the design and management to emphasize wildlife will only be successful if conflicts with pre-existing patterns of use are minimized and the public are kept fully informed of the reasons for change. The majority of wild life in towns is there incidentally, it is present because it is preadapted to the types of habitats that man creates for his own convenience and delight. But in parks with a permanent staff of gardeners the possibilities for manipulating the habitat to favour attractive plants and animals are considerable.

11.1 HABITATS WITHIN A PARK

The habitats provided by a park vary considerably depending on its age, size, function, topography and particularly the philosophy of its designer (Fig. 11.1). In every case mown grass containing scattered trees is the leading habitat often well supported by areas of woodland. Formal

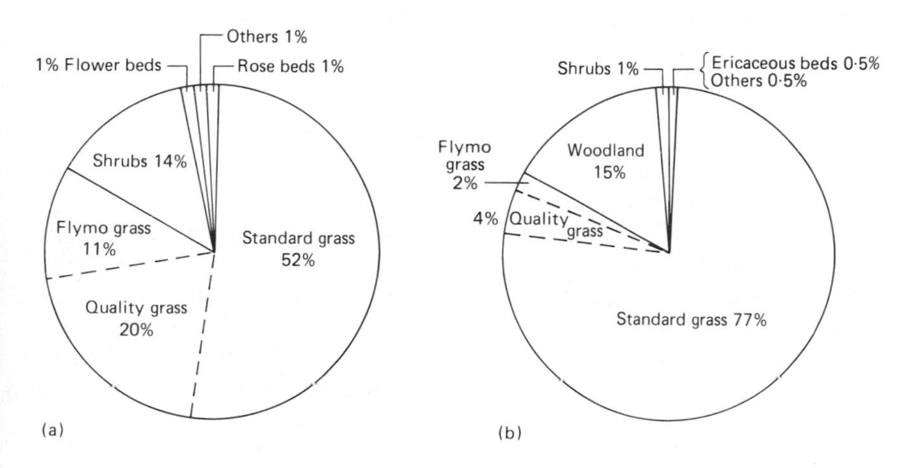

Fig. 11.1. Pie charts showing proportional representation of the habitats present in two types of city park in Sheffield. (a) prestigious, old-fashioned; (b) a district park combining active and passive recreation.

beds of roses, annuals, herbaceous plants and rockeries usually occupy a relatively small area as they are expensive to maintain. Shrubberies though initially costly are in the long term cheaper than mown grass to manage so are usually well represented.

The value of these habitats to wildlife depends on factors such as their size, plant species diversity, structure, availability of microhabitats and shelter, but in addition management is crucial. Operations that are particularly harmful are the use of herbicides, alterations to long-established management routines, undue emphasis on neatness and tidiness at the margins, and allowing water bodies to become hypertrophicated. The presence of large numbers of exotic plants, close mowing, disturbance by people and dogs, litter and vandalism are less detrimental to urban wildlife and may encourage it.

11.2 GRASSLAND

Grassland communities normally occupy between 75% and 95% of a park. It is the ideal surface for showing off buildings, flower beds, shrubberies and specimen trees, for walking over, lying upon or playing games on. Also it is easy to maintain and repair. Leaving aside very high-quality sports turf such as bowling greens, park keepers recognize three categories of grass: high standard, standard and occasionally mown. High standard grass is a fine-leaved bent-fescue turf mown once a week during the growing season at a height of 8 mm; clippings are boxed. Once or twice a year it receives a dressing of fertilizer to maintain vigour and promote good colour. In the past it was regularly aerated, scarified and also received a spring application of selective herbicide but finance now dictates that these operations are carried out more irregularly. Grass of this standard is universal in the ornamental core of parks, round entrances, and beside the main walks. With time the original sward composition alters with fescues becoming dominant on the drier sites. Overapplication of fertilizer leads to pearlwort (*Sagina procumbens*) and annual meadow grass becoming prominent, while, if nutrient levels fall, bryophytes such as *Calliergon cuspidatum, Eurhynchium praelongum* and *Rhytidiadelphus squarrosus* invade. One of the chief weed grasses which can stand close mowing (down to 5 mm) is Yorkshire-fog, while small-leaved forms of creeping white clover and yarrow may prosper. Not a great deal of wildlife can coexist with this intensity of management which is specifically designed to maintain the sown mix. In autumn, after mowing has ceased, older examples of this type of grassland which have developed a mat or 'sole' of organic matter may produce good displays of colourful fungi particularly fairy clubs (*Clavariaceae*) and wax caps (*Hygrophoraceae*), with mycorrhizal species under trees.

Standard grass is basically a rye-grass (*Lolium perenne*) sward. It is often sown as a mixture in the belief that by including small *Poa* and *Festuca* species a two-layered community will develop, increasing wearing capacity and reducing opportunities for weeds to invade, but in practice after a few years rye-grass is completely dominant. These grasslands are mown every fortnight during the growing season using machinery that does not collect clippings. Rye-grass has a wide ecological tolerance, only failing on soils that are heavily water-logged. It is the most hard-wearing grass in general use, easily beating commercial strains of crested dog's-tail (*Cynosurus cristatus*) which were once believd to possess a high tolerance of trampling. Numerous 'turf-type' cultivars of *L. perenne* are available; the current trend is to produce prostrate strains which require less mowing. Common grassland weeds tolerant of competition are frequently present in rye-grass swards where the sedge *Carex hirta* and moss *Calliergon cuspidatum* may become locally dominant, possibly as a result of soil compaction causing impeded drainage.

Occasionally mown grass is an infrequent community which may be missing altogether in the more central parks and when present is mostly associated with steep slopes or areas of tree planting. It offers much more in the way of variety than other categories with umbellifers, meadow buttercup, knapweed and a variety of legumes often present. Its value is well understood at Kew Gardens where areas of tall grassland are increasingly being left to develop for their interesting species composition and as a foil to close mown lawns. Management is by flail mowing or strimming once or twice a year. There are few parks that would not benefit from its inclusion.

This somewhat depressing review of park grassland is not the whole story. A grassland survey will usually bring to light small areas of turf containing a variety of attractive and interesting species. These are often found on banks or may be linked to areas of bulb planting. In such niches which are difficult of access to machinery so escape the most intensive mowing and chemical regimes (bulbs are sensitive to most herbicides, also areas with them are left unmown till mid-summer), an older type of grassland frequently persists containing species such as sweet-vernal grass (*Anthoxanthum odoratum*), field woodrush (*Luzula campestris*), heath bedstraw (*Galium saxatile*), bird's-foot trefoil (*Lotus corniculatus*), mouse-ear hawkweed (*Hieracium pilosella*) and cat's ear (*Hypochoeris radicata*). It is possible to develop an eye for these relic communities weighing up evidence such as slight changes in texture and colour of the sward, its relationship to topography, position relative to major facilities and re-designed areas, level of use, etc. Frequently, however, as with ancient meadows, survival is controlled by the human factors of idiosyncrasy and chance. One example I have encountered is a semicircle of grass sandwiched between rosebeds and a bowling green. It should never have escaped decades of herbicide and fertilizer application yet between weekly mowings large circular patches of a prostrate form of bird's-foot trefoil produce their flowers, a dwarf race of self-heal (*Prunella vulgaris*) is widespread, so too are field woodrush and heath bedstraw. Species density is 16 spp. m^{-2}.

The ecology of regularly mown town grasslands of known age has been investigated by Wathern (1976) and Wathern and Gilbert (1978) who studied 69 examples aged between 1 and about 240 years. Several types of variation were recognized most of which can be observed in any park of reasonable size that has had a varied history. New grassland produced by sowing species with high growth rates (*Lolium perenne, Phleum*) on to soils containing ample nutrients gives rise to monotonous swards. Initially these productive grasslands contain ruderal species introduced with topsoil (*Capsella bursa-pastoris, Matricaria perforata, Polygonum aviculare, Senecio vulgaris*), but they are rapidly eliminated by the first few mowings. A second group of species also found in young grassland is tolerant of mowing and survives much longer (*Cerastium holosteoides, Holcus lanatus, Poa annua, Ranunculus repens, Sagina procumbens, Trifolium repens*). Over many decades the sward structure and composition alters as the high-growth-rate species which dominated the young grasslands are gradually replaced by a wider diversity of plants, no one of which is dominant and all of which have intermediate to low relative growth rates (Grime and Hunt, 1975). In the oldest grasslands there is a return to a simple structure with one or two plants dominating a relatively species-poor sward. These older dated grasslands were mostly the lawns of large houses.

When data from the 69 stands were ordinated using coeffecients of interstand distance (Sokal and Sneath, 1963) it appeared that sward composition was controlled by one dominant factor which was expressed as a strong trend from immature to mature vegetation. Five types of species distribution were found which showed different responses to this factor.

1. A decline in frequency down the ordination diagram was shown by *Lolium perenne* and *Poa annua*. These species were ubiquitous in the most recently created grasslands but their frequency declined with increasing age and they were absent from the oldest examples.
2. An increase in frequency down the ordination diagram was shown by *Agrostis capillaris, Festuca ovina-rubra* agg. and *Galium saxatile* which were widespread and abundant in the older grassland.
3. A distribution perpendicular to the dominant factor was exhibited by *Trifolium repens*.
4. A nodal distribution. Certain herbs found in some of the older grasslands showed a nodal distribution; these included *Hypochoeris radicata* and *Luzula campestris*. On the ordination diagram they reached a maximum frequency in an intermediate position, then declined again.
5. No pattern of distribution. Certain species found in grassland of all ages showed no pattern in their distribution appearing tolerant of a wide range of conditions: *Cerastium holosteoides, Plantago lanceolata* and *Prunella vulgaris*.

When environmental conditions for each quadrat were plotted on the ordination diagram it was found that there was no relationship with either slope or aspect, only a poor correlation with age and soil pH, but soil nutrient status as determined by *Rumex acetosa* seedling bioassays (Rorison, 1967; Gupta and Rorison, 1975) fitted the nodal pattern described in (4) above. Knowledge of the ecology of the individual species tends to support the hypothesis that soil nutrient status could be the dominant factor. It also explains the unusual distribution of *Trifolium repens* which might be expected to act independently of soil nutrient status since it has the ability to fix nitrogen. Fertilizer experiments (Wathern, 1976) provided further confirmation of the importance of nutrient levels in the ecology of these man-made grasslands. Management which might have been considered equally important in determining grassland composition was largely standardized by only surveying sites believed to have a history of frequent close mowing.

The significance of these studies for town parks is that most can be shown to contain a series of grasslands the composition of which reflects the history of the site. Just as woodlands can be read like historical documents so too can town grasslands, and the archives needed for corroboration are likely to be no further away than the city library. A recent survey of a Sheffield park revealed that part of a bank that had been regularly mown and used by sunbathers since anyone could remember contained the moorland species mat-grass (*Nardus stricta*), heath grass (*Danthonia decumbens*) and harebell (*Campanula rotundifolia*) being a tiny remnant of Crookesmoor from which the district took its name. The bank is now valued as a historical feature and a mowing regime has been devised to show off the harebells. Not all parks contain such a gem, but most Victorian examples will have a lawn somewhere containing unusual species. The grassland survey showed that it is swards on neutral–base-deficient soil that support the highest diversity of species. The ultimate, so far, in the evolution of lawns can be seen in the grounds of country houses (many parks have originated from these). At Chatsworth, Derbyshire the Great Lawn sown by Capability Brown around 1760 has never been sprayed or fertilized, and deer grazing was replaced by mowing in 1833 one year after the invention of the lawn mower. Today this lawn is a series of grassland communities which reflect every change in the environment and the history of the garden (Gilbert, 1983b, c). Fifty-six angiosperms occur in the grassland including wildflowers not normally thought of as inhabiting lawns, e.g. tormentil (*Potentilla erecta*), milkwort (*Polygala serpyllifolia*), lady's bedstraw (*Galium verum*), yellow mountain pansy (*Viola lutea*), speedwell (*Veronica officinalis*), violet (*Viola riviniana*) thyme (*Thymus drucei*) and spring sedge (*Carex caryophyllea*). The owners have made a feature of this lawn with notices explaining its botanical interest; their thinking is ahead of that of most parks departments. The artificial diversification of grassland in parks should be attempted cautiously and only if the existing vegetation shows little scope for manipulation by other means. Drastic change could be looked on as interfering with the history and evolution of the site.

11.3 GROWTH RETARDANTS AND HERBICIDES

Parks departments are increasingly turning to growth retardants to check the growth of grass. Applications of maleic hydrazide, mefluidide and paclobutrazol are an acceptable and cost-effective means of reducing the growth of grass where mowing is difficult such as on banks, by path edges, where bulbs are growing, round obstructions, and wherever wet soil conditions prevent or delay mowing (Shildrick and Marshall, 1985). Many of these sites are exactly the places where swards with an interesting composition are found precisely because mowing is difficult. The long-term effects of maleic hydrazide have been investigated by Willis (1972, 1985) who found that smaller grasses like *Festuca rubra* and *Poa pratensis* increased at the expense of *Dactylis* and *Arrhenatherum*. Species diversity after some increase in the first 3 years declined a little, then fell again at the expense of *F. rubra* after about 20 years. Though Umbelliferae (*Heracleum sphondylium, Anthriscus sylvestris*) were strongly reduced, *Plantago lanceolata* and *Galium cruciata* were favoured by the spraying routine. The application of maleic hydrazide alone did not alter the contribution of broad-leaved herbs to the vegetation appreciably for the first 20 years.

The effects of a mixed application of maleic hydrazide + 2, 4–D, however, in addition to having the desired effect of keeping the grass height to about one-third that of the control plots quickly eliminated most of the broad-leaved species; only *Convolvulus arvensis* which became active after the spraying date, increased. Consequently if a neat uniform grass sward is required, the mixed spray is recommended by manufacturers. If care is not taken, parks where spraying is adopted as a method of managing amenity grassland could quickly become monotonous.

The use of sprays to encourage wildlife is in its infancy, as ecologists retain a deep suspicion of their use which dates back to the publication of *Silent Spring* (Carson, 1962). However, a single application of monocotyledon-targeted herbicide such as propyzamide can help dicotyledons to spread. A more sophisticated use is to discover which families show resistance to a particular herbicide, and use it to encourage them. For example, the Rubiaceae are resistant to simazine, the Solanaceae and Scrophulariaceae to lenacil, the Cruciferae to propachlor and so on. Perhaps one of their most appropriate uses is to help sown wildflower mixes establish as these are sown along with grass seed to provide a nurse. The catch is the grass often smothers out the flowers (Gilbert, 1985). In one field trial, plots of a wildflower–grass mix sown in May 1980 were treated once 8 months later with propyzamide (monocotyledon control) or glyphozate (monocotyledon and dicotyledon control) or a mixture of the two. Recording sward composition 3½ years later (Table 11.1) revealed that the sprayed plots supported 60%, 55% and 35% covers of the sown wildflowers while the controls contained only 8% and 5%.

Table 11.1. The survival of wildflowers in a wildflower–grass mix sown on to topsoil 5/1980, sprayed 1/1981 and recorded 6/1982 and 6/1984

	Control		*Propyzamide*★		*Glyphozate*		*Glyphozate and propyzamide*		*Control*	
	1982	*1984*	*1982*	*1984*	*1982*	*1984*	*1982*	*1984*	*1982*	*1984*
Total cover (%)	100	100	93	95	95	98	85	95	100	100
Cover-sown dicotyledons (%)	15	5	85	55	75	35	82	60	18	8

★ See text for specificity of the herbicides.

11.4 THE CHANGING WILD FLORA OF PARKS

This section is based on a long series of observations made on the flora of Hyde Park and Kensington Gardens in London, an area which, according to the principal field worker D.E. Allen (1965) 'has perhaps been investigated by botanists more minutely and over a longer period of time than any comparable area of the British Isles'. These two parks cover 265 ha, have been used intensively by the public for 300 years and today support a wild flora composed of native species growing alongside a large number of adventives. A detailed botanical study which included notes on frequency was carried out by Warren (1871, 1875) who recorded 181 flowering plants and ferns but he failed to refind 25 species mentioned in the earlier literature. In turn, 33 of his species have not been reported since and are assumed lost; many are marsh and aquatic plants which have vanished through habitat destruction. A few others such as *Molinia caerulea,* brought in with peaty soil for rhododendrons, and *Puccinellia distans,* from new laid turf, probably never established viable populations. Warren's list makes no mention of the modern weeds *Coronopus didymus, Chenopodium ficifolium, Epilobium angustifolium, E. ciliatum, Senecio squalidus, S. viscosus, Galinsoga* spp., *Matricaria matricarioides,* and others now considered common were rare in his day. The present weed flora seems largely to have replaced an older one in which *Euphorbia helioscopia, Ranunculus ficaria, Myosotis arvensis, Vicia cracca* and *Cerastium holosteoides* were leading species.

During World War II the park was used for allotments, camps, air-raid shelters and anti-aircraft emplacements which destroyed extensive areas of old turf, though this disturbance resulted in the appearance of several unusual plants. In the 1940s Kent (1950) carried out a second survey and during the 1958–62 period Allen (1965) made a most detailed

Fig. 11.2. Soft rush (*Juncus effusus*) and sheep's sorrel (*Rumex acetosella*) growing as weeds in an area of new planting to which a dressing of peat/bark mulch has been applied. Occasionally such introductions establish self-sustaining populations.

third one finding 309 taxa. He discovered a number of plants still in the stations where Warren had discovered them 90 years earlier; but more impressive were the changes since Kent's survey only a dozen years before; 17 of his species could not be refound and several annual grasses, e.g. *Hordeum murinum, Lolium multiflorum, Anisanthia sterilis* and *Bromus mollis* had become much scarcer. Conversely many of Kent's rarer

Table 11.2. Species accidentally introduced into planting beds with applications of peat–bark mulch

Common	*Occasional*	
Chenopodium rubrum	*Conopodium majus*	*Lathyrus pratensis*
Gnaphalium uliginosum	*Digitalis purpurea*	*Luzula multiflora*
Juncus effusus	*Dryopteris dilatata*	*Rorippa islandica*
Myosoton aquaticum	*Hypericum humifusum*	*R. sylvestris*
Rumex acetosella	*Juncus bulbosus*	*Spergula arvensis*
Stellaria palustris	*J. conglomeratus*	*Spergularia rubra*

Data supplied by J. R. Palmer.

species had become common, e.g. *Coronopus didymus, Epilobium hirsutum, Galinsoga* spp., *Medicago lupulina* and *Carex hirta*, so the vegetation of Hyde Park is clearly in a state of abrupt and dramatic flux. This even applies on an annual basis with *Lamium purpureum* being unusually abundant in 1960, *Epilobium ciliatum* in 1961 and *Arabidopsis thaliana* and *Gnaphalium uliginosum* in 1962. This led Allen to postulate that the timing of weeding by gardeners must be a significant biotic factor. It is possible that a strike among gardening staff falling at a particular time of year could significantly alter the composition of the flora of inner London. There is no reason to doubt that similar changes have been occurring within the flora of most parks since their inception.

An illustration of a new management operation which is currently being practised in parks (and elsewhere) is the use of peat/bark mulches to control weeds. I have observed that mulches often contain seeds of *Juncus effusus* and *Rumex acetosella* which germinate readily giving rise to well established plants (Fig. 11.2). J.R. Palmer (personal communication) has observed many species arising from applications of peat in south-east England (Table 11.2) while W.D. Campbell in Oxford has had the following from Somerset peat in his garden: *Filipendula ulmaria, Iris pseudacorus, Lysimachia nummularia, Stellaria palustris* and *Ranunculus sceleratus*. Only occasionally will such introductions establish viable populations, but where this occurs, perhaps at the edge of an ornamental lake or along a stream margin the equilibrium will have been disturbed and the accumulation of similar slight shifts throughout a park will over time affect its character. The movement tends to be of northern and western plants towards the south and east.

Only rarely is an area studied in sufficient detail to enable more than speculative statements to be made concerning the method of introduction of species. Allen's (1965) study of Hyde Park and Kensington Gardens, however, contains accurate information on this. He concluded that imported soil was the major source of propagules, particularly for species not forming part of the regular flora of inner London. He observed that soils of alluvial origin brought in species such as *Polygonum hydropiper* and *Alopecurus geniculatus*, the latter becoming well established. Other soil which came from light arable ground and was widely used for filling holes brought in *Scleranthus annuus, Aphanes microcarpa, Kickxia spuria, K. elatine* and *Anthemis arvensis*. Another source of introduced soil is that attached to the roots of shrubs and bedding plants; he suggested that *Montia perfoliata* and *Aegopodium podagraria* had arrived in this way. With the increasing use of container-grown material, this is likely to become a major route of dispersal for weeds of nursery gardens.

Renovation work provides further opportunities for plant dispersal. Gravel brought in for surfacing a path was responsible for the first record of sea plantain (*Plantago maritima*) from central London, while repair work to worn turf produced a temporary profusion of the meadow barley (*Hordeum secalinum*). For several years dredgings from the Serpentine strewn on adjacent beds produced crops of *Ranunculus sceleratus, Rorippa islandica* and *Juncus bufonius*.

A new major source of introduction in recent years is the scattering of commercial bird seed mixtures; certain London parks have vendors selling it on the spot. Allen attributed the presence of a dozen casuals to this pastime, viz., *Bupleurum lancifolium, Helianthus annuus, Salvia reflexa* and *Lolium rigidum*. They are concentrated at the edge of paths, in beds near paths and between cracks in paving stones. Discarded material from picnics are the probable origin of grape, strawberry, tomato and possibly young cherry, apple and raspberry plants. Some of these are also likely to be brought in by birds; where they occur under railings this is the most likely route.

Urban ecologists need to know a lot more about dispersal on footwear but apart from Clifford's (1956) paper little work has been done in this area. It is thought that *Matricaria matricarioides, Juncus tenuis* and possibly *Vulpia bromoides* entered Hyde Park and spread in this way. Species with fruit and seed bearing hooked spines, e.g. *Agrimonia eupatoria, Galium aparine, Arcticum* spp. and *Circaea lutetiana* must owe both their introduction and subsequent spread to transport on clothes and perhaps also on the coats of dogs. There is a further group of plants originally introduced for ornamental purposes which have run wild. Prominent here are *Oxalis corymbosa, Heracleum mantegazzianum* and *Reynoutria japonica* all of which have as firm a foothold in Hyde Park as many natives. In Brandon Hill Park, Bristol, *Carex pendula* has spread from the neighbourhood of pools where it was originally planted and is now a serious weed of plant beds.

11.5 LAKES

Possibly the most popular feature of the Royal Parks in London are their waterfowl. The birds include exotic species such as black swans, pelicans, red-crested pochard, mandarins and Carolina wood ducks which mix compatibly with native shoveler, pin-tail, gadwall, golden eye, pochard, tufted duck and mallard together with brent and barnacle geese. Only the more exotic species are pinioned. Features that make the habitat particularly favourable for wildfowl are an island and a border of grass and shrubberies along part of the margin to which neither people nor dogs have access. Given these conditions, many species will breed. Goode (1986) has pointed out how the coot population of St James's Park has become colonial in its habits quite unlike coot in more natural situations which tend to be aggressive towards one another. It is suggested that the abundance of food is responsible. A feature of these urban lakes is that the bird population is in a highly dynamic state. For example, in Battersea Park over the last few years there has been a massive build up of Canada geese and recent breeding by great-crested grebe and dabchick.

Partly because of their abundant wildfowl, many park lakes are not balanced ecosystems. St James's Park lake for example has a long history of dense algal growth, bad smells and the death of fish and ducklings (Pentelow, 1965; Wheeler, 1978). There is still much to be learnt about the ecology of shallow urban lakes lined with concrete. Reducing the factor 'guanotrophy' is a priority. Whitton (1966) who studied the algae in St James's Park lake over a number of years reported the regular occurrence of dense 'blooms' that coloured the water green. These were due to blue–green algae, the actual species of which differed from year to year, e.g. *Oscillatoria limnetica* (1961), *O. planctonica* (1962), *O. limnetica* and *Anabaena aequalis* (1964) and *O. agardhii* (1965). Very little is known about the factors affecting the size of these algal populations except that they are favoured by eutrophication. Despite numerous attempts at management, which have involved regular dredging, use of algicides, hand removal of algae, introduction of water plants and invertebrates, manipulation of the fish population and changing the water supply, fish deaths and algal 'blooms' have continued to characterize the lake.

By contrast, the Serpentine, which is deeper (to 8 m) and has a bottom that allows a certain amount of seepage which discourages anaerobic conditions, is better balanced with a diverse mollusc fauna and 11 species of fish. The fish population of park lakes is entirely artificial being due to stocking but carp, tench, perch, roach and gudgeon usually do well. Their presence produces a fourfold benefit; the promotion of angling, the maintenance of a more balanced ecosystem, they help control aquatic biting larvae, and provide food for fish-eating birds such as the great-crested grebe. Further information on the ecology of urban lakes can be found in Chapter 15.

11.6 MAMMALS

City centre parks are notable for grey squirrels which were first introduced into Britain in 1876. Though grey squirrels are also widespread in the countryside the extra food sources present in cities, where they hunt through litter bins and take food put out for birds, means that they maintain substantial populations in parks and large gardens. Before World War I the grey squirrel was so welcome an addition to our towns that in Hyde Park, for example, notices were put up to the effect that 'the public are earnestly requested not to molest the

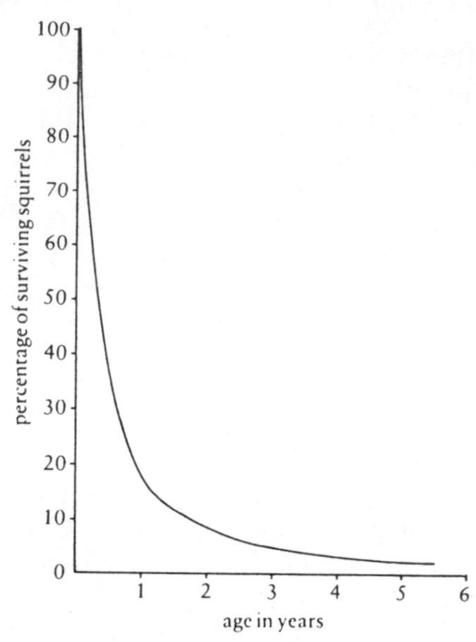

Fig. 11.3. Graph showing mortality in the grey squirrel plotted against age. Survival prospects after the first year are much brighter (Holm, 1987).

grey squirrels'. Forty years later many park authorities were desperately attempting to control their numbers because of the damage they were causing to trees. In Kew Gardens 4000 animals were destroyed and the species was officially banned from all Royal Parks. Three to five young are born in the spring though in their first year they suffer very heavy mortality from starvation, cats and dogs, and many are killed on the road. Males attempting to find mates in the spring, and young less-streetwise squirrels looking for somewhere to settle down, are particularly vulnerable to becoming traffic casualties. Survival prospects after the first year are much brighter (Fig. 11.3).

The grey squirrel is still spreading in the north of England where their increase has been sufficiently recent for naturalists to remember that 30 years ago red squirrels were the ones commonly seen in wooded parks. In Sheffield for example in 1955 the red was recorded from 33 parks and gardens but by 1965 it was known from only four localities within the city. The last urban sighting was made in about 1970, it having by then been completely replaced by the grey (Clinging and Whiteley, 1980). Attempts have been made to reintroduce the red squirrel to Regent's Park, London where its niche is now filled by the grey. Supplementary feeding was tried, to see if it would improve their chances of survival but the results to date have been disappointing (Holm, 1987). Parks in many American cities are frequented by flying squirrels which in the evening can be seen making glides 30–40 m long. The European flying squirrel visits parks in Finnish towns.

Other mammals that may be encountered in parks include hedgehogs. Morris (1966) observed that in central London where there are few private gardens large enough to support them their main populations are in the parks with sufficient records from Regent's Park, Kensington Gardens and Battersea Park to suggest breeding. Records include their nesting under sheds, planks and corrugated iron in addition to the more usual compost heap and bonfire piles. Certain parks, including Hyde Park, support rabbits; they often have coats of varying hue which suggests interbreeding with released tame ones. Rats, pipistrelle, Daubenton's and noctule bats, field voles, wood mice, fox and water shrew are all frequently or locally seen in parks but do not maintain major populations in this habitat. Other mammals such as stoat, weasel, deer, mole, pigmy shrew, brown hare and badger are not really urban mammals and if seen will most likely be isolated individuals.

11.7 BIRDS

There is a wealth of information on the birdlife of city parks. Those in London have been surveyed for many years so their bird populations are particularly well known. A list of these, their distance from the city centre and the approximate number of breeding species found in them is given in Table 11.3 (Hounsome, 1979). Though numbers correlate well with distance from the centre ($r = 0.72$, $P < 0.01$) the parks fall into two types within which there is no significant correlation with the city centre distance. They are the city centre parks (1–7) and the suburban parks (8–16). The former are unlikely to hold more than 30 species of breeding bird while the latter usually have between 40 and 50. The reason for this difference is that parks in the centre of towns suffer much more human disturbance, are usually smaller and are often more formal in their layout with large areas of mown grass and little in the way of wilder areas. One can also distinguish two subgroups within the first group; numbers 1–4 the central and royal parks, and numbers 5–7, the inner city municipal parks. The former tend to be more sympathetically managed for birds.

The requirements that birds seek from a park are mainly the provision of cover, food, nest sites and roosting sites. Cover is somewhere to be safe, somewhere to fly in the event of danger from predators or humans. The aerial hirundines and swifts rely on speed and agility in open flight for protection but most birds need cover in the form of vegetation. Vegetation also largely controls the food supply in terms of seeds, fruits,

Table 11.3. The relationship between numbers of breeding species of bird and the distance from the centre of London (both approximations). Data compiled from the *London Bird Report*, the *London Naturalist*, *Bird Life in the Royal Parks*, and Simms (1975)

Site	*Number of breeding species*	*Distance from Charing Cross (km)*
1. St James's/Green Park	22	1.5
2. Hyde Park/Kensington Gardens	29	3.0
3. Regent's Park/Primrose Hill	34	3.5
4. Holland Park	24	5.5
5. Clapham Common	17	5.75
6. Bishop's Park	16	7.25
7. Wandsworth Common	18	10.0
8. Wimbledon Common	46	11.25
9. Brent Reservoir	44	12.5
10. Richmond Park	56	12.5
11. Kew Gardens	39	12.75
12. Greenwich Park	27	14.0
13. Ham House	38	14.5
14. Osterley Park	46	15.5
15. Hampton Court	42	18.0
16. Bushey Park	47	18.0

Taken from Hounsome (1979).

phytophagous insects, litter organisms, etc. The most successful park birds feed on a combination of these and food provided by man. Specialized insectivores tend to be less common. Availability of nest sites is also a strongly limiting factor. Sparrows and starlings are able to use a wide variety of sites but other birds are more particular. Blackbird, robin, dunnock, song thrush and greenfinch prefer shrubs, bushes and low dense trees, while jay, magpie, missel-thrush, crow and wood-pigeon prefer to nest high up in trees. Tits, owls and woodpeckers favour holes in trees; a lack of suitable nest sites frequently limits their population size. Good roosting sites which afford protection against predators and the weather are vital to small birds in particular. Dense evergreen shrubberies in parks and islands on ornamental lakes form popular and safe roosts. The degree to which these requirements are met controls the bird population of any park. Studies which link the distribution of species to the various habitats within a park are scarce so

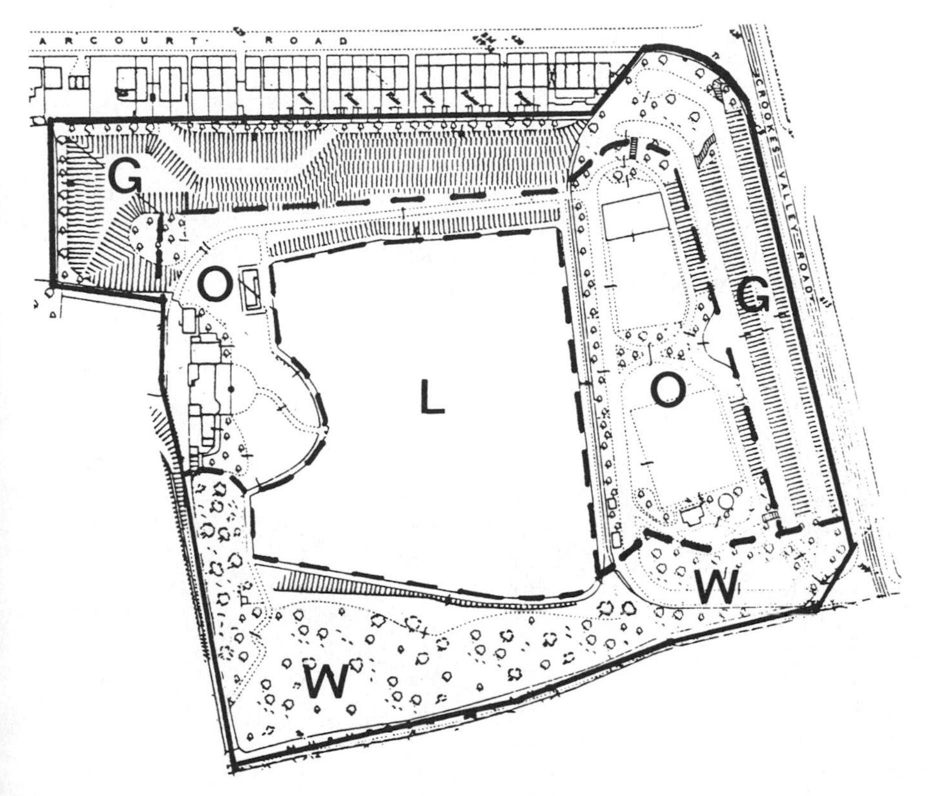

Fig. 11.4. Plan of Crookes Valley Park showing the habitats used during the bird survey. G, grassy banks; L, lake; O, ornamental gardens; W, woodland.

the detailed study by A. Bevington of a typical inner city park is of considerable interest.

Crookes Valley Park, 1.5 km from the centre of Sheffield contains the following habitats: a small deciduous woodland planted 80 years ago (1.5 ha); mown grassland with scattered trees (1.8 ha); ornamental gardens (1.3 ha) and a deep boating lake with vertical concreted sides (1.9 ha) (Fig. 11.4). During weekly visits made over 15 months the number of species present (the 'species count'), the number of birds present (the 'bird count'), and their activities were recorded. Birds flying over the area and pigeons were excluded; also species difficult to count accurately – house sparrows, starling, swift – were excluded from the 'bird' but not the 'species' count. The relative abundance and distribution of the 31 parkland species are shown on Table 11.4 from which the following observations can be made. All the common garden species except for bullfinch were present. Apart from greenfinch most small seed-eating birds such as goldfinch, linnet, redpoll and chaffinch were

Table 11.4. Relative abundance of birds within the main habitats present in Crookes Valley Park, Sheffield. See text for further details

Species	*Wood*	*Banks*	*Gardens*	*Lake*	*Total for Park*
Mallard	–	–	–	169	169
Black-headed gull	–	–	2	671	673
Collared dove	10	12	4	–	26
Great spotted woodpecker	2	–	–	–	2
House sparrow	√	√	√	√	–
Wood pigeon	51	24	44	–	119
Swift	√	√	√	√	–
House martin	√	√	√	√	–
Carrion crow	26	1	11	–	38
Magpie	18	43	9	–	70
Great-tit	19	8	4	–	31
Blue-tit	130	68	45	–	243
Wren	23	3	3	–	29
Missel-thrush	30	19	10	–	59
Song thrush	33	4	6	–	43
Redwing	1	–	–	–	1
Blackbird	125	69	55	–	249
Robin	37	20	13	–	70
Dunnock	34	8	14	–	56
Pied wagtail	–	4	3	13	20
Grey wagtail	–	–	–	1	1
Starling	√	√	√	√	–
Greenfinch	79	25	20	–	124
Goldfinch	5	5	1	–	11
Redpoll	–	8	–	–	8
Chaffinch	11	1	3	–	15
Linnet	–	–	1	–	1
Willow warbler	1	–	–	–	1
Spotted fly catcher	4	–	–	–	4
Rook	3	–	5	–	8
Blackcap	1	–	–	–	1
Total no. of records	643	322	253	854	
Without black-headed gulls				183	

Data from A. Bevington.

scarce, no siskin were recorded. Warblers were either absent or in the case of blackcap and willow-warbler present only during the spring migration. Only two species of tit were present.

The relative abundance of species is clearly controlled by the range of habitats present and their management. The scarcity of undisturbed dense cover, for example thick clumps of bushes which in Sheffield are discouraged for security reasons, can explain the absence of bullfinch and rarity of warblers. The shortage of fruiting weeds and other plants which is a characteristic of well-tended parks may explain the lack of variety of seed-eating birds, while the small amounts of birch and alder account for the virtual absence of winter flocks of redpoll and siskin. The special management of trees in public places which for safety reasons are removed before they grow old is responsible for the dearth of hole-nesting birds. Blue-tits have overcome this by using cavities in stone walls.

The species count was fairly constant at 12 spp. per visit for most of

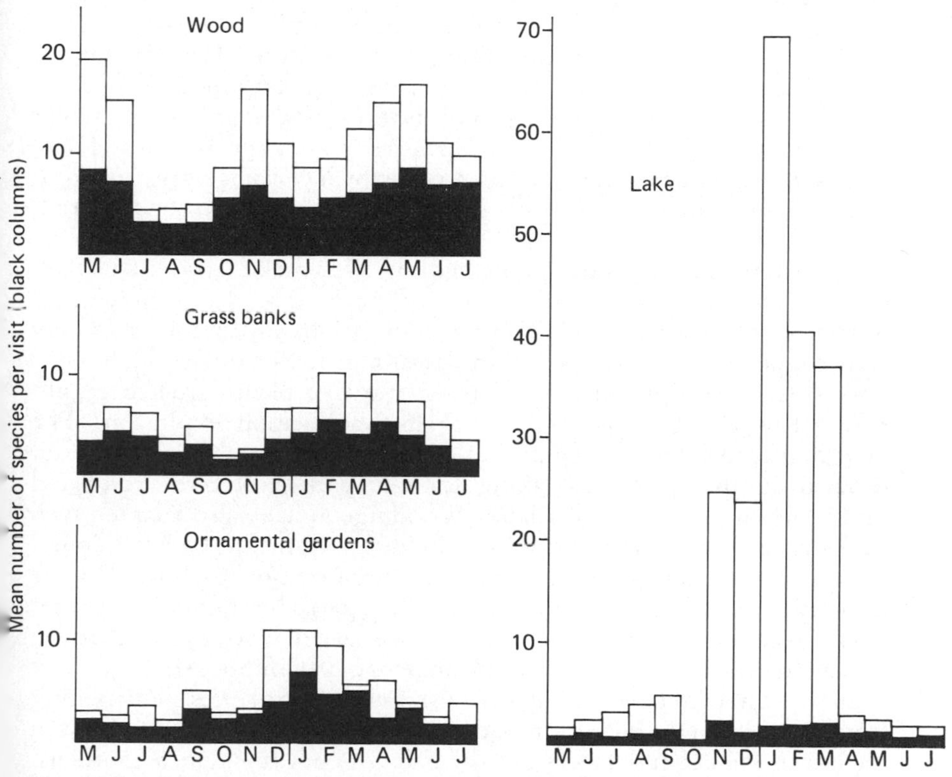

Fig. 11.5. Seasonal variation in the number of birds present and the number of species present in Crookes Valley Park, Sheffield (A. Bevington).

the year but dipped at the end of the breeding season (July–October). The distribution of bird numbers and species within the four habitats over the 15 months is shown in Fig. 11.5. This reveals that in terms of species density the small woodland, even though the shrub layer is almost wanting, is quite easily the best habitat in the park for birds; this is particularly true during the breeding season. It is followed by the grassland with scattered trees, then the ornamental gardens, while the boating lake which does not possess an island or any emergent vegetation is of minimum value to birdlife. During the breeding season the 'bird count' gave similar results, but for the rest of the year there was no discernible pattern except for the winter flock of up to 200 black-headed gulls that congregate on the lake once the boating season is over.

The avifauna of this park, controlled largely by its layout and management, shows interesting differences to that of residential areas. In particular certain species are absent due to the emphasis on safety, security and probably also high levels of disturbance by people and dogs. In one area where parks usually excel, waterfowl, Crookes Valley is disappointing showing that more than a simple sheet of water is needed to attract this group. During the period of study, as an experiment, considerable areas of grassland were left unmown but this did not increase its attractiveness to birds, the thrush family in particular preferring to feed on short grass. Most parks have a special feature; here extensive dry stone retaining walls down which curtains of ivy hang, to some extent compensate for the lack of cover, and conventional nesting and roosting sites.

In the 120 odd years since parks became a feature of large towns they have gradually become colonized by birds that were previously restricted to the countryside. Urban populations of carrion crow, jay and magpie have all increased dramatically in recent times. They need large trees to nest in and the maturing of earlier planting has been put forward as one possible reason for this population explosion. The increased availability of food must also be important as they scavenge on rubbish dumps as well as taking food in parks, off bird tables and feeding in large areas of grassland. Woodpigeon have also adapted well to the environment of parks; they provide an example of a behavioural change which usually accompanies the urbanization of birds. In the countryside it is a very shy bird, yet in town parks it is quite tame. Fitter (1945) recounts the story of a Scots gamekeeper who when asked on returning from a holiday in London what had impressed him most about the capital replied 'seeing the cushie-doos in St James's Park eating from a man's hand'. Initially woodpigeons began to nest on buildings in the city as well as in the trees but the habit did not catch on. Moorhens, starlings and gulls all started to increase in London parks at the end of the nineteenth century about the same time as woodpigeons. This movement of birds into cities is continuing as introduced species spread. Canada geese are now as much part of the London parks as any native bird with mandarin duck and ring-necked parakeets gradually advancing further and further into the capital. Only the availability of suitable nest sites seems likely to halt their spread.

Detailed information about urban birds is provided by Nicholson (1951), Summers-Smith (1963), Simms (1975) and Goode (1986). The picture these ornithologists paint is of a mobile avifauna continually adapting to the changing restraints and opportunities offered by the urban habitat. Reasons for this flux vary from lack of persecution (crow), cleaner air (house martin), cleaner water (water fowl), new habitats, and new introductions by chance (ring-necked parakeet) or intention (mandarin duck). Other species, for example gulls, have adapted perfectly to an urban lifestyle spending their winters commuting between copious sources of food on municipal rubbish dumps and around park lakes, and safe roosts on suburban reservoirs. Herring and lesser black-backed gulls have adapted more completely than other gulls

Fig. 11.6. Owing to supplementary feeding, the density of birds in town parks may be so great they they pose a threat to ornamental bedding displays; Kensington Gardens, London (O.W. Purvis).

and now breed in central London. Losses have included rooks which depend on farmland for their food and nightingales which require large thickets.

Birds are mostly welcome in parks though in Kensington Gardens notices requesting the public not to feed them in certain areas have been put up (Fig. 11.6). This is to protect bedding displays as sparrows, for example, are fond of marigold seeds and in searching for them often destroy the flower heads; pigeons are inclined towards begonias and their trampling also harms emerging bulbs. In the centre of Leicester starlings have selected areas of park shrubbery to form what is believed to be the largest roost in Europe; it involves over a million birds. Their guano has killed off large areas of woody planting; a research programme is attempting to discover how to disperse them. These instances apart, birds are welcome co-users of town parks.

11.8 INVERTEBRATES

The invertebrates of parks show many similarities to those found in suburban gardens, which are covered in Chapter 14. A few groups, however, are better represented in parks than gardens. These include grasshoppers which require areas of tall vegetation of the type which falls into the rough grassland category. In such habitats stridulating males of species such as the meadow grasshopper (*Chorthippus parallelus*) and the common field grasshopper (*C. brunneus*) can frequently be heard as summer advances. A surprising discovery in Regent's Park was a well-establihed colony of the rare Roesel's bush-cricket (*Metrioptera roeselii*) (Widgery, 1978). The colony is living in a 1 ha patch of coarse grassland containing young trees originally planted in winter 1975–1976. Till its planting and enclosure the site had been regularly mown and open to public access. Its presence, which consitutes a first record for Middlesex, is believed to be due to accidental introduction. Soil brought from Ongar, Essex, which is a good locality for the species, was used for filling the tree pits and it is thought likely that it contained eggs of the bush-cricket. The same locality contains the lesser marsh grasshopper (*Chorthippus albomarginatus*) which brings the list of grasshoppers recorded for this central London Park to four.

A group of solitary bees collectively known as mining bees (here taken to include the genera *Andrena* and *Halictus*) are often conspicuous in parks due to their habit of digging burrows in areas of short turf underlain by sandy soil. *Andrena fulva*, the tawny mining bee, is one of the most conspicuous; often several hundred individuals can be seen digging burrows, each nest entrance surrounded by a conical mound of fine excavated soil. The first heaps are made of soil pushed up by the bees that have emerged from last year's burrows, but soon the females pair and can be seen sinking new shafts. Each leads down about 15 cm and contains a number of brood chambers which she furnishes with eggs and balls of honey-moistened pollen gathered locally. One unusual feature of the mining bees is that they return to the same patch of ground where they were born so that villages of up to 2000 burrows can be found together. In parks many of the burrows get seriously disturbed and some of the bees killed by the first passages of mowing machinery. Sometimes wasp-like homeless bees (Nomadidae) can be seen hanging about the burrow entrances waiting a chance to enter and lay their eggs in the cells.

About a dozen species of butterfly are regular visitors to parks though the national trend is for those in the south to be considerably richer than ones in the north. As adult butterflies are nectar feeders, encouraging their presence is entirely compatible with ornamental horticulture; both endeavours aim to provide a succession of flowers, but butterflies prefer certain species above others. Among the best butterfly plants are the spring flowering *Aubretia* and honesty which provide for the four hibernating species and early hatching orange-tips. As summer progresses lavender, red valerian and *Hebe* spp. take over as the main butterfly flowers of a park, then from mid-July buddleia comes into its own. By choosing a range of varieties and pruning judiciously its flowering can be extended into October, though at the end of summer Michaelmas daisy and *Sedum spectabile* are also very attractive to late butterflies. It is rare for parks to provide more than a feeding station for butterflies. Small white, large white (nasturtium, cabbage) and small tortoiseshell (nettles) are the nearest we have to urban species but many of those seen in parks will have originated from colonies breeding in the urban fringe.

Little detailed work appears to have been carried out on other invertebrate groups. Davis (1982) sorted through 39 soil litter samples from the well-wooded Holland Park, London. From them he identified nine species of millipede, four woodlice and three pseudoscorpions. These comprised common woodland and litter species not adapted for disturbed habitats; species often present in parks and gardens but with no special affinity for them; and an assortment of soil and compost species which appear to flourish in artificial habitats. These included the rare millipede *Cylindroiulus vulnerarius* which is native to Italy but has spread north in association with man and the very rare *Microchordeuma gallicum* which also has synanthropic tendencies. The principle here is that if any habitat is examined in sufficient detail unusual species will be found which in towns are likely to be introduced, culture-favoured organisms not present in the surrounding countryside.

Chapter 12

ALLOTMENTS AND LEISURE GARDENS

Allotment development perceived as clusters of semiprivate rectilinear plots, monotonous in layout and often unattractive in appearance, is a feature of most towns and cities in Britain. The average provision in urban areas is around 0.4 ha (1 acre) per thousand head of population, though in 1967 the railway town of Wolverton, Bucks achieved a record 6.0 ha (15 acres) per thousand population. By way of contrast, a number of London Boroughs and certain towns in South Lancashire have little, if any, allotment land. Its importance as a habitat lies in the rich and varied communities of annual weeds which are constantly present in all worked plots, the strange secondary successions which commence once holdings become abandoned, the frequent occurrence of stable supplementary habitats; and a few species that have been identified that are commoner in allotments than anywhere else. There is some overlap with gardens.

12.1 BACKGROUND

The important literature on allotments is mostly associated with Professor Harry Thorpe who, in 1965, was appointed Chairman of a Government Departmental Committee of Inquiry into Allotments (Thorpe, 1969). His report, though now dated, is a mine of information. This was followed by detailed case studies of Birmingham (Thorpe, 1975; Thorpe *et al.*, 1976, 1977). As the major events that have shaped this habitat are particularly well documented for that city, it will be taken as an example.

By 1750 Birmingham was encompassed by a belt of flourishing

Fig. 12.1. Distribution of 'guinea gardens' in Birmingham 1824 together with the detailed layout of two areas.

allotments popularly known as 'guinea gardens', as the annual rent was often one sovereign. Also known from Coventry, Nottingham, Sheffield, Southampton and Tyneside these 'small gardens' were for the recreational enjoyment of the middle classes who cultivated them for both flowers and produce, added a summer house and often dug a well (Fig. 12.1). After 1830 the land was mostly sold for development, but they persisted longer in certain cities, such as Newcastle upon Tyne, where certain inner city place names such as Regents Gardens have their origin in this former land use. If historical research were undertaken, allotments that have been under continuous cultivation for over 200 years could probably be identified.

Their successor, the urban allotment, was the result of a completely different social concept. These were originally provided by charity, so poorly paid labourers could supplement their income. After the *Allotment Acts* of 1887 and 1890, local authorities were obliged to provide them by law, but many sites were only temporary; for example, of the 45 established in Birmingham by 1889, only five remained under cultivation in 1970. Nationally, interest boomed during World War I, fell off in the depression, then reached an all-time peak during the 1940 'Dig for Victory' campaign when 1.5 million plots covering 57 000 ha (143 000 acres) were being worked (Fig. 12.2). A rapid decline in interest followed peace so, by 1949, about a third of all plots were uncultivated. Despite much land being relinquished, the decline lasted until the early 1970s when whole sites had returned to nature. An inquiry in Sheffield

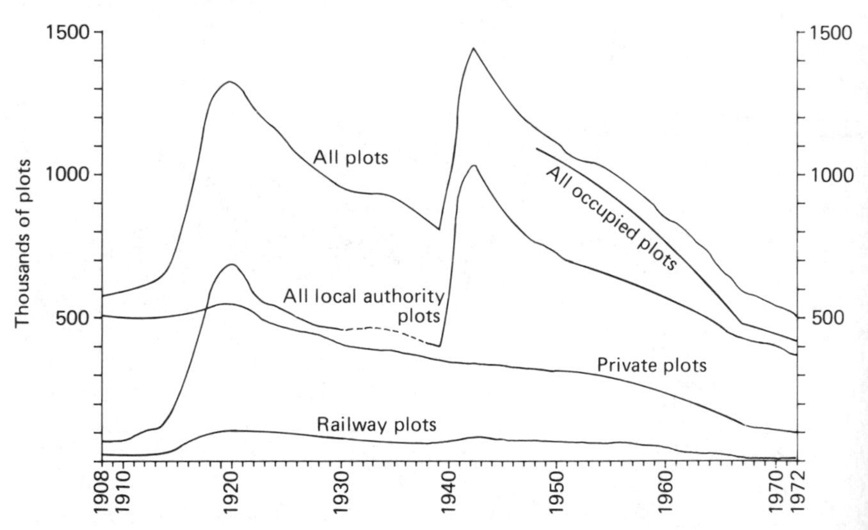

Fig. 12.2. The number and types of allotment plot in England and Wales 1908–72 (Thorpe, 1975).

revealed that by 1966, 40 ha (100 acres) of former allotment land lay derelict. Subsequently a large proportion of this was sold for housing, converted into public open space or thickened with extra planting and left to grow into woodland.

Following the 'Thorpe Report' (1969), allotments suddenly achieved political significance and the decline was halted by spending capital money on site development and amenity provision, often to a very high standard. Thus the Leisure Garden Movement was launched, whereby selected allotment sites were redesigned to incorporate more varied layouts with 10–20% of the area being given over to tended grassland, trees, summer houses, toilets, pavilions, children's play, ornamental planting, surfaced roads, car parks, etc. This philosophy has only made a significant impact in a few towns such as Birmingham, Bristol, Cardiff and Coventry, but is still the dream of most allotment officers.

The last year for which allotment statistics relating to England and Wales were compiled was 1978 when 488 186 plots covering 21 000 ha were in existence. Today the great majority of sites remain traditional in layout, 0.8–4.0 ha (2–10 acres) in size and situated in the middle to outer zones of our cities. Their location is the result of chance rather than planning. Space use in the average plot is 70% vegetables, 10% flowers and 4% fruit, the remainder being uncultivated, paths, huts, etc. Owner motivation tends to be relaxation, recreation and leisure rather than productivity. In a lecture given in 1979, Thorpe suggested that leisure gardens could be used for the teaching of ecology. The main habitats are cultivated ground, compost heaps, temporary buildings, paths, sometimes hedges, and often derelicts plots; they have largely been ignored by ecologists.

12.2 CULTIVATED PLOTS

The soil in allotments is dug over at least once a year, a factor which precludes most biennial and perennial plants, but provides suitable conditions for over 50 species of annual. The prevalent and sophisticated used of herbicides on farmland means that allotments are now a major refuge for many formerly more widespread species of cultivated ground. Three groups of annuals can be recognized. Summer annuals have their main period of germination in spring and early summer, having spent the winter as dormant seeds. They are commonly natives of more southern areas. Examples are the sun spurge (*Euphorbia helioscopia*), petty spurge (*E. peplus*), dwarf spurge (*E. exigua*), red-shank (*Polygonum persicaria*), scarlet pimpernel (*Anagallis arvensis*) and, in allotments at Swansea, fluellen (*Kickxia elatine*). In contrast, winter annuals germinate in the autumn so are in full leaf during the winter; many have a basal rosette of leaves. They include dove's-foot cranesbill

(*Geranium molle*), hairy bitter-cress (*Cardamine hirsuta*), ivy-leaved speedwell (*Veronica hederifolia*) and, in southern towns, Canadian fleabane (*Conyza canadensis*). Yet other annuals will germinate and flower at any time of the year, provided the temperature is sufficiently high and soils moist, e.g. chickweed (*Stellaria media*), shepherd's purse (*Capsella bursa-pastoris*), groundsel (*Senecio vulgaris*), annual meadow-grass (*Poa annua*).

Germination characteristics, whether quasi-simultaneous, continuous, or the commonest, intermittent, e.g. shepherd's purse, smooth sow-thistle (*Sonchus oleraceus*), common speedwell (*Veronica persica*), need to be determined before the ecology of allotment weed assemblages can be understood, as they exert a control over the seed bank. Weed assemblages are so sensitive to the pattern of cultivation that it is possible to distinguish lists made from early and late potato crops.

A few perennials with underground stems thrive in allotments where they are, by common consent, regarded as the most troublesome weeds. They include common horsetail (*Equisetum arvense*), bindweed (*Con-*

Table 12.1. The commonest weeds of regularly cultivated allotments, determined from a nationwide survey by the author

The ten most abundant and widespread weeds	
Agrostis stolonifera	*Poa annua*
Capsella bursa-pastoris	*Senecio vulgaris*
Chenopodium album	*Sonchus oleraceus*
Epilobium ciliatum	*Stellaria media*
Euphorbia peplus	*Veronica persica*
Slightly less abundant and widespread	
Elymus repens	*Lapsana communis*
Convolvulus arvensis	*Matricaria perforata*
Equisetum arvense	*Plantago major*
Euphorbia helioscopia	*Ranunculus repens*
Lamium purpureum	*Taraxacum officinale*
Abundant only in certain areas	
Conyza canadensis	*Solanum nigrum*
Geranium molle	*Stachys arvensis*
Mercurialis annua	*Urtica urens*
Polygonum persicaria	*Viola arvensis*

volvulous arvensis), couch-grass (*Elymus repens*), coltsfoot (*Tussilago farfara*) and ground elder (*Aegopodium podagraria*). All can regenerate from the tiniest rhizome fragment, provided it contains a node. Even when eliminated from plots they quickly reinvade from refugia beside paths, huts or among bushes. It is noticeable that these rhizomatous weeds are commonest on clay soils which are utterly impossible to clear by hand. Horsetail is resistant to all commonly available herbicides and grows so vigorously that where it gets a hold, plots have to be abandoned.

In a nationwide survey of allotments the weeds in Table 12.1 were found to be the most abundant. No strong regional patterns were discernible, variation within towns tending to be as great as between them. Soil type exerted the greatest influence with annual mercury (*Mercurialis annua*), small nettle (*Urtica urens*), gallant soldier (*Galinsoga parviflora*) and black nightshade (*Solanum nigrum*) reaching unusual abundance on deep fertile well-drained soils, while *Anagallis arvensis,* corn spurrey (*Spergula arvensis*) and field pansy (*Viola arvensis*) pick out less-fertile more-mineral substrates. Many local authorities offer a free soil pH-testing service advising that a pH of around 6 is optimal. In the north of Britain allotment soils may have a pH as low as 4.5; these support rather specialized weeds such as marsh cudweed (*Gnaphalium uliginosum*), sheep's sorrel (*Rumex acetosella*) and creeping soft-grass (*Holcus mollis*). Field woundwort (*Stachys arvensis*) formerly a locally frequent weed of ploughed fields is becoming increasingly scarce in that habitat and is now probably commoner in allotments than anywhere else. Several annual species of *Euphorbia* and *Viola* may be starting to take on a similar distribution.

In a survey of 59 urban allotments distributed over six sites in Hertfordshire, Willis (1954) recorded 58 different weeds; their order of abundance was similar to my own findings. Certain species such as *Anagalis arvensis*, scentless mayweed (*Matricaria perforata*), *Geranium* spp. and *Urtica urens* were prevalent on some sites, almost absent on others. After comparing the species in uncultivated plots, by paths and along fence lines with those in cultivated areas, he concluded that with the possible exception of *Taraxacum officinale, Convolvulus arvensis* and *Sonchus* an allotment holder's main source of weed seeds were the plots of his fellow diggers.

A phenomenon first noticed in Sheffield, but since observed elsewhere, is the occurrence of bird-seed aliens on allotments where pigeon lofts are permitted. Manure and other residues emanating from the lofts are piled up and later dug into the cultivated area of the holding. Since pigeon fanciers do not keep their plots well weeded (hence their restriction to nine out of the 67 allotment sites in Sheffield), stray bird seed from the cotes frequently germinates and grows to flowering by

Table 12.2. Composition of three seed mixes commonly fed to pigeons in Sheffield and the appearance of the species in allotments

	1	*2*	*3*	*4*
Sorgum (*Sorgum bicolor*)	+	+	+	·
Maize (*Zea mays*)	+	+	·	√
Wheat (*Triticum aestivum*)	+	+	+	√
Barley (*Hordeum vulgare*)	+	·	+	√
Oats (*Avena sativa*)*	+	·	·	√
Italian rye-grass (*Lolium multiflorum*)	+	·	+	√
Canary grass (*Phalaris canariensis*)	+	+	·	√
Common millet (*Panicum mileaceum*)	+	·	·	√
Darnel (*Lolium temulentum*)*	·	·	+	·
Rice (*Oryza sativa*)	·	·	+	·
Bristle-grass (*Setaria pumila*)*	·	·	+	·
Garden pea (*Pisum sativum*)	+	+	·	√
Broad bean (*Phaseolus vulgaris*)	+	·	·	√
Cultivated flax (*Linum usitatissimum*)	+	·	+	√
Knotgrass (*Polygonum aviculare*)*	+	·	·	√
Pale persicaria (*P. lapathifolium*)*	·	+	·	√
*Arbutilon theophrasti**	·	+	·	·
Hemp (*Cannabis sativa*)	·	·	+	·
Opium poppy (*Papaver somniferum*)	+	·	·	√

Column 1, general feed; 2, young bird mix; 3, conditioner; 4, mature plants recorded in the allotments. Seed in the general feed is whole, in the young bird mix it is crushed, and in the conditioner it is mixed with an aromatic oil.
* Present only in small amounts.

late summer when potato patches in particular are gay with flax and poppies. Table 12.2 gives the composition of three types of pigeon feed and details those species recorded in the allotments of Sheffield's pigeon fanciers. The populations rely on repeated introduction for their persistence. There is also a possibility of unusual aliens in plots cultivated by Asians where crops such as coriander, mustard, fennel and other curry herbs are grown, but none have been noted (if the sown species are disregarded).

Certain local authorities do not allow any livestock to be housed on their sites, while others tolerate tame rabbits, poultry and bees. Beekeepers are required to erect a solid 2.5 m-high fence round their hives to ensure the bees return on a high flight path so as not to cause a hazard to plot owners. The value of the bees as pollinaters for legume crops must be considerable. Regulations usually allow only two hives per site.

12.3 SECONDARY SUCCESSIONS

The decline of interest in allotments has resulted in most sites containing abandoned plots, so a mosaic of successional stages is often present. After 2–3 years without cultivation a variable Yorkshire-fog (*Holcus lanatus*)–creeping bent (*Agrostis stolonifera*) grassland usually establishes containing scattered patches of taller species such as docks, thistles, nettles and hogweed (*Heracleum sphondylium*). Woody plants may establish at this stage, e.g. broom, gorse, willow, sycamore. By 4–6 years, tall herbs have become the physiognomic dominants with *Holcus lanatus* and *Poa trivialis* confined to the understory and *Agrostis* and *Rumex* declining. Many plants of garden origin are now prominent such as Michaelmas daisy (*Aster novi-belgii*), shastri-daisy (*Leucanthemum maximum*), golden-rod (*Solidago canadensis*), goat's rue (*Galega officinalis*), horseradish (*Armoracia rusticana*), mints (*Mentha* spp.), fennel (*Foeniculum vulgare*), lupin (*Lupinus polyphyllus*), tansy (*Tanacetum vulgare*), soapwort (*Saponaria officinalis*) and cultivated brambles like *Rubus* 'Bedfordshire Giant'. Large grasses such as false oat (*Arrhenatherum elatius*) and cocksfoot (*Dactylis glomerata*) increase as do the woody species. After 10 years areas have the appearance of open scrub with brambles, raspberries and even bracken on the more acid sites, but still contain persistent garden relics; in wet areas rushes (*Juncus effusus, J. inflexus*), tufted hair-grass (*Deschampsia cespitosa*) and great hairy willow-herb (*Epilobium hirsutum*) are common.

To find plots that have lain derelict for over 10 years, sites that have been totally abandoned need to be sought. The invasion of woody plants tends to be negligible once a dense ground cover has established, so from that point the progression to woodland largely depends on the growth of young trees already present and especially on the presence of hedges. Hedges and bramble patches remain 'open' which allows tree establishment to continue for longer than in grassland. The young woodland is variable in composition, but ash, sycamore, elder and goatwillow are usually present, together with hawthorn and oak, both of which have some ability to establish in dense vegetation. The constant presence of woody exotics not normally present in British woodland separate these young allotment woods from all others, e.g. privet (*Ligustrum ovalifolium*), lilac (*Syringa vulgaris*), *Cotoneaster frigidus*, bullace (*Prunus domestica* subsp. *insititia*), various cultivars of soft fruit and *Spirea salicifolia*. It is not yet possible to say how persistent these garden relics will be, but it should be noted that a number are shade tolerant and reproduce vegetatively.

12.4 BIRDS

Allotments in cultivation, especially if there is a small amount of cover in the form of hedges and bushes, support quite high densities of territorial insectivorous birds such as song thrush, blackbird, robin, wren and dunnock. Seed eaters like greenfinch, house sparrow and wood pigeons, and omnivorous starling and magpie will also be present. All these species feed on the allotments and nest in nearby cover, their density being controlled by nest site availability. If large enough, and laid out as leisure gardens, real field species such as meadow pipit and even skylark and partridge may be attracted. Once derelict plots start to appear, new seed eaters such as goldfinch, redpoll, bullfinch and linnet move in, while the open-ground feeders decline. When whole sites are abandoned hedges become overgrown and scrub increases so that other birds are favoured. Garden warbler, reed bunting, collared dove, jay, mixed winter flocks of blue, great, long-tailed and occasionally marsh-tits plus goldcrest often make use of scrub-infested allotments, especially if a few trees are present. If there are good berry sources, winter thrushes and continental blackbirds will be present.

A study in Poland (Luniak, 1983) concluded that the avifauna of worked allotments was little diversified when compared with that of other green urban space; though fair numbers of birds were present in the breeding season they dropped off in winter. Reasons advanced for this were the homogeneity of the habitat, intensive human presence combined with lack of cover, lack of medium-sized and large holes for breeding sites, and less anthropogenic food than in parks. After allotment areas had been established for 6–8 years, full density was attained but species such as redstart, golden oriole, collared dove and icterine warbler were not regularly present till after 10–20 years. Compared with wooded parks there was a reduction in tree crown nesters, while flocks of rook, jackdaw and woodpigeon were rarer. Conversely there is a higher percentage of fruit eaters in winter, e.g. blackbird, fieldfare, waxwing, and a higher density of breeding blackbird, redstart, tree sparrow and great-tit. The territorial relationships of birds in well-cultivated allotment land has not been investigated.

12.5 SUPPLEMENTARY HABITATS

Many small allotment sites provide only a narrow range of habitats, viz., dug plots, paths, compost heaps and temporary buildings. However, around half the 50 sites examined contained significant additional areas. Thorpe *et al.* (1977) found that a third of the sites in Birmingham incorporate some water feature, usually a stream, but sometimes a spring, pond or marsh; nationally 18% of sites are affected by bad drainage. Elsewhere old hedgerows have been incorporated into layouts to provide internal divisions or planted to create shelter. In Sheffield for example the standard plot is approximately 10% larger than in the rest of the country to allow hedges of privet to be grown as wind breaks.

Overshading and soil depletion, often caused by a thoughtlessly sited line of poplars, can result in adjacent plots becoming derelict. These then become much richer in wildlife, particularly if diversified by rubbish dumping and bonfires. Large isolated trees such as the huge oaks incorporated into Walkers Heath Leisure Gardens, Birmingham, encourage woodpeckers on to the site and are quite compatible with gardening if located on the northern boundary or near to a central pavilion, toilet block, etc. These additional habitats increase the wildlife potential of sites and can be beneficial to plot holders. For example in towns where magpies have started to spread, damage from woodpigeons, which have a fondness for brassicas, has declined dramatically as the magpies control their numbers through egg predation.

Allotment tenancy agreements promote uniformity by restricting the scope for plot holders to create niches for wildlife. In Sheffield, hedges can only be of privet or thorn (though in practice a blind eye is turned to *Rosa rugosa*), the sinking of wells is prohibited even on sites without water, nor can minerals be quarried by enterprising miners. Fruit trees are discouraged as it is believed they encourage trespass and timber sheds must be of an approved pattern. Regulations in Birmingham differ in that livestock (including pigeons) is not allowed, hedges are discouraged, fruit trees are welcomed, and there is a particular clause forbidding the placement of nightsoil or putrid fish upon the allotment. On certain sites skips are provided for rubbish, elsewhere it must be burnt or composted. The range of supplementary habitats is, therefore, to some extent controlled by the local authority.

12.6 FOSTERING WILDLIFE ON ALLOTMENTS

Certain forms of wildlife are incompatible with well-tended allotments. These include woodpigeons, grey squirrels, wild rabbits, slugs, snails, insects such as greenfly, blackfly and cabbage white butterflies, most perennial plants and tree belts which cast a heavy shade. Despite action taken to discourage these, a lot of incidental wildlife is present. Annual

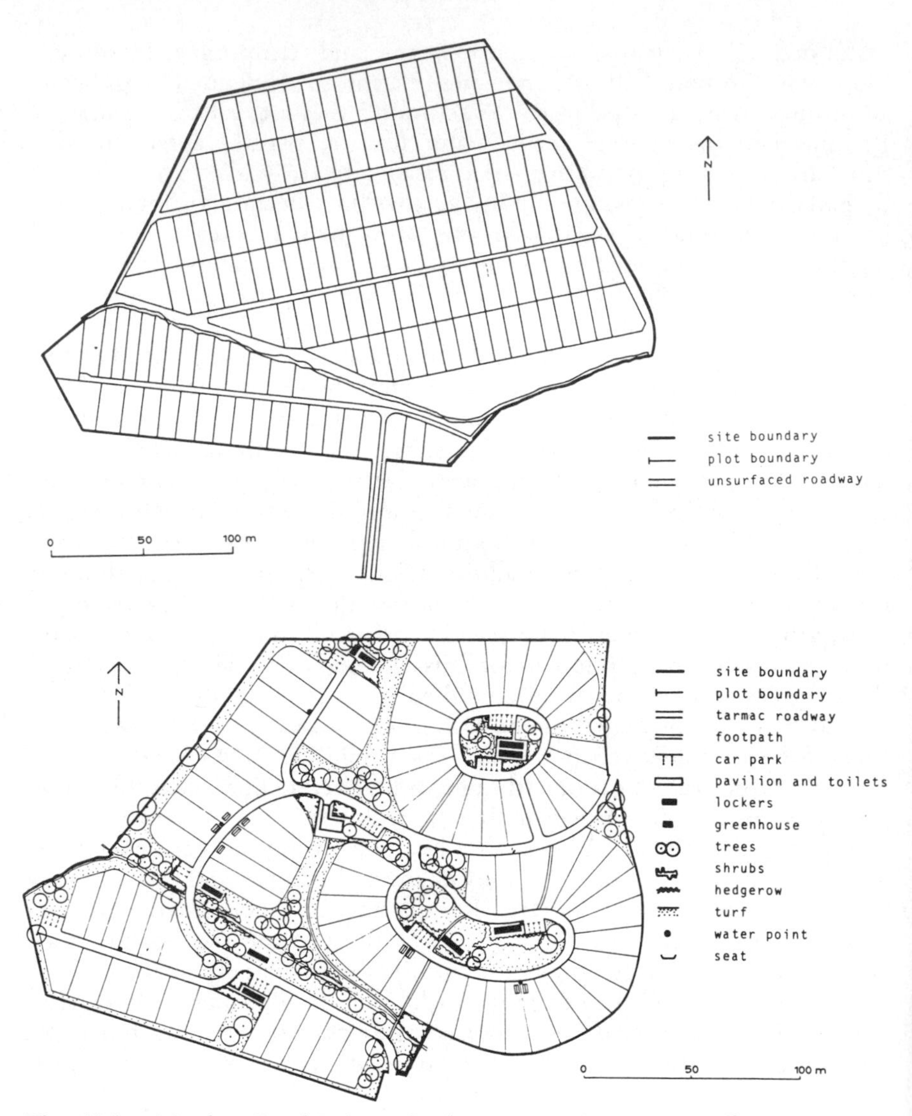

Fig. 12.3. Meadow Road Leisure Garden, Birmingham before (upper plan) and after improvements carried out in 1975. The new layout provides considerable additional areas of wildlife habitat.

weeds that accumulate seed banks are perfectly adapted to the conditions, mobile species such as birds will always be prominent, foxes will continue to breed under the huts and invertebrates to form rich communities in the numerous compost heaps (see page 262) which are made by well over half the owners.

The Leisure Garden Movement which seeks to devote up to 20% of each site to general amenity offers the greatest opportunity for enhancing the natural history interest of allotments. Incorporating existing site assets such as streams, hedges and rough grassland into new or redesigned layouts, as has been achieved unintentionally at many sites in Birmingham, provides a highly satisfactory pattern of cultivated plots, tended open space and informal wildlife areas (Fig. 12.3). Locally, simple habitat creation within the wildlife/tended open space zone can, by allowing long grassland to develop, planting shrubs as cover for birds or introducing 'insect trees' such as alder, birch, oak and willow, further enhance its value. Bordsley Green Leisure Garden, Birmingham is internationally famous for the way it was redesigned in the early 1970s to provide attractive facilities and a diverse layout. It is also, unintentionally, an example of how wildlife can be incorporated as an additional and completely harmonious amenity. Designating long derelict allotment sites as local nature reserves is a further way of changing the use of council-owned land to fit the aspirations of modern city dwellers.

Chapter 13

CEMETERIES

All towns contain cemeteries. These can vary in size from 1 to 50 ha and in age from the second quarter of the nineteenth century to modern. In general, the earlier a cemetery was established the better it is for wildlife today. The very best originated as private commercial ventures, and most cities have a 'Victorian cemetery' which falls into this category. In the last few years cemeteries have become a major source of interest to architects, social historians and latterly urban ecologists. One reason for their popularity undoubtedly lies in two books written by James Curl, *The Victorian Celebration of Death* (1972) and *A Celebration of Death* (1980). These vividly describe the history of cemetery design, the rarefied pleasures of funerary architecture and the strange melancholy of investigating old graveyards. Urgency is added to his revelations by the fact that many of the examples he describes are being deliberately destroyed in the interests of administrative convenience and current fashion. As a reaction to this official vandalism, increasing numbers are being surveyed by voluntary groups with a view to formulating management plans which safeguard their special features. One reason for this conflict of interests lies in the steep rise in crematoria burials which has taken place since 1940. Today, two-thirds of the population choose this method of disposal, so the traditional concept of burial grounds is of declining relevance, as crematoria increasingly get priority funding at the expense of traditional cemeteries.

In the 150 years since the first Metropolitan cemeteries were laid out they have developed a unique ecology related to land use and management. Relic grassland, heathland and woodland often survive within their boundaries, contrasting with the secondary communities of sycamore–bramble–ivy which invade neglected areas. Lichens on stonework, herbicide-resistant communities around kerbs, mown grassland on the verges, gravel in the grave spaces and the now mature structure planting, together produce a varied but integrated vegetation capable of explanation in terms of anthropogenic factors.

13.1 BACKGROUND

Cemeteries

By the end of the eighteenth century city churchyards were having difficulty coping with the dead of an expanding population. Bodies were piled so high that the churches appeared to stand in pits and the ground was so frequently disturbed that no grass would grow on it. Attempts to deal with this state of affairs were resisted by the church in order to safeguard its income from burial fees. Eventually, however, as part of the general movement towards hygiene in towns, helped by a growing desire for a family grave as a mark of affluence, large cemeteries were developed. The first was in Liverpool (1825) rapidly followed by Edinburgh, Belfast and Glasgow (1832) which is judged to have the most spectacular necropolis in Britain. Between 1832 and 1847 Parliament authorized the establishment of commercial cemeteries in London at Kensal Green (1833), Norwood (1836), Highgate (1839), Nunhead (1840), Abney Park (1840), Brompton (1840) and Tower Hamlets (1841). As cholera raged, more and more cemeteries and burial boards were set up so by 1850 every town and city had acquired a cemetery of some sort.

Right from the start the influential landscape architect John Loudon (1843) championed the Garden Cemetery Movement in particular by writing a book entitled *On the Laying Out, Planting and Management of Cemeteries, and on the Improvement of Churchyards*. In it he argued that at very little extra expense the new cemeteries should be made into botanic gardens to improve the moral sentiments and general tastes of all classes, more especially the great masses of society. At Abney Park cemetery, of which he particularly approved, the trees and shrubs were all labelled and it was reckoned to be the most complete arboretum in the neighbourhood of London. He advocated chalky or gravelly soils because their good drainage would aid decomposition and preferred conical trees and shrubs to allow a free circulation of air and reduce shading. Poplars and willows due to their associations with water were considered inappropriate. The recommended trees in order of preference were Italian cypress (*Cupressus sempervirens*), Irish yew (*Taxus baccata* 'Fastigiata'), third equal upright yew (*Taxus baccata* 'Erecta'), Swedish juniper (*Juniperus communis* 'Suecica'), and tall juniper (*Juniperus excelsa*).

Also highly recommended were *Thuja orientalis,* holly, cedars and firs due to their form and the belief that they promoted feelings of solemnity and grandeur; he had seen all these doing well in cemeteries on the continent. They were to be deployed mainly along the roads and walks with only a few among the graves. He found much cemetery planting objectionable and criticized the design for Norwood – deciduous trees in belts and clumps – as being in the style of a common pleasure ground. His book recommends some deciduous planting including pendulous varieties (trees of sorrow), shrubs, wall plants and spring bulbs as being compatible with scythed grass.

During the period 1850–1900, hundreds of local authority cemeteries were established. Being funded by public money they did not need to make a profit for shareholders nor was it necessary to attract custom with prestigious layouts, fashionable architecture or imaginative treescapes. For this reason their design and planting rarely matched that of the private cemeteries. Their siting appears almost random at what was the city limits at the time of their designation but a few were linked to a nearby fever hospital or the end of a tramline. What is now relic countryside has often been protected within their boundaries.

Fig. 13.1. Neglected cemeteries turn over to woodland. The ivy is *Hedera helix* 'Caenwoodiana' (I. Smith).

While a large permanent staff maintained local authority cemeteries to 'park' standards the private ones fell on hard times due to high overheads and falling custom resulting from a declining interest in the celebration of death. Once a section in a private cemetery had filled up, management was neglected and it turned over to woodland (Fig. 13.1). Many struggled on till a shortage of new grave spaces or bomb damage forced them into liquidation. Their fate then varied from being acquired by development companies as building land, cleared to provide public open space, taken over as going concerns by local authorities or being managed by groups of 'friends'.

By comparison, burial in the subsidized run of the mill public cemeteries remained cheap and relatively popular. After 1940, however, money was increasingly diverted to crematoria with the result that staffing levels were gradually run down. For example at the very large (43 ha) Witton Cemetery, Birmingham the cut was from 50 to 10 men. The response to this – lawn conversion – is still being implemented. It involves removing kerbs round the graves to leave lines of headstones between which grass is sown. Management then involves either mowing or spraying with a growth retardant together with herbicide application as granules or a spray round the base of the memorials. Where mowers cannot get, a strimmer maintains the neat appearance.

For some time local authorities have been promoting lawn sections in their cemeteries where no grave mound or kerbing is permitted, and grave spaces are turfed or sown as soon as possible after interment to ease maintenance problems and promote uniformity. Headstones of only a limited size and of a limited number of materials are allowed. Cultivated ground is not permitted, though, to relieve the starkness, rows of hybrid tea rose bushes may be planted between the graves by the authorities. Certain towns, for example Weston-super-Mare, only permit horizontal tablets so the entire area can be maintained by gang mowers. The concept of cemeteries as semiwilderness areas is now true only of a few formerly private examples but the older parts of the public ones still contain much of interest.

Crematoria

The 250 crematoria in Britain handle half a million cremations annually which generate 400 tons of ash. They are profit-making concerns and most cities regard them as status symbols, their managers attempting to further increase business through superb ornamental planting, tasteful facilities and the highest possible standards of maintenance. Loudon would have been delighted as they conform in many ways to his concept of garden cemeteries with their close mown grass, well stocked rose beds, highly tended shrubberies and abundant ornamental trees. While they have only a limited value for wildlife, one of the consequences of cremation is that a proportion of the older cemeteries are now neglected, almost forgotten wildernesses within the urban fabric.

Churchyards

A Local Government Act of 1972 enables the Church of England to transfer to local authorities responsibility for the maintenance of vegetation, paths and walls in churchyards which have been formally closed. Through this provision an increasing number of city churchyards are being handed over. To reduce maintenance costs they tend to get cleared, levelled, sown, mown and sprayed, which has catastrophic effects on their wildlife. Birmingham City Council for example maintains over 30 but asks that they are in a reasonable condition before handover. In the long term, local authorities see old burial grounds as having a potential for quiet recreation and possibly also the disposal of cremated remains and commemorative tree planting. Very few see them, or parts of them, as being managed for nature conservation.

13.2 STRUCTURE PLANTING

All cemeteries contain a structural framework of trees planted when they were first laid out. These are now large often free-growing specimens which have been allowed to attain their full natural shape unaffected by lopping, topping or having their canopies raised. The structure planting usually defines the boundary, the main circulation routes, and may also be concentrated round entrances and chapels. Planting in the burial areas is less common, but scattered trees occur often representing the last remnants of clumps, the rest of which have been removed to create more grave plots.

A survey of over 50 cemeteries in England and Wales has shown that ash, lime, plane (in the south) and sycamore are easily the most frequently planted trees. Regularly present but in far smaller numbers are Austrian pine, beech, birch, holly, horse chestnut, poplars, robinia (in the south) and yew. Following dutch elm disease the status of elms is difficult to determine – they were certainly once a major feature of many cemeteries. The above trees are amongst the most regularly planted in towns generally so at first sight cemetery planting is not particularly distinctive. Slightly more unusual mature species encountered include Italian alder (*Alnus cordata*) at Church cemetery, Nottingham; a huge willow-leafed pear (*Pyrus salicifolia*) and avenue of silver maple (*Acer saccharinum*) at Witton, Birmingham; narrow-leaved ash (*Fraxinus angustifolia*), maidenhair tree (*Ginkgo biloba*) and cut-leaved broad-leaved lime (*Tilia playtphyllos* 'Laciniata') at Nunhead, London; red maple (*Acer *rubrum*), manna ash (*Fraxinus ornus*), Spanish oak (*Quercus* × *hispanica*), swamp cyprus (*Taxodium distichum*) at Kensal Green, London; and box elder (*Acer negundo*), cut-leaved alder (*Alnus glutinosa* 'Laciniata'), Swedish birch (*Betula pendula* 'Dalecarlica') at Norwood, London. There are only four Bhutan pines (*Pinus wallichiana*), three Austrian pines (*P. nigra*), one Indian bean tree (*Catalpa bignonioides*) and a group of Spanish oaks (*Quercus* × *hispanica*) left of interest from the 'arboretum' days at Abney Park and only a hint at most other cemeteries. The 16 sites surveyed in Sheffield and Rotherham produced only one tree of note, a large silver pendant lime (*Tilia petiolaris*) in the formerly private General Cemetery.

Arnos Vale Cemetery in Bristol provides a rare example of mature structure planting dominated by columnar evergreens as recommended by Loudon (1843). Set up in 1837 it is still run by a private cemetery company and is something of a period piece; memorials of any type can still be erected and management tends to be minimal. Here the planting is dominated by Irish yew, common yew, Sawara cypress (*Chamaecyparis pisifera*) and other cypresses (*Cupressus* spp.), while large pines, cedars of Lebanon (*Cedrus libani*) and a monkey-puzzle (*Araucaria araucana*) provide ornament. These are supported by evergreen shrubs such as box and juniper, the whole climbing up a steep hillside to produce a landscape classified in Grade 2 by English Heritage. Other cemeteries dominated by evergreens are Danygraig, Swansea (1856), where 24% of the mature trees are holm oak (*Quercus ilex*) and the Morriston also in Swansea which has been entirely planted with pines.

An analysis of the survey data shows that the following trees are unexpectedly abundant in cemeteries when compared with surrounding parks and gardens. Turkey oak (*Quercus cerris*), birch, false acacia (*Robinia pseudoacacia*), variegated sycamore (*Acer pseudoplatanus* 'Lute-ovirens') and yew. A Turkey oak overhangs Karl Mark's grave in Highgate East. Beech and native oaks were less common than expected, possibly an air pollution effect; poplars and large willows are only locally present as their widespreading shallow roots cause fatigue to grave-diggers. It is, however, difficult to generalize; the 3 ha Warstone Lane Cemetery in Central Birmingham contains 27 railway poplars (*Populus* × *canadensis* 'Regenerata') and just one other tree, a London plane. Cultivars of common holly and Highclere holly (*Ilex* × *altaclarensis*) are a feature of the older private cemeteries which makes them atractive places to holly enthusiasts who can expect to find at least five, and exceptionally up to 12, different ornamental taxa per site.

A feature that distinguishes the best cemetery planting from all other types is the abundance and variety of weeping trees. These appealed to the Victorians who called them 'Trees of Sorrow' or in Germany 'Trauerbaume' which translated means tree of grief or mourning.

Loudon (1843) thought these particularly appropriate, despite their spreading nature, and went to considerable trouble to search out suitable varieties. A quarter of the trees he recommended are pendulous. These include eight evergreens and 22 deciduous species. Most cemeteries established in the nineteenth century contain Trees of Sorrow. Two out of three support the weeping ash (*Fraxinus excelsior* 'Pendula') with its widespreading mound of divergent pendulous branches, and, prior to dutch elm disease, the domed Camperdown elm (*Ulmus glabra* 'Camperdownii') and weeping wych elm (*U. glabra* 'Pendula') with its stiffer habit were equally common. Weeping beech (*Fagus sylvatica* 'Pendula') including the superb weeping purple beech (*F. sylvatica* 'Purpurea pendula') are present in about 20% of cemeteries and the silver pendant lime in 15%. The elegant weeping holly (*Ilex aquifolium* 'Pendula') was encountered only twice. Other trees with a natural pendulous habit such as silver birch and the deodar cedar are popular and still being planted. Weeping trees are usually deployed in prestigious areas such as near chapels and at major path intersections. For maximum effect they should be used singly but the General Cemetery, Nottingham contains over 50 arranged in groups of four and five.

It might be expected that trees with religious associations would be well represented in churchyards and consecrated sites like cemeteries. Such trees include the Abraham's oak (*Quercus coccifera*) under which Abraham pitched his tent, weeping willow (*Salix babylonica*) the tree on which the Israelites hung their harps when they sat down to weep by the river of Babylon, Judas-tree (*Cercis siliquastrum*) on which Judas hanged himself, and various trees associated with manna such as the manna ash (*Fraxinus ornus*) and tamarisk. These were uniformly rare in the 50 sites examined.

Much of the structure planting in our major cemeteries is over 130 years old and replacement planting has begun. During the survey the age of trees was estimated and they were placed in the following classes: mature (over 60 years old), intermediate (20–60 years old) and young (less than 20 years old). Few fell into the intermediate age class and these tended to be similar in species composition to the mature group, but there had been a lot of recent planting, particularly in cemeteries run by local authorities. This new wave of activity, the first serious tree planting since the burial grounds were established, makes no attempt to perpetuate existing patterns. It is largely composed of small ornamental trees, often hybrids and cultivars chosen for special leaf or flower effects. The genera *Malus, Prunus* and *Sorbus* provide the flavour of this new planting. It will gradually change the character of cemeteries from rather austere, gloomy places to something approaching that of parks with profusely flowering or coloured leaved material eventually taking over from the ashs, limes and sycamores. Though small numbers of traditional species such as laburnum, gean (*Prunus avium*) rowan and ash are being planted they are subordinate. Moor Cemetery (1832) in Rotherham has gone furthest in acquiring new taxa; during the last five years over 180 have been introduced including such ornamentals as *Acer palmatum*, *A. rubrum*, *Betula papyrifera*, *Camellia japonica*, *Ginkgo*, *Magnolia stellata*, *Prunus avium* 'Plena' and *Pyrus salicifolia*. This reflects a new attitude to this burial ground which is being converted into the botanical garden which Rotherham at present lacks.

13.3 SHRUBS

Shrub planting forms only a minor element in the design of most cemeteries where it is used mainly to soften entrances and buildings. Victorian cemeteries, however, often contain a random scatter of evergreen shrubs which were planted on graves by mourning relatives. The field evidence suggests this must have been quite popular, perhaps as an alternative to placing wreaths, but no literature references to the practice, which has now all but died out, have been found. The most

Table 13.1 Evergreen shrubs formerly planted on graves; these still survive in neglected cemeteries where they have historical interest

Common	Rare
Aucuba japonica	*Catalpa bignonioides* (decid.)
Euonymus japonicus	*Chamaecyparis lawsoniana*
Ilex × *altaclarensis*	*C. pisifera*
Ligustrum ovalifolium	*Erica carnea*
L. ovalifolium 'Aureum'	*Escallonia macrantha*
Taxus baccata	*Olearia* × *haastii*
	Osmanthus delavayi
Frequent	*Rosmarinus officinalis*
Buxus sempervirens	*Spirea* spp.
Ilex aquifolium	
Juniperus spp.	Ground cover
Prunus laurocerasus	*Hedera helix* cultivars
	H. canariensis
Occasional	*H. colchica*
Cryptomeria japonica	*Hypericum calycinum*
Lonicera nitida	*Vinca major*
Mahonia aquifolium	*V. minor*
Skimmia japonica	

popular species were the Victorian favourites *Aucuba japonica, Buxus, Euonymus japonicus, Juniperus, Ligustrum* and *Taxus,* which are easily kept under control by clipping, but over 20 species (and many more cultivars) including the Indian bean tree (*Catalpa bignonioides*) and *Osmanthus delavayi* have been recorded (Table 13.1). The best places to see them are in the now overgrown London cemeteries such as Highgate and Nunhead where they form an understory in areas which have reverted to woodland. Today local authorities discourage the planting of anything other than grass on grave plots so they have historical interest.

13.4 THE GROUND LAYER

The expectation of cemetery visitors is for a neat and tidy appearance which usually means one thing – short grass. In practice there is usually only sufficient labour to achieve three to six cuts a year so, in early summer, the grass grows quite tall between treatments, though increasingly growth retardants are being employed to combat this. Around entrances and chapels higher horticultural standards are maintained, including grass cut 26 times a year, edging, the introduction of spring bulbs, roses and summer bedding. A wider range of ground cover is found in the general burial areas, its detail depending largely on the management regime which in turn depends on the type and arrangement of the memorials. Three broad categories can be distinguished.

Lawn cemeteries

This type of layout covers considerable areas of local authority cemeteries and has been the only arrangement permitted in most new burial areas for the last 15–20 years. It fulfils the modern requirement of being cheap to maintain, but is mostly ecologically dull (Fig. 13.2). Mowers with their blades set at about an inch pass regularly between ranks of uniform gravestones standing in short grassland. The turf usually contains much rye-grass together with mowing-tolerant and trampling-resistant species such as creeping buttercup, daisy, dandelion, ribwort plantain and white clover. Where disturbed by recent interments annual meadow-grass is abundant. The composition of the grassland differs little from that of a recently established lawn.

Lawn conversion

A different vegetation occurs where lawn conversion has taken place in older parts of the cemetery where soils have become acid due to leaching and lack of disturbance. Here the grassland is much less fertile so is

Fig. 13.2. A modern lawn cemetery, cheap to maintain but ecologically dull.

frequently dominated by finer leaved species such as fescues and bents or on heavier soils Yorkshire-fog (*Holcus lanatus*) and rough stalked meadow-grass (*Poa trivialis*). Associated speices include sheep's sorrel (*Rumex acetosella*), heath bedstraw (*Galium saxatile*), yarrow, cat's ear, field woodrush (*Luzula campestris*), sweet vernal grass (*Anthoxanthum odoratum*) and occasionally (seen three times) the sedge *Carex nigra*. Many of these species are locally abundant and have quite showy flowers so between mowings lawn conversion areas can look extremely attractive. Thought could be given to varying the time of cutting to take advantage of this. Owing to the uneven ground, caused by subsidence when coffin lids collapse, mower blades frequently scour the tops of hillocks. Such niches become occupied by communities of bryophytes and lichens in which *Polytrichum* spp. and *Cladonia* spp. are prominent. These are calcifuge (acid-loving) species which emphasize that cemeteries contain some of the most acid infertile soils to be found in towns. Lime and fertilizer are never applied to the general burial areas.

Close-set graves with kerbs

The majority of cemeteries are still dominated by close-set graves in which the grave space is outlined by substantial raised kerbs often enclosing an area of shallow stone or glass chippings lying over a hard base. Alternatively the kerbs outline soil cultivated as a small garden or left for natural species to colonize. This type of layout persists in areas where expensive memorials occur which relatives are unwilling to see meddled with. Managers find these areas difficult to maintain to a high standard as the 45 cm gap between the graves is too narrow for a mower. Once scythed or 'hooked' they are now more often blanket treated with a growth retardant or residual herbicide. This can leave the area looking messy so as an alternative they get worked over with a strimmer which has revolutionized cemetery maintenance since 1980. I have seen many solutions tried in an attempt to cut labour in this type of layout ranging from eliminating all vegetation using chemicals to blanket planting with *Cotoneaster* 'Skogholm'.

Where herbicide is used only sparingly, areas with close-set graves tend to support a tall-herb/grass community which between mowings can look most attractive and provide something of the atmosphere of a country churchyard. The dominant grass is usually false oat (*Arrhenatherum elatius*) but may sometimes be Yorkshire-fog (*Holcus lanatus*) and include others such as hairy oat (*Helictotrichon pubescens*), sweet vernal-grass (*Anthoxanthum odoratum*) and creeping fescue (*Festuca rubra*). Growing amongst them are a wide variety of tall herbs typical of neutral grassland, e.g. cow parsley (*Anthriscus sylvestris*), knapweed (*Centaurea nigra*), ox-eye daisy (*Chrysanthemum leucanthemum*), St John's-wort (*Hypericum perforatum*), clovers and vetches. The more acid sites have foxgloves. The most diverse communities encountered were at Arnos Vale, Bristol, where the sward contains abundant bed-straws, cowslips, primroses, germander speedwell and yellow oat-grass (*Trisetum flavescens*) in addition to many legumes and Umbelliferae. This richness was related to the alkaline soil and regular, but low-key, maintenance.

By contrast regularly sprayed cemeteries carry a coarse unattractive vegetation. Spraying encourages deeply rhizomatous species such as couch grass (*Elymus repens*), bellbine (*Calystegia sepium*), bind-weed (*Convolvulus arvensis*), horsetail (*Equisetum arvense*) and coltsfoot (*Tussilago farfara*) which, owing to their avoidance mechanisms, increase to weed proportions. The rampant bellbine at Beckett Street, Leeds is a legacy of past spraying. Hairy sedge (*Carex hirta*), in 10% of cemeteries, also belongs to this group. Another strategy that provides some protection against herbicides is rapid growth from seed which enables *Epilobium* spp. and *Senecio* spp. to multiply on recently sprayed areas.

A walk around any cemetery will show woody species particularly ivy, bramble and raspberry establishing at the foot of memorials. This is a result of their use as bird perches, in fact grave plots are one of the few niches where I have regularly observed ivy seedlings at the cotyledon stage. The establishment of these woody plants is helped by the spraying as it reduces competition, and, once well grown, they show resistance to many chemicals.

13.5 GRAVE PLOTS

A feature which distinguishes cemetery vegetation from all other is associated with the actual grave plots which in theory are privately owned. It was customary, particularly pre-1940, for small gardens to be created on graves; these were regularly tended by relatives. With changing fashion and as people moved away, these tiny gardens became neglected and while most of the plants succumbed to competition a few of the more aggressive became naturalized so that today cemetery vegetation, when not closely mown, is characterized by populations of garden relics. Evergreen shrubs have already been mentioned in this connection (Table 13.1). Francesca Greenoak (1985) has identified a number of plants linked by their name to the Christian faith and asserts that these are particularly common in burial grounds. They include snowdrops, lady's mantle (*Alchemilla mollis*) and star-of-Bethlehem *Ornithogalum* spp. These are certainly present in many neglected cemeteries but the most widely naturalized garden species are lesser meadow rue (*Thalictrum minus*), London pride (*Saxifraga* × *urbinum*), daffodil (*Narcissus pseudonarcissus*), hart's-tongue fern (*Phyllitis scolopendrium*), asparagus (*Asparagus officinalis* subsp. *officinalis*) and, in shady localities, ivy, periwinkle, rose of sharon (*Hypericum calycinum*) and creeping Jenny (*Lysimachia nummularia*). Individual cemeteries often contain a garden escape of unusual abundance or one which is nationally rare, e.g. *Montia sibirica* (Beckett Street, Leeds), *Euphorbia uralensis* (Norwood), *Geranium phaeum* (York), *Lathyrus latifolius* (Kensal Green).

A further niche, unique to cemeteries, though with parallels in Highland Britain, is provided by the thin (1–3 cm) layer of chippings overlying a solid stone base which often occupies the rectangular area enclosed by the kerbs. The chippings are usually of granite or limestone, though recently coloured glass has become popular. The main characteristics of this niche are drought, low nutrients, intermittent disturbance and low competition, ideal for stress-tolerant plants. Several garden escapes are particularly common in this microhabitat, for example the succulents wall-pepper (*Sedum acre*), white stonecrop (*S. album*), English stonecrop (*S. anglicum*), rock stonecrop (*S. forsterianum*), *S. reflexum, S. spurium*. In nature these species occur on scree, rocks, cliff ledges, shingle and wall tops so are well suited to this urban habitat where they are also favoured by possessing a tolerance to many of the common herbicides used to prevent rank growth developing on grave plots. Other species of garden origin regularly encountered on the chippings are a beautiful orange hawkweed from central Europe called fox and cubs (*Hieracium aurantiacum*), the strict calcicole bloody cranesbill (*Geranium sanguineum*) and thrift (*Armeria maritima*).

Though many of the above species are natives, most of the propagules will have originated from gardens or adjacent graves, as areas of chippings are rarely, if ever, deliberately planted up. A further ecological group found on grave gravel are annuals of dry disturbed places such as herb robert (*Geranium robertianum*), wild pansy (*Viola tricolor*), sheep's sorrel (*Rumex acetosella*), pale toadflax (*Linaria repens*), Oxford ragwort (*Senecio squalidus*), the grass *Bromus mollis* and in the south Canadian fleabane (*Conyza canadensis*). Once the hard base which underlies the chippings gets cracked through age or ground subsidence a new group of strongly rhizomatous but less drought-tolerant species take over, e.g. the grasses *Agrostis capillaris, Festuca rubra, Holcus mollis, Poa pratensis,* or colourful legumes such as meadow pea (*Lathyrus pratensis*), bird's-foot trefoil (*Lotus corniculatus*) and vetches. Where shrouded by dense ground vegetation or trees, communities of bryophytes typically dominate.

13.6 RELIC COMMUNITIES

Cemeteries are normally established by enclosing land at the outskirts of towns, but urban growth has ensured that most now lie well within the built-up area. A number still contain, protected within their boundaries, relics of the countryside as it existed many years ago. Large isolated trees, often in an irregular line but considerably older than the structure planting and unrelated to layout, can be correlated on old maps with former hedgerows or parkland planting. Numerous examples could be quoted but perhaps the most famous is the large cedar of Lebanon at Highgate West, a relic of Lord Ashurst's garden which formerly occupied the site.

Remains of woodland are not uncommon. In Sheffield the last fragment of Burngreave Wood which features prominently on Fairbank's 1793 map of the city, but long thought to have disappeared, was recently discovered in a corner of Burngreave Cemetery (1860) as a few oaks with a ground flora of bluebell, wavy hairgrass, foxgloves and creeping soft-grass (*Holcus mollis*). Similar remnants of oak woodland representing an outlier of the Great North Wood persist on a bank in Norwood Cemetery, Lambeth. The best example encountered during the survey was at Lawnswood, Leeds where an intact ground flora consisting of ramsons (*Allium ursinum*), bluebell (*Hyacinthoides non-scripta*), wood anemone (*Anemone nemorosa*), hairy woodrush (*Luzula pilosa*) and enchanter's nightshade (*Circaea lutetiana*) persists among the gravestones.

Relic grassland is less easy to recognize as many of the species are more mobile and others can get mown out. At Wisewood, Sheffield a wet grassland containing drifts of lady's smock (*Cardamine pratensis*), large bird's-foot trefoil (*Lotus uliginosus*) and blinks (*Montia fontana*) clearly represents pastures from the days before efficient drainage was employed. At Lawnswood a wet area with hard rush (*Juncus inflexus*), pale sedge (*Carex pallescens*) and square-stemmed St John's-wort (*Hypericum tetrapterum*) is probably relic. A drier grassland occurs at Norwood and Nunhead but it is difficult to know whether species such as meadow cranesbill (*Geranium pratense*) and hoary plantain (*Plantago media*) have a relic status.

No such difficulties attend an appraisal of the heathland communities encountered in 10% of the cemeteries visited. Danygraig Cemetery, Swansea (1856) contains several extensive areas of heather (*Calluna vulgaris*) growing with the moorland grasses *Nardus stricta, Molinia caerulea, Aira caryophyllea,* and various calcifuge lichens (*Cladonia furcata*) and bryophytes (*Polytrichum piliferum*). Other Swansea cemeteries are occupied by dense swards of *Molinia,* while bilberry, crowberry (*Empetrum nigrum*), wood sage (*Teucrium scorodonia*) and much wavy hairgrass are preserved at Wards End less than 2 km from the centre of Sheffield. An unusual fragment was discovered in The Rosary, Norwich (1838) which may represent an outlier of the formerly extensive heaths which lie north of the city. It contains a number of species rarely encountered in urban grassland, e.g. heather (*Calluna vulgaris*), wood speedwell (*Veronica montana*), barren strawberry (*Potentilla sterilis*) and violet (*Viola riviniana*), together with the moss *Pseudoscleropodium purum.*

13.7 SUCCESSION

Neglected cemeteries turn over to woodland quite quickly, as the varied microtopography provides plenty of germination sites for the heavy seed rain. Once management ceases, tall grasses such as false oat (*Arrhenatherum elatius*), meadow foxtail (*Alopecurus pratensis*), cocksfoot (*Dactylis glomerata*) and at Nunhead, tall fescue (*Festuca arundinacea*), increase together with rosebay willowherb, ragwort, thistles and docks. Soon ivy and bellbine start to scramble over the graves. Then the initially subordinate brambles begin to form mounds which eventually coalesce to cover large areas penetrated only by occasional paths leading to graves which are still being tended. All this time tree seedlings have been getting established, even in bramble patches so within ten years, flourishing almost even-aged stands of saplings are present everywhere. The composition of this young woodland depends very much on local seed sources but ash and sycamore are always leading species, hawthorn, horse chestnut and willows are normally present, and in the south Norway maple, Turkey oak and holm oak can be abundant. Lime rarely regenerates, but holly and yew are often frequent.

The succession passes through a dense pole stage which eliminates most of the previous grassland species, then the canopy gradually opens out over a rather limited shade-tolerant ground flora. This is normally dominated by common ivy (*Hedera helix*) and the extremely vigorous

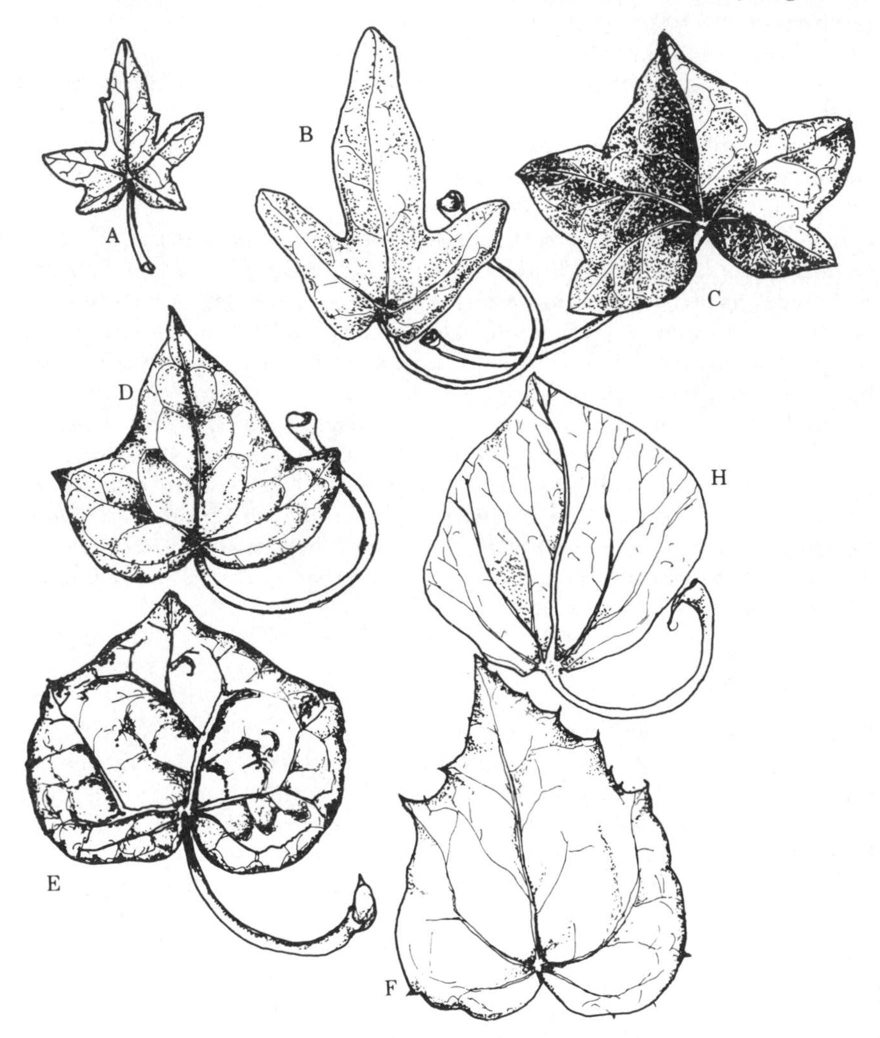

Fig. 13.3. Species and cultivars of ivy commonly found in overgrown cemeteries. A, *Hedera helix*; B,C, *H. helix* 'Hibernica'; D, *H. algeriensis*; E, *H. colchica*; F, *H. colchica* 'Dentata'; H, H. aff. *algeriensis* (A. Rutherford).

Irish ivy (*Hedera helix* 'Hibernica') which being disinclined to climb forms a dense ground layer. Morphological differences between these two ivies and various other cultivars common in burial grounds are illustrated in Fig. 13.3. Other species that give character to this woodland are survivors from grave plantings such as periwinkle (*Vinca* spp.) and creeping Jenny (*Lysimachia nummularia*) together with typical mobile colonizers of secondary woodland, e.g. lords and ladies (*Arum maculatum*), cow parsley (*Anthriscus sylvestris*), celandine (*Ranunculus ficaria*) and various ferns. Other areas remain as bramble for a long time but eventually it is only along the rides that grassland and woodland edge species survive. Kensal Green which is composed of a mosaic of neglected, half-managed and tended areas has a particularly rich flora that includes 313 native and naturalized angiosperms (Latham, 1984). Nunhead, Highgate and Abney Park are fine examples of cemeteries that have turned over to secondary woodland. A number of others have had the succession reversed by teams of unskilled labourers clearing the woody growth so not all towns now have good examples. Overgrown cemeteries are among the most important sites for nature conservation in urban areas. The following descriptions outline the range of variation in composition, management and wildlife interest they have to offer.

13.8 MANAGEMENT OPTIONS

Nearly as many management options exist as there are cemetery managers, so only a few of the sites visited which were under a regime favourable to wildlife can be described. Traditional management as tended open space is not considered. A common objective should be the preservation of any original structure planting, both as an example of high Victorian fashion and for its value to hole-nesting birds. The historical importance of relic and naturalized grave planting should also be recognized and no attempt made to eliminate exotics as they are part of the land use and help to make cemetery vegetation distinctive. Occasionally excessive dominance by Japanese knotweed, bramble, horsetail or *Calystegia sepium* may need to be checked. In general, low levels of maintenance carried out regularly produces the richest wildlife; Arnos Vale Cemetery, Bristol, is a magnificent example of this.

The Nature Reserve – Rawmarsh High Street Cemetery, South Yorkshire

This 2 ha local authority cemetery (1856) is rarely visited by the public as it is full up. Increasingly reluctant to maintain it to traditional standards the superintendent approached Rotherham Museums Department and the Yorkshire Naturalist Trust for advice because he was interested in managing it for nature conservation. As a result, since 1983 the cemetery has received just one cut a year in late summer. This is done with strimmers, the cut grass being raked off and removed by tractor. The operation, which takes six men a month, has reduced maintenance costs by 80%. The cemetery is now a mass of daffodils in spring and cow parsley in June, while in late summer large beds of bramble, nettles and ivy which are left unmown form a feature. Since 1983, dead elms have been left standing and there are plans to kill other trees by ring barking where they are too dense. Mature ash, horse chestnut and laburnum are well represented, but there is little shrub cover for birds. A small fragment of old pastureland is present which contains *Succisa pratensis, Sanguisorba officinalis* and *Potentilla erecta*; the museum has recorded several unusual insects. Perhaps in these early days the cemetery's main contribution to the area is the atmosphere of naturalness it provides; later new species will arrive in response to the sympathetic management.

Non-intervention – Nunhead, London

In 1975 the Borough of Southwark bought the long-neglected Nunhead cemetery for £1. They had grandiose plans to divide it into three sections with 10 ha remaining as a working cemetery, 10 ha managed as a nature reserve and 4 ha converted into a park. This plan has only been partly executed with the consequence that most of the site has seen little maintenance for nearly 50 years. A dense secondary woodland of sycamore, Norway maple, oak, ash and elm covers the site, its composition depending on the distribution of seed parents. The trees stand over a vast undulating bed of ivy which in places rises up over the gravestones and elsewhere climbs trees to the top. Spindly regeneration only 1 m high, yet over 20 years old, is kept in check by the low light; there is dead wood everywhere. *Aucuba, Buxus, Euonymous* and *Ligustrum* form a sparse evergreen shrub layer. At the margins of paths blackberry forms an almost impenetrable barrier and a few woodland herbs such as dog's mercury, creeping Jenny and male-fern occur.

Where this woodland has been cleared the glades quickly fill up with bramble or Japanese knotweed, though a few have maintained themselves as tall herb meadows with big clumps of rosebay willowherb, Michaelmas daisy, hoary ragwort (*Senecio erucifolius*), zig-zag clover (*Trifolium medium*), tall fescue (*Festuca arundinacea*), tufted hair-grass (*Deschampsia cespitosa*) and locally traveller's joy (*Clematis vitalba*). This type of vegetation also extends along the margins of the main tracks. Throughout the site the original structure planting provides an older age class of trees. The plant diversity, the varied structure of the vegetation and the size of the cemetery suggest the site will support a rich fauna;

foxes, owls and bats have been reported as breeding. Nunhead probably contains the largest area of the most advanced successional stage of any cemetery in Britain.

The Highgate approach – Highgate West, London

Highgate, an internationally important cemetery, is cared for by a voluntary group called the Friends of Highgate Cemetery. Since 1975 they have been managing its 7 ha in a way that preserves the historical interest and romantic quality while, at the same time, enhancing the flora and fauna; indeed wildlife is an integral part of the Highgate experience. Until well into the twentieth century the cemetery was kept meticulously and the structure planting lopped to maintain the views out over London. Eventually, however, sycamore and ash regeneration started to take over, and, although this was half-heartedly cleared, in time the coppice regrowth got out of hand and an intimate overgrown woodland character prevailed. Today the site is largely occupied by a dense young to middle-aged secondary woodland within which the old structure planting survives (Fig. 13.4).

Fig. 13.4. Highgate Cemetery has, through neglect, turned into a unique urban woodland where introduced plants outnumber native species (J. Gay).

The Friends are working to a management plan which by zoning the site into a 'wildlife refuge' and a 'sensitive area' walks the knife-edge between non-intervention which eventually leads to dereliction of the artifacts and the single-minded preservation of historical interest which would completely destroy the intimate atmosphere. The general aim is to encourage ecological diversity by introducing as many locally native trees, shrubs and ground-layer species as possible. At present the distinctiveness of the soil types is masked by sycamore, but this will gradually be placed by birch, oak and pine on the Bagshot Sands with a more diverse woodland dominated by ash, oak and cherry on the lower ground which is underlain by clay. The use of trees and shrubs attractive to insects is being given priority. Currently the richest herbaceous communities are along path sides which will be developed as wood margins; elsewhere small glades are being created. Between 1976 and 1982, 83 herbaceous plants and 25 woody ones were introduced by seed, turf transplants or as whips. Today a walk at any season shows a wide range of introduced species such as primrose, lords and ladies, teasel, bistort, broom, greater celandine, foxglove, honesty, lady's mantle, fox and cubs, snowdrops, wood spurge, balm and violets. As this is an entirely artificial site these introductions are quite appropriate. The area should be seen as a botanical garden for wild plants which previously grew in the Highgate area.

The wildlife zone is a continuous area which occupies half the cemetery. Access off the paths is discouraged by barrier planting of prickly shrubs to increase the refuge value and reduce pressure on fragile sites, such as the pond. A central flowery meadow with the atmosphere of a country churchyard was developed in 1985. Within the sensitive zone lie the catacombs, most of the more impressive tombs and buildings, and the largest trees. Here, apart from gradually replacing the sycamores and implementing the general wildlife prescriptions, little management is required. Path margins are on the whole being opened up to reveal glimpsed views of memorials among the trees and a dense network of small informal paths will be allowed to develop. Areas of 'Victorian gloom' currently dominated by evergreens will be reinforced.

Bird life is already quite varied with wren, blackbird and robin abundant, goldcrest common, spotted flycatcher, blackcap, nuthatch, tree-creeper and hedge sparrows nesting on ivy-covered headstones, tawny owl roosting, woodpeckers regularly seen and so on. The total breeding population is around 80 pairs.

Managing overgrown cemeteries – Abney Park, London

This 13 ha private cemetery was laid out in 1840 as an arboretum.

Though many of the trees were later removed to make way for additional graves, several Bhutan pine (*Pinus wallichiana*), swamp cypress (*Taxodium distichum*) and Spanish oak (*Quercus* × *hispanica*) survive. Between 1939 and 1979 the area received little maintenance, and sycamore, ash and Japanese knotweed took over as the cemetery moved towards becoming a dense species-poor woodland. A recent management plan with the objectives of providing for passive recreation, education, burials, nature and memorial conservation through a system of zoning is being implemented. Though 28% of the mature trees are sycamore, a tolerant view is taken of its presence because of the food chain it supports (Fig. 9.6); consequently it is only cleared where glades are required. By contrast, ways are being sought of totally eliminating Japanese knotweed.

Since 1979, 37 birds have been recorded, of which the 10 commonest breeding species are (in order) blackbird, woodpigeon, starling, wren, blue-tit, robin, greenfinch, song thrush, great-tit and stock dove. A comparison with the results of the National Garden Bird Survey showed that seven species seen only rarely in gardens nationally were present at Abney, while a comparison with two local woods showed that one contained 10 and the other 13 of Abney's top 20 birds. The conclusion was that the bird life of this neglected cemetery showed features typical of both gardens and woodland. Practical points to remember when managing a cemetery for birds are that it is said all species eat elderberries; alder and birch encourage redpoll; crabapple and rowan retain their fruit well so are good for thrushes; goat willow which flowers early attracts insects which in turn attract chiff-chaff, willow warblers and tits. Ivy is particularly valuable as it provides nest sites for goldcrest and dunnock, autumn food for insectivorous birds and spring food (berries) for blackcaps, etc. Fastigiate conifers provide cover for nesting goldfinch and also for greenfinch near the top and dunnock lower down.

Ornamental horticulture – Moorgate, Rotherham, South Yorkshire

Most cemetery managers receive only a very basic training in horticulture but the superintendent of Moorgate Cemetery, Rotherham (1846) holds a Kew diploma. Since 1978 he has greatly enriched the exceedingly dull Victorian structure planting (50% sycamore, 20% holly) by introducing over 180 trees and shrubs of high horticultural value. These are mostly deployed in groups of five and show a great diversity of form, leaf colour and seasonality; in 8 years this 6.5 ha cemetery has been turned into a botanical garden. The new planting includes 19 varieties of *Acer*, 11 Rhododendrons and spectacular species such as the dove-tree (*Davidia involucrata*), honey-locust (*Gleditsia triacanthos* 'Sunburst') and several Magnolias which, if they grow well, will be worth making special visits to see.

The trees are shown off against a background of highly maintained landscape. For the last 7 years the herbicide simazine has been regularly applied among the graves; this has encouraged a spectacular growth of Crassulaceae (stonecrops) on grave chippings and, where used on soil, deeply rhizomatous vetches, peas and convolvulus have increased together with willowherbs and creeping cinquefoil (*Potentilla reptans*), all of which benefit from the absence of competition from more susceptible species. This approach, using the methods of ornamental horticulture, has the merit of encouraging visitors thereby reducing vandalism. The general structure and diversity has incidental wildlife benefits.

Chapter 14

GARDENS

Compared with the rest of Europe, England and Wales have an unusually high proportion of houses with private gardens, e.g. England and Wales, 78%, Belgium, 70%, Holland, 56%, West Germany, 49%, France, 32% (Evenson, 1979). Provision in Scotland is much lower than in England and Wales as there the housing contains a higher proportion of tenements and flats without private gardens. Though exact data are hard to come by, it is thought that in Britain there are over 15 million gardens which in England and Wales cover 400 000 ha or 3% of the land surface. In towns residential areas may extend over 60–70% of the total built-up area.

The high level of garden provision in England and Wales can be attributed to several historical factors which are discussed in detail by Kellett (1982). The stable political system in this country meant that town defences were not so vital as on the continent where interstate and intercity rivalries were common. Consequently English towns were not nearly so densely populated in the pre-industrial era; early maps clearly show burgage plots and gardens widespread within them. So in England towns tended to maintain a rural tradition of single family houses with garden plots while on the continent land shortages for safe urban expansion resulted in tenement living. Throughout the industrial revolution single family dwellings continued to be the norm even though garden space was restricted. The major influence that reinstated the private garden to its former prominence stemmed not from government initiatives but from the 'garden city' movement pioneered by Ebenezer Howard (1898) and Raymond Unwin (1909). They were a powerful influence in the town-planning field promoting the idea of 12 dwellings per acre (28 ha^{-1}) by practical example in schemes at Letchworth and Hampstead. The influential report of the Tudor-Walters Committee (1918) on working-class housing officially recognized the importance of gardens and from then on their provision was nearly universal, usually at both back and front. Low densities and generous gardens were also the norm in the early new towns. Public sector building in the 1960s and 1970s reversed this trend but was unpopular as it curtailed such activities as cultivation, sitting out, children's play and drying washing. Gardening, one of the most intensive forms of land management known, remains a national characteristic of the English and Welsh.

One of the keys to the richness of the garden ecosystem is its variety. A mosaic of minihabitats is provided by lawns, shrub beds, rockeries, old fruit trees, vegetable patches, hedges, walls, the house, flower beds, ponds and compost heaps. The range of ecotones present, the selection of plant species which produce copious supplies of nectar and pollen, and the results of pruning which provides a succession of young growth makes gardens a particularly rich habitat for insects. Work in an African tropical garden (Owen, 1971) suggested that it was richer in butterfly species per unit area than nearby primary and secondary forest. This was interpreted as an expression of its patchwork nature of sun and shade, the abundance of flowers and the fact that in some species new larval food preferences were evolving to exploit plants of cultivated origin. So while the centre of cities are simplified by urbanization, the opposite is happening in the suburbs.

14.1 VEGETATION

The structure of gardens

It might be thought that the species range and the physical structure of garden vegetation was infinitely and continuously variable, but this is not so. Schmid (1975) in his study of Chicago, Illinois recognized two basic types of residential neighbourhood. The first, designated 'open landscape' was characterized by small ornamental trees which never overshaded the houses or streets. Privacy was obtained by large plot size but no house was hidden by the trees which provided ornament and shade. Layout was relatively formal with lawns showing off both trees and buildings. By contrast 'closed landscape' presented vegetation as its main visual feature, large trees and shrubs separated and screened property, being arranged in clumps and belts often standing over unmanaged herbaceous vegetation. Closed landscapes were the habitat of the upper social strata being found only in the wealthy suburbs.

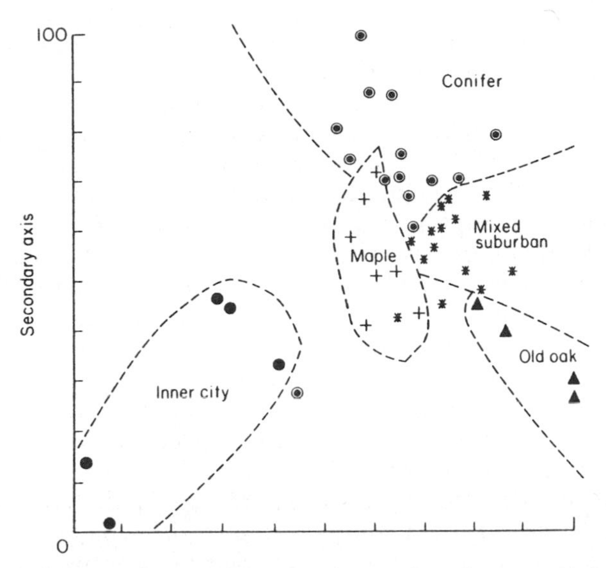

Fig. 14.1. Ordination of vegetation occurring on front lawns at 44 sites in Arkon, Ohio; five assemblages are identified (Whitney and Adams, 1980).

A more sophisticated analysis of vegetation in residential areas of Akron, Ohio was carried out by Whitney and Adams (1980). They recorded ornamental and shade trees occurring in front gardens at 44 sites dispersed throughout the urban area. Each site was uniform regarding features such as age of home and socioeconomic background, this information being extracted from census returns. Ordination of the data by means of reciprocal averaging enabled five major community complexes to be recognized on the basis of their more representative or characteristic species (Fig. 14.1).

The old oak complex comprised large extensively wooded gardens dominated by an overstorey canopy of *Quercus velutina*, *Q. rubra* and *Q. alba* together with *Fraxinus americana*; native species predominated. This was more or less equivalent to Schmid's closed landscape. A mixed suburban complex characterized post-1945 housing, where diverse cultivars of *Acer saccharinum*, *A. saccharum*, *A. rubrum*, *A. platanoides* and *Quercus palustris* mixed with *Betula pendula*, *Liriodendron tulipifera* and *Liquidambar styraciflua*. A conifer complex was associated with modest single-family interwar residences; here *Thuja occidentalis* and *Picea pungens* were ubiquitous interspersed with *Picea abies*, *P. glauca*, *Juniperus* spp. and *Taxus* spp. Many were present as foundation plants at the base of buildings and tree density was low to very low. The maple complex was linked with prosperous turn-of-the-century middle-class housing and also some post-war areas. An inner city complex contained fast-growing sometimes self-sown trees tolerant of urban stress, e.g.

Ailanthus altissima, Morus alba, Ulmus rubra; fruit trees were also a characteristic feature of the inner city.

When cultural variables were matched with each of the complexes, taste, fashion and the relative availability of plant material were found to be the major factors influencing the composition of the woody plants in front gardens. With regard to income and house value, the old oak and mixed suburban complexes were the hallmark of the affluent and prestigious neighbourhoods. The conifer complex picked out working-class areas where landscape is not a great concern, while the inexpensive fast-growing trees of the inner city reflect a casual attitude. The maple complex echoed the tastes of an era around the turn of the century when exotics, often of European origin, were in vogue.

When the 44 sites were plotted on a map of Akron the various complexes were seen to be arranged in an irregular concentric pattern which fits Burgess's five-zone theory of urban expansion (Burgess, 1925). He suggested that a city is composed of the central business district (non-residential); a zone in transition which is an area of deterioration having at one time contained prosperous residences but now being invaded by commercial interests (inner city and maple complexes); a zone of workers' homes (conifer complex); a residential zone of prosperous single-family housing (old oak and mixed suburban complexes); and the commuter zone (old oak and mixed suburban). A similar pattern of vegetation related to socioeconomic factors has been reported from Ann Arbor, Michigan (Detwyler, 1972).

Producing a classification of gardens based on floristic and structural criteria is one of the more pressing needs in urban ecology, as they are the major wildlife habitat in towns. As most properties are privately owned remote sensing techniques coupled to ground truth surveys may produce the initial breakthrough. A preliminary reconnoitre of Sheffield using Landsat Thermatic Mapper imagery on four wavebands showed very complex patterns even where housing types appeared uniform on maps, and there were additional complications such as allotments registering as large gardens. Combining remote sensing with ground truth surveys the following types of front garden vegetation have been recognized in Sheffield: the account is not exhaustive.

Large gardens: pre–1920

Massive multilayered boundary planting in which freely growing evergreens predominate: *Ilex aquifolium, I × altaclarensis, Taxus, Prunus laurocerasus, P. lusitanica, Ligustrum,* Rhododendrons, *Aucuba, Hedera,* all frequent, often as cultivars; usually mixed, occasionally of a single species. Forest trees, *Tilia, Pinus, Acer, Ulmus, Fagus* normally stand over the evergreens. The planting screens large lawns with specimen trees and island shrub beds. This closed landscape was created for the wealthy of a previous era.

Terrace housing; pre–1920 gardens < 4 m wide

About 50% are bordered by a 1–1.5 m tall clipped hedge of *Ligustrum ovalifolium* frequently var 'Aureum'. Privet is occasionally replaced by *Fuchsia magellanica, Cotoneaster simonsii, Mahonia, Fagus,* etc. The majority contain a small bed of hybrid tea roses; the only other woody species commonly grown for ornament are *Hydrangea macrophylla, Cotoneaster* spp., *Fuchsia* and dwarf conifers. This semiopen controlled vegetation contains very few trees. Lawns are absent. This type of front garden vegetation characterizes low-income group housing where low clipped privet hedges were formerly a status symbol.

Detached and semi-detached housing, pre–1920, gardens 4–10 m wide

Mature lopped forest trees (*Tilia, Betula, Populus,* copper beech, *Fraxinus*) are frequently seen with supporting smaller trees also mainly natives (*Crataegus monogyna, Ilex aquifolium* or old favourites *Syringa vulgaris, Laburnum anagyroides, Prunus cerasifera* 'Pissardii'). Poorly maintained hedges may be present. They are equivalent to the zone of deterioration in Ohio. There was only a limited range of plants available when the gardens were originally laid out.

Detached and semi-detached housing, 1920–1960, gardens 4–10 m wide

A large selection of ornamental shrubs is present in combination with blossom trees (*Prunus, Malus, Laburnum* × *watereri, Sorbus* and fastigate conifers). Properties are divided by hedges, often of beech, and the road frontages are normally of specimen shrubs. The centre of the garden may be occupied by a lawn, rockery, roses or a flower bed. These gardens represent middle-class decorative horticulture, strongly influenced by the garden city movement.

Detached and semi-detached housing, post-1970, gardens 4–10 m wide

Open lawns are a feature, often to the pavement edge, containing an island bed of dwarf conifers, small ornamental shrubs including heathers, or specimen trees and shrubs in grass. Striking 'garden centre' plants predominate: *Pyrus salicifolia* 'Pendula', *Acer platanoides* 'Drummondii', *Cytisus × kewensis, Berberis darwinii, Viburnum farreri, Salix caprea* 'Pendula', *Salix matsudana, Juniperus virginiana* 'Skyrocket', *Prunus* 'Amanogawa', plus diverse small conifers.

It would be counterproductive to recognize a large number of garden types but as a start it appears that within the field of almost continuous variation represented by front garden vegetation a number of noda can be identified. In Sheffield their distribution seems to be controlled by garden size and age of property both of which are linked to socioeconomic factors such as income. Techniques for identifying the plant assemblages and the factors controlling them are well known; it would considerably benefit urban ecology if further studies were made in this area.

The nature of garden plants

The word garden comes from the Hebrew meaning 'a pleasant place' and many of the plants grown in domestic gardens emphasize this character by being highly ornamental. Originally they were chosen from the more useful or showy of the native species but these were soon supplemented by large numbers of exotics as plant collectors scoured the world for horticultural material. Later, plant breeders working on both native and exotic elements produced a wide variety of hybrids and cultivars so that the range of spectacular plant material available is now very large indeed. Hillier's (1983) *Manual of Trees and Shrubs*, for example lists 32 cultivars of the native shrubby cinquefoil (*Potentilla fruticosa*), 20 of beech (*Fagus sylvatica*), 25 of ivy (*Hedera helix*), 27 of yew (*Taxus baccata*) and 10 of elder (*Sambucus nigra*). The speed with which introduced species are adopted is well illustrated by the flowering currant (*Ribes sanguinium*). Seed was sent by David Douglas from Oregon to this country in 1826; by 1840 it was in almost every suburban garden and the Horticultural Society considered its introduction alone had covered the entire cost of Douglas's 3-year expedition. Today 11 cultivars and one hybrid are on offer.

If the nature of cultivars is examined it becomes apparent that they consist, in the main, of plants ill-equipped to survive outside gardens. Many trees have a pronounced pendulous or fastigiate habit, shapes that would not allow persistence in closed canopy forest. Even more disadvantageous is selection for dwarfism indicated by the frequency in catalogues of names such as Compacta, Compressa, Densa, Globosus, Humilis, Minima, Minor, Nana, Pygmaea, Procumbens, Prostrata and Repens. These plants with shapes and statures that deviate markedly from a norm that has evolved over millions of years are mostly perpetuated and maintained only by the combined partnership of the horticultural trade and gardeners.

Other cultivars are distinguished by red, purple, golden, yellow or variegated foliage. There has been little recent research on the anatomy and physiology of coloured leaves but older botanical works such as Rabinowitch (1956) contain a little information. In 'Purpurea' leaves the colour is due to the combined effect of reddish water-soluble anthocyanin pigments in the cell sap and the green colour of normal chloroplasts which impart a bronze or purple colour to the leaves. The anthocyanin is often limited to the vacuoles of the epidermal cells where its concentration and persistence may be such that the cultivars are known as 'Nigra' as in *Prunus cerasifera* 'Nigra', or alternatively in 'Rosea' of the same species it gradually breaks down as summer advances so the canopy returns to green. Most such cultivars arose as sports and are perpetuated vegetatively. The yellow colour of 'Aurea' leaves is due to a low content of the green to blue–green pigments chlorophyll *a* and *b* which allow yellow carotenoids to become dominant. In one analysis the chlorophyll *a* and *b* content of ordinary green *Ulmus* leaves was 0.56% dry weight while 'Aurea' leaves of the same species contained only 0.05% of these pigments. This implies a reduction in photosynthetic capacity, and gardeners are aware that variegated and yellow-leaved varieties are less hardy than green ones. Fig. 14.2 demonstrates the lower photosynthetic efficiency of leaves of the 'Aurea' variety of elder. Despite their lack of persistence in nature and the theoretical reasons for regarding them as crippled plants, it is

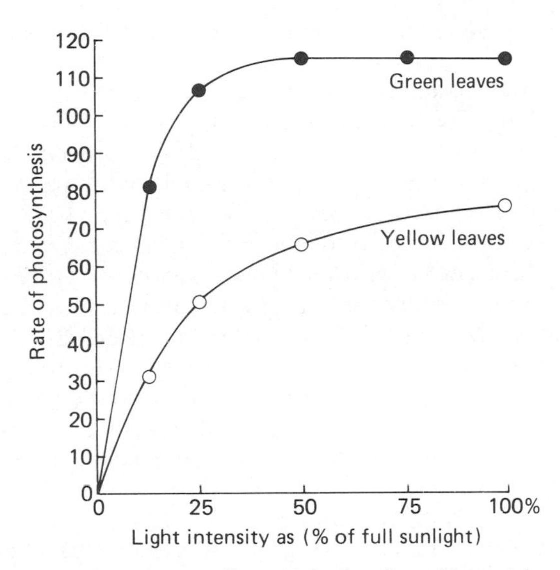

Fig. 14.2. Photosynthetic rates of green and yellow 'Aurea' leaves of *Sambucus nigra*; the low concentration of chlorophyll in the 'Aurea' leaves reduces their photosynthetic efficiency at all light intensities (Willstatter and Stoll, 1918).

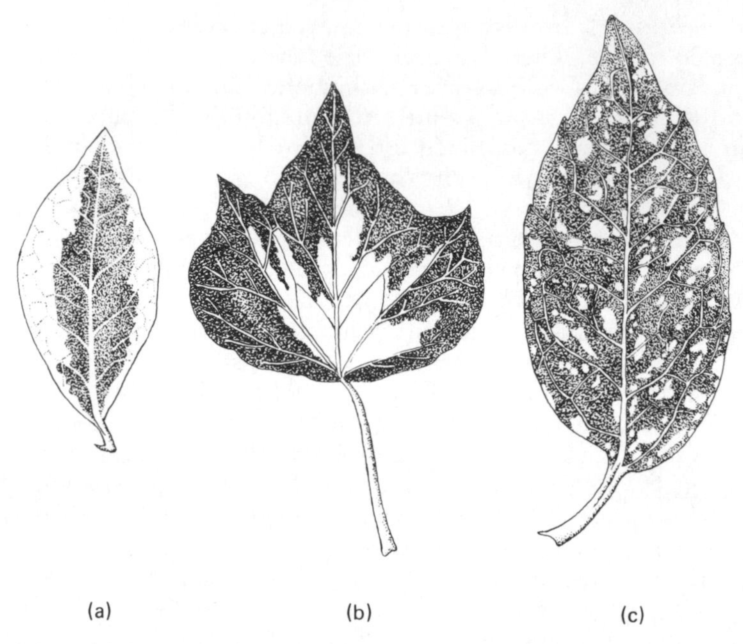

Fig. 14.3. Three types of variegated leaf. (a) Marginal (*Ligustrum ovalifolium* 'Aureum'); (b) central (*Hedera helix* 'Gold Heart'); (c) Mosaic (*Aucuba japonica* 'Gold Dust').

anomalous that golden varieties of elder show considerable vigour in the garden habitat.

With variegated foliage only part of the leaf is not green. The pattern of variegation can take several forms (Fig. 14.3). Holly cultivars show many types with the white margins of 'Silver Queen' containing no plastids and no chlorophyll, while the yellow margins of 'Golden Queen' contain plastids but they do not produce chlorophyll *a* or *b*. Causes of variegation can be nutrient deficiency, the presence of a virus, or infectious chlorosis (*Ligustrum vulgare* var *aureo-variegatum*) but the most usual type is a chance somatic mutation manipulated by breeders to create artificial chimeras. For a discussion of these see Timson (1959). From observations made on a large number of variegated privet bushes it has been estimated that their chlorophyll content varies between 10 and 70% (mean 40%) of that of normal green leaves. Yellow and white areas do not produce starch, and the reduced efficiency means that such plants do not normally thrive in the shade.

Selection for vivid autumn colours, together with other traits, such as silver or grey foliage associated with hairiness or wax deposits are less harmful to survival prospects. Disadvantageous characteristics of flowers are doubling in which stamens are turned into petals and gigantism which can render nectar unavailable; both reduce chances of pollination and the setting of ripe seed.

This section has emphasized how gardens are largely stocked with poorly competitive, slow-growing and crippled plants often living out-side their normal climatic range. For the majority, gardens are their only habitat with perpetuation of the cultivar by seed impossible; they only persist due to constant intervention, and often reintroduction, by man.

Weeds of cultivation

The weed flora of cultivated areas has many similarities to that described for allotments even down to the regular occurrence of bird seed aliens. Domestic gardens, however, are less frequently completely dug over so contain a higher proportion of rhizomatous weeds. Tutin (1973) has provided an insight into the weeds of a 0.5 ha garden in Leicester which he kept under observation for 25 years during which 95 species appeared. A number of distinct weed communities were present, their dynamic aspects are highlighted.

For example, while the vegetable garden was dressed with farm-yard manure it contained an abundance of *Chenopodium polyspermum, C. rubrum* and *C. ficifolium* but after manuring ceased they gradually disappeared and there was a parallel decline of *Atriplex hastata, A. patula* and *Polygonum convolvulus*; these are all annuals which require high soil fertility. Not all species responded in this way, however, for within this part of the garden the rhizomatous perennials *Aegopodium podagraria, Elymus repens* and *Campanula rapunculoides* occupied more or less the same area over the 25 years despite efforts to get rid of them, while many annuals with a wide edaphic amplitude such as *Bromus sterilis, Geranium robertianum* and *Myosotis arvensis* also maintained fairly stable populations in their own small areas. Extinctions included the once abundant *Oxalis corniculata*, while new arrivals were *Arctium minus, Dipsacus fullonum, Epilobium ciliatum* and *Senecio squalidus*. Following deep cultivation, the rare corn buttercup (*Ranunculus arvensis*) appeared, the most likely source being previously buried seed. A surprising feature was the way very local native British species such as *Elymus donianum* and *Hierochloe odorata* became extremely aggressive weeds. *Poa annua* was present as three genetically distinct variants, one perennial, another a tall annual, and the third a dwarf annual; each showed a distinct preference regarding fertility and soil moisture.

Patterns such as these can be confirmed in many gardens though they have rarely been put on record. Owen found at least 80 weeds in her 0.6 ha garden. The reduced competition resulting from cultivation operations releases plants from one of the chief restraints on their ecological amplitude, as Tutin discovered with *Elymus donianum,* normally a plant of base-rich mountains. Weeds with very different requirements in the wild are often found growing side by side in a semistable community, the common link being an ability to withstand attempts at eradication. For example, spontaneous *Vicia sepium* (ungrazed grassland), *Prunella vulgaris* (grazed damp grassland), *Aegopodium podagraria, Geum urbanum* (woodland), *Epilobium ciliatum, Equisetum arvense,* (ruderal), *Lamium purpureum* (nitrophile), *Bellis perennis* (sun lover) and *Ranunculus ficaria* (shade lover) can co-exist in the same border over a long period. The garden bed characterized by disturbance and with many opportunities for aliens to be recruited into the weed flora is a typical urban habitat.

Lawns

The composition of lawns is controlled by a wide range of factors amongst which its age, height and frequency of cut, soil fertility, use of weed killers and initial formulation of the sown mix are important. Most can be regarded as productive vegetation subject to frequent defoliation which Grime (1979) equates with disturbance. Turf grasses respond in two ways to frequent mowing, *Agrostis capillaris* and *Festuca rubra* v *commutata* are stimulated to produce a very large number of small tillers, while *Lolium perenne* S23 responds with a rapid almost vertical regrowth of the damaged leaves. The current trend is for turf grass breeders to produce lawn cultivars of *Lolium* that respond to mowing like *Agrostis* and *Festuca,* e.g. *L. perenne* 'Lorina' and 'Loreta'. The prostrate growth helps to suppress weeds. The average domestic lawn is bumpy and cut only when the grass has reached a height of about 6 cm with mower blades set at 2.5 cm. Under this regime a wide range of species survive unless selective weed killers are employed against them. Lawn specialists probably present nowhere else in the garden include *Achillea millefolium, Hypochoeris radicata, Leontodon autumnalis, Lotus corniculatus, Luzula campestris* and *Senecio jacobaea,* which occur alongside species equally common in cultivated ground, e.g. *Bellis perennis, Ranunculus repens, Taraxacum officinalis, Trifolium repens.* Age is a crucial factor, this aspect is dealt with in the chapter on parks.

Garden species of suitable growth form may sow themselves into lawns and thrive there. The best known example is *Veronica filiformis* introduced to Britain as a rockery plant in the early nineteenth century from its home in the mountains of Caucasus and Asia Minor (Bangerter and Kent, 1957). For a century it behaved as a model garden plant, then between the wars it became noted as 'a beautiful but rampant' weed of lawns. This speedwell is now widespread in lawns, where its slender creeping stems which root at the nodes equip it well for survival and vegetative dispersal (ripe seed is rare) in mown grassland. It is spreading

Table 14.1. The most constant (column 1), the most widespread (column 2) and the most prolific (column 3) species of macrofungi recorded from a lawn in Bristol over a period of 7 years

	1	*2*	*3*
Calocybe carnea	·	+	+
Clavaria fragilis	·	+	+
Clavulinopsis corniculata	·	·	+
C. helvola	·	+	+
C. pulchra	·	+	·
Conocybe tenera	+	+	+
Coprinus friesii	+	+	+
C. lagopus	+	·	·
C. plicatilis	+	+	+
Corticium fuciforme	·	+	+
Cuphophyllus niveus	+	·	+
Deconia spp.	+	+	+
Dermoloma pseudocuneifolium	·	+	+
Entoloma lampropus	·	+	+
E. sericeum	·	+	·
Galerina clavata	+	+	+
G. laevis	+	+	+
Hemimycena mairei	+	+	+
Hygrocybe conica	·	+	·
Marasmius graminum	·	+	+
M. recubans	·	+	·
Mycena flavoalba	+	+	+
M. olivaceomarginata	+	+	+
M. swartzii	+	+	+
Panaeolus ater	·	·	+
P. foenisecii	+	+	+
Sclerotinia trifoliorum	·	+	·
Tubaria autochthona	+	·	+
T. pellucida	·	+	·
Total – 29 species	14 spp.	24 spp.	22 spp.

Column 1, occurring in at least 6 years out of 7; column 2, recorded from at least four lawn areas out of five; column 3, producing > 100 sporophores, 1972–1976. Taken from Bond (1981).

in a similar fashion throughout Europe. Other garden plants which at times invade lawns are the double daisy (*Bellis perennis* 'Flora-plena'), lady's mantle (*Achemilla mollis*) and chamomile (*Chamaemelum nobile*). It is just possible that the chamomile lawn at Buckingham Palace represents a native population.

Little attention has been paid to the macrofungi occurring on lawns but a detailed survey of a large garden at the edge of Bristol (Bond, 1981) provides some insight into the factors controlling them. The recorded area of 380 m^2 of lawn in five blocks was laid out in the 1920s, since when regular mowing and removal of cuttings has left the soil impoverished. The lawns were surveyed several times a month from 1972 to 1978. The entire collection comprised 86 species (67 Agaricales, nine Aphyllophorales, ten other groups) with annual species totals ranging from 23 to 58 which led Bond to suggest that regular recording for 5–7 years will account for about 90% of the species likely to be present. Differences in productivity between the years both in species and total number of fruit bodies depended on meteorological conditions. Table 14.1 gives details of the most constant, widespread and prolific species. Of the 10 254 sporophores for 1972–76 the most abundant, *Galerina clavata* and *Galerina laevis,* together contributed 34%. At the other extreme nearly half produced not more than 10 fruit bodies over the 4 years. Three-quarters of the sporophores appeared in the months September, October and November which contrasted with an orchard sward in the neighbourhood where there was a preponderance of summer fruiting species (Bond, 1972).

Shade and moisture content of the soil were important in determining the detailed distribution of species, both being favourable. Not all the fungi were typical lawn or even grassland species; some (*Coprinus lagopus, Mycena flavescens, M. stylobates, M. tenerrima* and *Typhula erythropus*) occurred on bare soil, amongst moss or on miscellaneous detritus; others were clearly adventive from adjacent garden features, such as herbaceous border (*Agrocybe erebia*), clump of ferns (*Collybia confluens*), shrubbery, etc. (*Agaricus* spp.; *Laccaria laccata, Calvatia excipuliformis*). *Armillaria mellea* and *Coprinus micaceus* were from unidentified tree roots or buried wood. Other species were associated with particular trees around the boundaries, e.g. *Mycena mucor* on fallen beech leaves, *Tubaria furfuracea* from woody beech detritus, *T. autochthona* exclusively from buried *Crataegus* haws, *Clitocybe fragrans* under yew and *Inocybe brunnea* under hazel. Fungi of special habitats within the 'lawn' community included *Cordyceps militaris,* from buried lepidopterous larvae, and two small agarics, *Marasmius calopus* and *Mycena integrella,* which were found only on the dead outer leaves of plantain rosettes. These categories of fungi comprised about a third of the total species but contributed only some 6.5% of the total sporophores.

Garden walls

Several synoptic accounts of wall vegetation have been prepared, that by Segal (1969) being particularly exhaustive. He points out that some of the best-developed mural communities in Europe are to be found in towns in parts of Britain and France where relative humidity remains fairly high. A number of species have so taken to the habitat that they now have their headquarters on walls, e.g. ivy-leaved toadflax (*Cymbalaria muralis*), wallflower (*Cheiranthus cheiri*), snapdragon (*Antirrhinum majus*), red valerian (*Centranthus ruber*), pellitory-of-the-wall (*Parietaria diffusa*) and yellow corydalis (*Corydalis lutea*) (Fig. 14.4). Other species which could be added for lowland Britain are the ferns wall-rue (*Asplenium ruta-muraria*), common spleenwort (*A. trichomanes*) and black spleenwort (*A. adiantum-nigrum*). Other ferns such as hartstongue, male fern and polypody are excluded from the list as they remain stunted on walls, growing much better in woodland.

The colonization of walls is favoured by age, the presence of lime-mortar, any aspect other than south, exposure to rain, and verticality. Most true wall species are only found on vertical walls; as the angle of inclination decreases an ever-widening range of common species colonize.

Fig. 14.4. Yellow corydalis (*Corydalis lutea*), one of eight species which in Britain have their headquarters on urban walls.

A survey of the flora associated with 650 walls in Essex (Payne, 1978) included 278 urban garden walls from which 150 species were recorded. Among the most abundant native plant species it was no surprise to find common garden weeds such as *Mercurialis annua, Epilobium montanum, Euphorbia peplus, Veronica hederifolia, Poa annua, Sonchus oleraceus* and *Senecio vulgaris* heading the list. A further example of the importance of local seed sources is that 30–40% of the garden walls supported the garden escapes snapdragon, wallflower and ivy-leaved toadflax which were not present on the remaining 372 walls in other parts of the area. Garden walls therefore hold distinctive communities, even among mural vegetation, and, owing to the regular presence of well-established colourful garden escapes, they have a high amenity value.

14.2 ANIMALS

Gardens are first-rate habitats for two groups of animals. The presence of cover, a wide range of feeding opportunities, water, nest sites and roosting sites attracts a rich and varied population of birds. It has been estimated that suburban gardens support a denser bird fauna per unit area than deciduous woodland. The other group, on which the bird life is partly dependent, are insects drawn by the wide range of larval food plants, abundant nectar and pollen sources, opportunities to parasitize other insects, presence of accessory habitats such as compost heaps and rotting fruit, all in a sunny and sheltered setting. Recording carried out by Owen (1983) between 1972 and 1978 in her 0.6 ha Leicester garden has quantified this richness. She listed 323 lepidoptera, including more than a quarter of the large family Noctuidae and a third or more of the butterflies, hoverflies and bumblebees on the British list which visited her garden together with six out of seven social wasps (Vespidae) and a quarter of the ichneumon flies. This latter group is not particularly well known and included eight new to Britain plus two undescribed species. Extrapolating from these well-studied groups it has been suggested that sooner or later a third of the British insect fauna will visit this Leicester garden and probably a similar proportion of spiders. Support for this view comes from a study of beetles in a garden at Blackheath, London where between 1926 and 1973 over 700 species were recorded. One reason for these extensive lists is that long-term recording has shown that there is no such thing as a normal year; the fauna changes continually with arrivals, departures, sporadic appearances and mass migrations depending on the activity of predators, parasites, events further down the food chains, the weather and unknown factors.

Mammals

Residential areas with large gardens provide an optimum habitat for a number of mammals. An analysis of records of hedgehog (*Erinaceus europaeus*) for example shows that it is predominantly a suburban specialist inhabiting allotments, parks and particularly gardens. Distribution maps based on road casualty records reveal a distinct hedgehog zone in suburban and commuter districts while in rural areas occurrences are far less frequent (Massey, 1972; Howes, 1976). It has still to be satisfactorily demonstrated why hedgehogs should exhibit this synanthropic tendency. The ready availability of daytime and breeding retreats, hibernacula and articifial feeding may all be advantageous and outweigh the hazard of busier roads. Analysis of droppings from garden sites show their food includes beetles, earthworms, earwigs and caterpillars but few woodlice. Their density has been estimated at three to five family groups per 1 km^2 in suburbia with a home range of 20–30 ha for males and 10 ha for females (Herter, 1965; Plant, 1979; Morris, 1983).

Since World War II foxes (*Vulpes vulpes*) have become common in many British cities. Work by Harris suggests that while the general density of foxes in lowland Britain is one per 1 km^2 in towns such as Bristol it is four per 1 km^2, and even higher in Cheltenham. There have been many explanations for their colonization of suburbia. One, that it was due to a lack of food in the countryside in the wake of myxomatosis is attractive, but the invasion pre-dated the advent of this disease of rabbits. Harris (1986) believes that isolated fox populations trapped in rural enclaves by rapid interwar suburban development were the first to discover that gardens offered a habitat in which they could thrive. Urban foxes are fussy about the area in which they live being common only in certain types of town or city. They reach their greatest

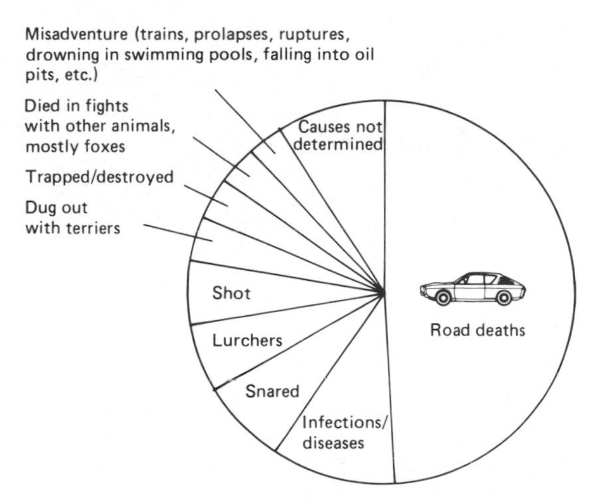

Fig. 14.5. Causes of mortality in foxes in Bristol (Harris, 1986).

abundance in big uniform owner-occupied tracts of residential housing built in the 1930s when land prices were low. Their preference is for a quiet garden (no dogs) about 30 m long with a shed, hedges or other cover to lie up in and people who like them and perhaps also feed them. They are almost indifferent to the presence of cats but big strong dogs kill their cubs.

Each year up to 60% of a town fox population dies. About half the deaths are road casualties (Fig. 14.5) and a quarter are deliberately killed in snares, with lurchers, or shot. This means the life expectancy of urban foxes is short, ranging from 14 months in London, to 18 months in Bristol. The high death rate has a significant effect on fox social structure. Foxes normally live in stable family groups and pair for life, but with partners regularly dying, dog foxes may get replaced by their sons, vixens by their sisters or daughters and non-breeding vixens attached to the family group are rare. There is little opportunity for the large stable family groups found in the countryside to develop. An additional result of the high mortality is that the home ranges of foxes, which overlap anyway, are less exclusive and always changing. The size of ranges varies from 25 to 40 ha in privately owned, interwar suburbs where supplementary feeding is available to over 100 ha in less favourable areas such as industrial or council-rented property. The diets of urban foxes living in the inner and outer parts of London have been investigated by analysing the stomach contents of 571 animals (Fig. 14.6). Scavenged food comprised the bulk of the diet with wild birds, mammals, earthworms and insects forming much of the rest. Inner city animals consume a higher proportion of food scavenged from bird tables,

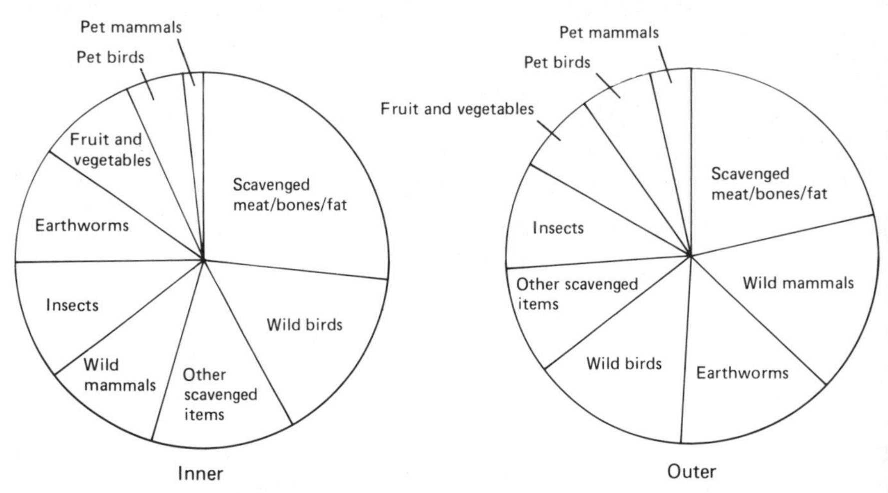

Fig. 14.6. Food of foxes living in the inner and outer parts of London (Harris, 1986).

dustbins or specially put out for them, but fewer earthworms, domestic pets and wild animals. Their diet is very much omnivorous with fruit and vegetables forming a significant part of their food in the autumn.

The work of Harris (1986) and Macdonald (1987) on urban foxes is some of the most detailed and sophisticated in the field of urban ecology. It has been pursued with such vigour because of the fear that if rabies is reintroduced to this country foxes will become infected and pose a major health hazard. The current epidemic that is advancing west across Europe at a rate of 30 km per year is spread primarily by foxes; where their numbers have been successfully controlled the incidence of the disease has fallen dramatically.

A third distinctly suburban mammal is the pipistrelle bat (*Pipistrellus pipistrellus*) which, though it has a wide distribution, for preference selects modern semidetached houses, flats and garages in which to establish its summer breeding colonies. In May pregnant and non-

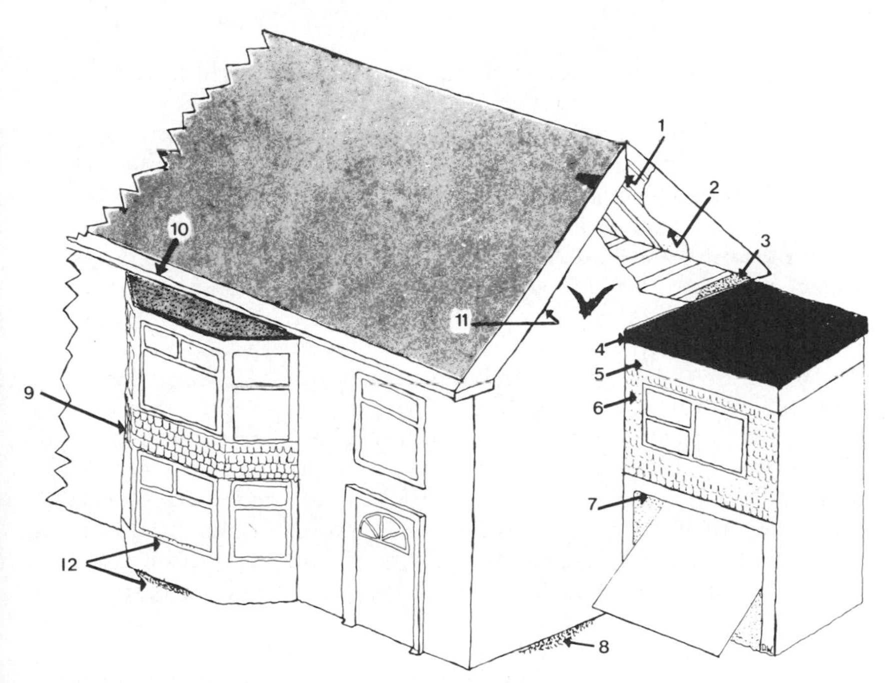

Fig. 14.7. Roosting sites favoured by pipistrelle bats on a modern house: 1, roost in gable apex; 2, roost under internal roof felting; 3, bat droppings in loft adjacent to exit holes; 4, under extension roof felt; 5, behind fascia boards; 6, behind hanging tiles on extension; 7, above garage ceiling; 8, droppings below exit holes; 9, behind hanging tiles above bay windows; 10, between soffits and walls; 11, behind barge boards; 12, droppings on ground and window sill from roost in hanging tiles (Whiteley, 1984).

breeding females establish nursery colonies under hanging tiles, behind barge and fascia boards, between soffits and walls or in the apex of gables (Fig. 14.7). Colonies may average about 100 adults. Each female gives birth to a single baby in June–July and by late August the entire colony leaves the home to spend the winter in trees before returning often to the same roost. Feeding flights are carried out mostly near to water, in open woodland or over gardens.

Other garden mammals are the grey squirrel (see Chapter 11), wood mouse, occasionally the common shrew and in well-vegetated gardens, bank voles. It should be noted that moles and badgers are not yet urban animals though there are indications from Bristol and Essex that the latter is starting to move into suburbia in the same way that foxes did 40 years ago.

Birds

Suburban gardens are believed to support the highest density of breeding birds of any habitat in Britain. Evidence for this comes from studies in London where Batten (1972) reported a total density of nearly 600 pairs per 1 km^2 (minus house sparrows) in an area of 30-year-old interwar housing, whilst in Ealing, Ferguson (Simms, 1975) recorded 641 pairs per 1 km^2, also minus house sparrows. The regular breeding species of suburbia are mostly woodland birds which find the mosaic of habitats resemble the richest of wood margin ecotones. A study of blackbirds in the Oxford area revealed that in woodland surrounding the city their density was one pair per 2.6 ha whilst in the suburbs they were present at 7–8 pairs per ha. Outside the breeding season the numbers of birds using a garden is larger than might be expected. Ringing has shown that while in a well-stocked garden the maximum number of blue-tits or greenfinches present at any one time might be a dozen, over a thousand different individuals of each species probably visit the site during a winter (Glue, 1982). For example, rural blue-tits come in from woodland to search for food and each autumn a large influx of starling, blackbird, song thrush and others arrive from the continent.

Returns from the *Garden Bird Survey* being conducted by the British Trust for Ornithology have identified the 15 species shown in Table 14.2 as the most frequently recorded in gardens during autumn and winter 1987/88. The list is similar to results produced by the BTO's *Garden Bird Feeding Survey* (Glue, 1982) and a recent European garden bird survey. Eventually differences in the composition of the garden bird fauna across Britain and Europe will be identified. A long-term benefit of such surveys are that changes in the behaviour of species will be quantified. During recent decades there have been regular additions

Table 14.2. The most frequently recorded birds seen in gardens in Britain during autumn and winter 1987/1988. The order of frequency given here varies slightly depending on the region and weather

Blue-tit	Greenfinch
Blackbird	Wren
Robin	Song thrush
House sparrow	Collared dove
Starling	Coal-tit
Great-tit	Magpie
Dunnock	Black-headed gull
Chaffinch	

Data are from BTO/BASF survey.

to the avifauna of suburbia. For example, the bullfinch started to become common in gardens in the 1950s. A suggested explanation for this phenomenon is that about then the birds made the discovery that the buds of cultivated fruit trees are of a higher nutritive value than those of many wild trees and could substitute for their traditional diet of seeds. During the severe winter of 1962/63 siskins in Surrey started feeding on peanuts and fat at bird tables where they had previously been very rare visitors. The habit spread and the siskin is now a regular garden bird. Other opportunists which are extending their range into gardens include sparrowhawk, kestrel, collared dove, magpie, carrion crow, tawny owl, greater spotted woodpecker, blackcap and reedbunting; the trend is certain to continue. A possible eventual recruit is the Syrian woodpecker (*Dendrocopos syriacus*) which is very much a bird of man-made habitats. It is slowly spreading west across Europe.

The reason gardens accommodate so many birds is because they provide for the basic needs; food, cover, water, nest sites and roost sites. Areas that are dug over expose invertebrates suitable for song thrush, missel thrush, starling, robin and dunnock while the same areas, when allowed to grow weeds or cultivated for flowers, are attractive to greenfinch, goldfinch, bullfinch and house sparrow. Lawns are particularly favourable to thrushes, blackbird, starling, pied wagtail, crow and sparrows. Areas of tall grass provide less favourable conditions. A further natural food source is the aerial plankton which builds up over gardens. Swifts find it worth flying many thousands of kilometres from tropical Africa to nest and feed in our cities, where they can guarantee to find the estimated one million food items required to raise a brood. Some ecologists recognize a 'swift zone' in our cities where there is a combination of buildings for nest sites and gardens to provide a suitable density of flying insects.

For centuries householders have put out food for birds, prompted during post-war years by the BBC, which in times of severe weather followed its news bulletins with requests to provide food and water for them. The BTO *Garden Bird Feeding Survey*, which has been running since 1970, has a great deal of information on the food preferences of garden birds. For example, it is now known that greenfinch and great-tit prefer peanuts, chaffinch and collared dove mixed seeds, magpie and blackbird bread, long-tailed tit and greater spotted woodpecker fat, while black-headed gulls prefer meat and bones (Glue, 1982). There are, however, certain objections to the supplementary feeding of garden birds. One is that the balance of species is altered, with those maintaining artificially high populations able to compete unfairly. House sparrow and starlings may become nest site competitors or nest stealers from great-tits and greater spotted woodpeckers respectively, or the tits may occupy all available nest sites so that when the migrant redstarts and pied flycatchers return they are forced to build in unsuitable sites. Information on drinking habits was also collected during the BTO survey. It was discovered that birds feeding on tree seeds are particularly regular visitors to drink at bird baths because of the low moisture content of their food. The insectivorous wren uses them chiefly for bathing.

Most gardens provide a wide range of natural nest sites. Open-nesting birds like blackbird, thrushes and finches are attracted by thick hedges, shrubs and trees, the spinier and denser the better to combat predators. Tall conifers such as Lawson or Leyland cypress are favoured by collared dove, dunnock which nest near the base, and goldfinch and greenfinch which nest near the top. In contrast many small birds such as warblers and buntings nest fairly close to the ground. Another group prefers crevices in walls or buildings – robin, wren, pied wagtail – while others require holes and can be helped by erecting nest boxes. More research needs to be done on nesting success in the face of predation by cats, squirrels, magpies and crows coupled with high human disturbance.

One of the main determinants of the bird species visiting a garden is its surroundings. This section has concentrated on the avifauna typically present in large blocks of suburban housing but an estate adjacent to woodland might well attract into its gardens, jay, tree creeper, nuthatch, hawfinch, goldcrest, lesser whitethroat, chiff chaff, tawny owl and many others as occasional visitors.

Garden ponds

One in ten gardens contains a pond. It is widely appreciated that they provide a major refuge for the common frog, toad, and palmate and smooth newts (see Chapter 15). They are also the only habitat in Britain that has successfully been colonized by the midwife toad (*Alytes obstetricans*) a native of central and southern Europe. They occur in three areas of Britain, the best-known colonies being around Bedford. These originated in a consignment of ferns and water plants delivered to a nursery garden from France and around 1922 were introduced into a number of local gardens. In 1947 five toads and a dozen tadpoles were taken from Bedford to Rotherham and almost forgotten, but have recently been rediscovered as a very strong colony in the suburb of Woodsetts (Ely, 1985). The only other successful introduction was into the garden of Terry's, the chocolate firm in York, where the toad is still present after 55 years. It is not known what factors limit their spread, but most introductions fail.

Insects

The insect class is by far the largest in the animal world and with the possibility of one-third of the British insect fauna visiting a single garden it is difficult to known how best to demonstrate interrelationships within the group particularly as larval stages often have very different requirements to the adults. A selection of the better-studied orders has been chosen to illustrate the ways in which insects use gardens.

As garden vegetation is often selected for the abundance and longevity of its flowers, insects foraging for nectar and pollen are a major feature of the habitat. Different groups of insects to some extent partition out the flowers showing a preference for species with a certain structure or colour, though a more complete explanation of choice is provided if the concept of community interactions that are at times competitive are included. This places a premium on efficiency of foraging so that flower abundance and density influence choice together with feeding hierarchies in which the larger and more abundant species such as the hoverflies *Eristalis* spp. and *Syrphus* app. are dominant.

A very wide range of adult hoverflies are found feeding on flowers in gardens. Among the commonest being *Episyrphus balteatus, Melanostoma mellinum, M. scalare, Metasyrphus corollae, Platycheirus albimanus, Scaeva pyrastri, Sphaerophoria scripta* and *Syrphus* spp. Records from a malaise trap operated between 1972 and 1982 in the Owen's Leicester garden have shown how the assemblage varies from year to year. The number of hoverfly species caught fluctuated from 61 in 1973 to 37 in 1977 with five different species ranking as commonest in different years. Each year, those with aphid-feeding larvae formed the majority of the catch but in the drought year of 1976 genera with herbivorous larvae were common and in 1973 and 1978 species with larvae feeding on decaying organic matter (*Eristalis* spp.) were prominent. In addition to flowers, which may be visited for pollen (*Melanostoma, Episyrphus*) or in the case of larger-bodied species for pollen and nectar (*Eristalis, Metasyrphus*), honeydew (*Xylota*) and rotting fruit (*Eristalis*) are food sources. *Volucella* spp., the larvae of which scavenge the nests of wasps, bees and ants, may be frequent in gardens close to woodland. The hoverfly population of gardens is often influenced by migration from the countryside which may result in large influxes of *Syrphus vitripennis, Metasyrphus corollae, Episyrphus balteatus* and *Scaeva pyrastri*.

For much of the year bumble bees are the most conspicuous garden insects. About eight species regularly occur in gardens in the south of England, rather fewer further north. They differ in seasonality, feeding behaviour and nest sites so they coexist satisfactorily. First to emerge from hibernation are *Bombus pratorum* queens which may be on the wing in March while *B. lapidarius* does not emerge before late May. At the other end of the year *B. agrorum* and *B. terrestris* are still on the wing in October. All eight nest and hibernate in gardens usually underground but sometimes in clumps of grass or an old bird's nest. Flower-foraging preferences are controlled by tongue length with *B. lucorum* and *B. terrestris* having the shortest tongues (10 mm) while *B. ruderatus* and *B. hortorum* have tongues 20 mm or over and only feed at flowers with long corolla tubes or spurs.

After hoverflies and bees, lepidoptera are the most noticeable group of flower-feeding insects. Marking and recapture has shown that most garden butterflies are actively on the move flying several kilometres a day and using gardens as refuelling stations. Only orange-tip and the whites are likely to breed there. The longer a garden is observed the longer the list of butterflies visiting it will become. The Owens recorded 21 species in 7 years with rarities like silver-washed and marsh fritillaries, and the white-letter hairstreak being seen on single occasions. Garden butterflies reflect those of the countryside surrounding the town and also the scale and extent of northward migrations of species such as red admiral, painted lady and clouded yellow.

Many insects that feed at flowers have larval stages that eat the leaves, stems and occasionally the roots of garden plants so their complete life cycle can take place in residential areas. Some of the most abundant garden moths are common everywhere as their caterpillars are polyphagous; these include the large yellow underwing, dot, bright-line brown-eye, knotgrass, angle shades and hebrew character. Native woody species such as poplar, apple, hawthorn, lime, privet and various willows support many lepidoptera but it is those with larvae that feed on exotic plants which are the true garden specialists. Jennifer Owen recorded 33 species of moth feeding on 52 species of alien plant in her garden. Spotted dead-nettle (*Lamium maculatum*) from continental northern Europe supported eight, the purple leaved cherry-plum

(*Prunus cerasifera* 'Pissardii') from the Middle-East nine, and *Buddleia davidii* from the Far East eleven (Owen and Whiteway, 1980), so moth caterpillars are well able to exploit the garden flora. Where specialist feeders are involved considerable range extensions have been achieved. The juniper pug and juniper carpet are starting to be recorded on ornamental junipers and other conifers, *Lobesia littoralis*, the larvae of which feed on thrift, is turning up in gardens well away from its usual coastal habitats and the scarce tissue (*Rheumaptera cervinalis*) which feeds on the rare and decreasing wild barberry owes its continued existence in Britain to having transferred on to cultivated species of *Berberis*.

Robbins (1939), writing about the moths of his London garden in the 1880s considered that it maintained a small group of species that were not easily come by in rural districts and that these represented a town garden fauna. This is still being added to. Delphinium and monkshood are the food plant of the golden plusia (*Polychrysia moneta*), a moth first recorded in Britain in 1890 since when it has spread through the country by utilizing the plentiful supply of its food plant in gardens. A similar history is shown by the varied coronet (*Hadena compta*) first seen in Kent (1948) from where it spread rapidly through the gardens of southern England feeding on sweet William (*Dianthus*) and Blair's shoulder-knot (*Lithophane leautieri*) collected in the Isle of Wight (1951) and now spreading rapidly by feeding on the flowers and leaves of Monterey cypress in parks and gardens (Owen and Duthie, 1982). One of the latest examples is a tiny plume moth (*Stenoptilia saxifragae*) discovered in 1970 feeding on the rockery plant mossy saxifrage and now quite widespread in gardens. The artificially high plant species diversity of gardens favours them as breeding sites for large numbers of lepidoptera; while gardening flourishes, nothing can prevent this habitat from becoming even richer through the phenomena outlined above.

Ground-living arthropods and molluscs

These species live on the ground and are therefore more dependent on local habitat conditions than are flying insects. Davis (1979) studied the group in 15 private gardens in London employing pitfall traps operating for 4 weeks in early summer. They caught 33 species of beetle, five of centipede, eight of millipede, seven of woodlice, 59 of spider, two of harvestmen and two of pseudoscorpion. Few species were ubiquitous; for example, among the beetles none occurred in all gardens and only five in half or more. The commonest and most widespread ground-living invertebrates were the woodlice *Oniscus asellus, Philoscia muscorum* and *Porcellio scaber,* the centipedes *Lithobius forficatus* and *L. microps*, and the harvestman *Oligolophus spinosus* but variability seemed to be the chief characteristic of the lists. Many of the species are considered synanthropic and a few of the centipedes and millipedes were local to rare in the context of London. The best predictor of faunal diversity was the proportion of land occupied by gardens, parks, etc. within a 1 km radius of the site. Off-site influences therefore largely determine the diversity of ground-dwelling arthropods in the area studied.

A survey carried out in gardens at Harpenden in Hertfordshire recorded extraordinary densities of slugs; at certain seasons 20–80 individuals were moving about and feeding on every square yard during the night (Barnes, 1949). Despite removing between 10 000 and 17 000 slugs a year for 4 years from a single garden no reduction in the population was observed. The survey found that the slug fauna in adjacent gardens varied considerably in a way that was far from clear, though topographic control of moisture may have been important. In one garden it was estimated that slugs were eating 1.5 cwt of material a year, and their biomass was equivalent to two plump rabbits. The good news is that they are quite indiscriminate in their feeding, eating whatever lies in their path whether it be a young plant or dogs' faeces; only 10% of their diet is prized by the gardener for food or pleasure.

The ecology of compost heaps would repay detailed study as, in addition to holding large slug and earthworm populations, they are breeding sites for the larvae of numerous diptera including the soldier flies *Sargus bipunctatus, Microchrysa polita* and *M. flavicornis*. These rather lazy insects often make their way from gardens into houses where their striking blue–green metallic coloration renders them noticeable.

14.3 DISCUSSION

Cultivated gardens cover a larger area in towns than any other land use, yet they are among the most fragile of ecosystems. Composed principally of 'crippled' plant cultivars and species growing beyond their natural climatic and edaphic ranges they are maintained solely by the energy of gardeners who toil to reduce competition, manipulate soils and introduce species. Any relaxation in this intricate and costly management regime results in widespread extinction leading to simplification. The consequence of this management is usually an extremely diverse mosaic of habitats ranging from alpine cliff ledge (rockery) to Mediterranean mattoral (herb garden) to deciduous wood margin (shrub bed) in which a total of 200–250 plant species may coexist on one property; as plants are the base of food chains, associated herbivores are numerous. Since Elton in his authoritative work *The Pattern of Animal Communities* (1966) wrote off gardens as tending towards biological deserts, it has been recognized that many insect species have successfully adapted to feeding on introduced plants which have enabled considerable extensions in range to be achieved. One of the most familiar examples is the orange-tip butterfly which has become a member of the garden fauna since its larvae took to feeding on garden arabis (*Arabis caucasia*) and honesty (*Lunaria annua*).

A further misconception is that to be rich in wildlife, a substantial part of a garden needs to be wilderness. Nothing could be further from the truth. Gardens managed along conventional lines, not aping more ancient parts of the countryside, are a community in their own right. Jennifer Owen's well-studied suburban garden in Leicester is full of interesting species including so many rare ones and others new to science that a good case could be made for scheduling it as a site of special scientific interest. Gardens provide a perfect example of how the native fauna whether mammal, bird, amphibian, insect or mollusc can adapt to changing land use. Introduced species with specialized requirements also find them an acceptable habitat. The long-standing English passion for gardening may be the salvation of much wildlife, with residential areas providing important refuges from which it can later recolonize the countryside.

Chapter 15

RIVERS, CANALS, PONDS, LAKES, RESERVOIRS AND WATER MAINS

Urban water provides some elegant examples of how man, by carrying out his normal day-to-day activities and making himself comfortable in his environment, can comprehensively alter an ecosystem. As usual, certain species are eliminated by urban influences, others are strongly favoured by them while a few groups appear unaffected. Hydrologists are able to provide a broad picture of the physical and chemical aspects of water flow in urban areas (Douglas, 1983; Hall, 1984; Lazro, 1979), but there are only patchy biological data to complement this. Reviews of urban aquatic ecosystems are provided by Hynes (1960) and Whitton (1984), while at a natural history level Teagle (1978) and Kelcey (1985) have collected a certain amount of information. Given the 'state of the art', a fully integrated treatment of this habitat is not possible, so the general principles of urban hydrology will be outlined followed by a range of examples which illustrate the special ecological relationships that characterize urban rivers, canals, lakes, ponds, reservoirs and water mains.

15.1 URBAN HYDROLOGY

The most important factors are stream flow, channel modification and water quality; these need to be understood in an urban context before biological aspects can be dealt with.

Stream flow

In towns the natural circulation of rain water is modified by the

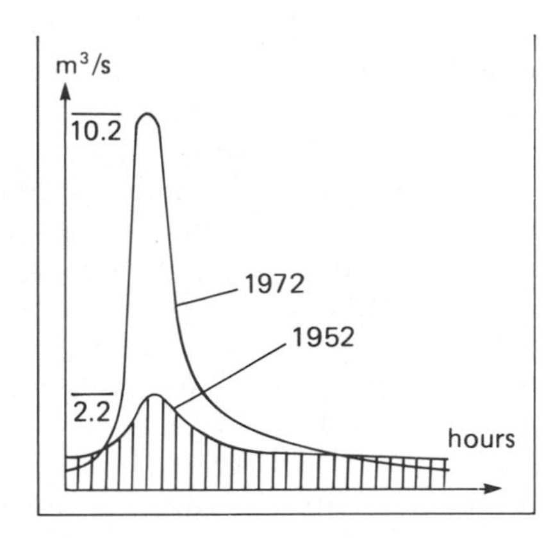

Fig. 15.1. Development of run-off into the Korsch river near Stuttgart, West Germany, following urbanization of the catchment (Kienle and Luz, 1977).

impervious nature of many surfaces which increases run-off. This is efficiently collected up in gutters and drains, then discharged into the nearest water course which consequently shows a pattern of high peak flows of short duration. Their 'flashy' character is further accentuated by the fact that the impervious surfaces result in a loss of recharge to the ground water-table, which means base flow also decreases; as a result water courses are subjected to periods of alternate severe scouring followed by drying out. This has been quantified for the Korsch river near Stuttgart (Kienle and Luz, 1977) where a 20-year study showed that after the built-up area increased from 7% in 1958 to 20% in 1978, peak flow volumes were five times higher, and setting in much sooner after a downpour, while at the same time, the natural low water flow decreased by half (Fig. 15.1). Of additional significance are the findings of Hollis (1977) who discovered that at Harlow New Town storm peak discharge was more severe in summer, when organisms are at their most active, than during winter. One benefit to urban aquatic life, however, is that to reduce the possibility of flooding, an increasing number of storm water drains no longer discharge directly into streams, but instead flow into specially constructed basins known as balancing lakes. This new aquatic habitat is now incorporated into most developments; for example, Milton Keynes has twelve varying in surface area from 3 to 40 ha. Many are designed with nature conservation in mind.

Channel modification

In the past, stormwater management has involved the destruction of almost every feature valued by biologists: luxuriant aquatic vegetation, meanders with sand bars, islands, pools and riffles, quiet bays, trees and shrubs along the banks. All these are regarded by river engineers as causes of flooding and their removal, usually taken to its logical conclusion around towns, results in wide straight rivers running in deepened beds between high uniformly strengthened banks. Alternatively, streams may be removed from the landscape altogether by culverting. Recently, in the light of experience gained abroad and the activities of pressure groups, river engineers are starting to incorporate wildlife and landscape considerations into their schemes. The destruction of trees and shrubs growing beside water courses is particularly contentious, as increased light levels tend to accelerate weed growth which itself can create a flooding hazard. Ecologists have recently pointed out the advantages of half-shade, which allows some plant growth, rather than the large amounts found in open streams (Dawson and Haslam, 1983).

As an alternative to channelization, 'natural' river engineering is advocated for at least suburban areas whereby wildlife conservation and natural beauty are enhanced (Newbold *et al.*, 1983; Lewis and Williams, 1984). These techniques involve the construction of low flood banks set back from the river which allow for overflow on to washlands used for recreation or nature conservation. Meanders can often be retained by cutting a short flood relief channel across their neck which takes water during the highest peak discharges. Where land uses press in on the river corridor, flooding can be alleviated by constructing a two-stage channel with a composite cross-section. The deep main channel is flanked by a wider, shallower one, the flat bottom of which is known as a berm. If possible, berms should be cut in one bank only, or on alternate banks so that only 50% of the bank community is destroyed along any length. To increase habitat diversity and avoid a canalization effect, the angle of slope of the banks and the width and alignment of the channel and berm should all be varied. Working from one bank only is a valuable technique often acceptable to engineers; it has been successfully employed along the Ford Brook, Walsall. The untouched bank acts as a reservoir from which species can colonize the rest of the river. These types of natural river engineering can cost less than 5% of an equivalent channelization scheme, but need to be linked to a programme of balancing lake provisions.

Water quality

As water passes through urban areas, a wide range of pollutants are discharged into it which profoundly affect its ecology. The most important is organic matter originating from untreated, or only partly treated sewage as this causes the water to become deoxygenated in the following way. The often finely divided organic residues, either in suspension or coating the river bed, become so densely clothed with bacteria, fungi and predatory protozoa, that these absorb most of the oxygen from the water, causing anaerobic conditions in which only a few specialized organisms can exist. The amount of bacteriologically degradable organic matter in water can be measured indirectly in a laboratory using the biological oxygen demand (BOD) test, but water boards also employ biologists who assess dissolved oxygen by sampling stream invertebrates, many of which are very sensitive to this factor. Once the organic matter has been broken down, the populations of decomposers decline and oxygen levels recover, particularly if riffles and weirs are numerous. In practice, this self-purifying process rarely gets a chance to run its course as further outfalls discharging organic matter prevent a full recovery. The freshwater classification used in England and Wales ranks water into four categories from good quality (class 1A and 1B) to bad quality (class 4). Though the different qualities relate to the potential use of the water, they also reflect their biological richness. Much of the worst-quality water is associated with urban areas (Department of the Environment, 1986).

A second form of pollution is greatly enhanced nutrient levels, especially inorganic forms of phosphorus and nitrogen. Water entering towns may already carry a considerable burden of nutrients originating from agricultural land, but large extra amounts are added by sewage treatment plants most of which are only designed to remove organic matter and other suspended solids. Highly nutrient-enriched water is known as eutrophic to separate it from moderately enriched mesotrophic and nutrient-poor, oligotrophic waters. Even in the absence of man's activities, water in lowland rivers is naturally enriched to the level where bream, rather than trout, are the dominant fish, but levels are excessively enhanced by urbanization. Eutrophication has its greatest effects in lakes and reservoirs where it promotes a massive algal growth of either plankton, which causes the disorder known as 'green water,' or filamentous species (*Cladophora, Spirogyra, Ulothrix*) which cover the entire surface with a mat of 'blanket weed.' When the algae die, their decay can produce deoxygenation. Some lakes are rendered eutrophic by resident duck populations or roosting gulls. Nutrient loading from urban storm water drainage is variable but does not differ too much from Owens (1970) findings in the Great Ouse basin of 0.9 kg of P ha^{-1} $year^{-1}$ and 9.5 kg of N ha^{-1} $year^{-1}$. The source of these nutrients is uncertain; once thought mainly to be due to fertilizers and animal droppings, a significant fraction is now known to be leached from vegetation.

A third widely applied criterion of water quality is the content of suspended solids (SS). A high level adversely affects aquatic ecosystems, but to kill adult fish the rarely recorded level of 20 000 mg l^{-1} must be exceeded. At much lower concentrations, subtle changes occur such as niche simplification, the blanketing of spawning grounds, or the rendering of parts of the stream bed anaerobic; the stream bed may be rendered too unstable for plants to colonize. Siltation and turbidity not only reduce diversity but result in declining productivity across all trophic levels (King and Ball, 1967). Shallow light penetration is particularly damaging to communities of submerged aquatic plants which need at least half a day's bright sunlight to thrive.

Toxins, an extremely variable group of contaminants, originate mainly from industry. Among the commonest are the heavy metals lead, cadmium, zinc, copper, nickel and mercury, and cyanides, phenols, pesticides and herbicides. In coal-mining areas, ochrous mine water and saline mine water can be discharged untreated into watercourses. Other industries have their distinctive toxins such as bleach and mothproofing agents from textile works. Possibly the most harmful is un-ionized ammonia which is discharged in the 'liquor' from coking plants. Three in the Sheffield–Chesterfield area are responsible for one of the longest stretches of 'dead' river in Europe.

Water quality in an urban area is largely controlled by the efficiency of sewage treatment, the nature of trade effluents and features such as whether storm water (rainwater) and foul water (domestic sewage, bath water, washing water, certain industrial effluents, etc.) are conveyed in single or separate systems. In older cities, foul and storm water are carried in the same pipe and both treated at the sewage works. A disadvantage of this system is that the Victorian trunk sewers are now seriously overloaded and at only twice the dry weather flow, i.e. during almost every rain storm, large volumes of very dirty water, including raw sewage, are discharged into the nearest river via various overflow pipes. The concept of separate systems whereby foul water goes to the treatment plant and storm water is discharged direct into watercourses was evolved to avoid this nuisance. Since 1950, all new towns in the UK have adopted this concept in the belief that storm water was less of a pollution hazard than the overflow from a combined system. Work by Wilkinson (1956) on a post-war separate-system housing estate in south-east England demonstrated that BOD loadings were reduced by 60% and suspended solids increased by 650% of those that would be expected if both were treated to Royal Commission standards (20 mg l^{-1} BOD, 30 mg l^{-1} SS) in a combined system. So lower BOD is traded against much higher SS loadings.

15.2 RIVERS

Rivers form important mixed-habitat corridors as they run through towns, and where canalization is not too severe or water pollution too oppressive, they can be the richest of all urban wildlife sites. Particularly diverse are islands, lateral shoals and the channel edge; these are unbelievably complex as conditions alter with each variation in water depth, flow rate, substrate type, slope and aspect. In addition, the intricate pattern of zones is not stable, an irregular erosion cycle related to floods being superimposed. Restrictions on the full development of aquatic wildlife along urban stretches of water courses are gross pollution, scour, unstable substrates, disturbance by anglers and dominance by aggressive alien plant species. Where the river is confined by high walls or buildings, bankside and bank top vegetation is replaced by simple mural communities and the channel is severely scoured.

Vegetation

Mid-channel and channel edge zones

Under conditions of clean water, full light and only gentle scouring, the mid-channel is occupied by rooted, submerged, streamlined plants while floating-leaved species predominate where flow velocities are lower. At the channel edge robust emergent plants form a fringeing reedbed with a border of floating-leaved species in deeper water and perhaps duckweed occupying quiet bays. This idealized picture is severely modified where gross pollution is found in urban rivers. Seriously polluted watercourses contain either no aquatic vegetation or sparse fennel-leaved pondweed (*Potamogeton pectinatus*). The Rother is an example of a river with no aquatics, the grossly polluted water flows between banks thickly clothed with tall herbs, the lower limit of which are controlled by scouring. Accidental very poisonous pollution episodes are less damaging than continuous moderate pollution since they rarely kill the underground parts of plants so the vegetation recovers within about 3 years unless spillages recur.

Continuous mild pollution is the norm in towns but its effects go largely unrecognized. First there is a decline in species diversity. At a clear water site, say a bridge, 7–10 (–16) aquatic plant species might be expected covering a good deal of the bed, but at a typical town site only half this number would be present and showing a reduced total cover. *Potamogeton pectinatus* which is actually favoured by pollution forms dense stands in the bed of such rivers (Fig. 15.2). Species tolerant of sewage and industrial pollution, but not favoured by it are curled pondweed (*Potamogeton crispus*), bulrush (*Schoenoplectus lacustris*), bur-reed (*Sparganium erectum*), unbranched bur-reed (*S. emersum*) and monkey flower (*Mimulus guttatus*). Slightly less tolerant are flowering rush (*Butomus umbellatus*), reed sweetgrass (*Glyceria maxima*), duckweed (*Lemna minor*) and yellow waterlily (*Nuphar lutea*). Indications that a

Fig. 15.2. A typical urban river with the channel full of fennel-leaved pondweed (*Potamogeton pectinatus*) and the banks lined with Japanese knotweed (*Reynoutria japonica*). River Don, Sheffield (I. Smith).

river is becoming cleaner and also less eutrophicated are the presence of water-plantain (*Alisma plantago-aquatica*), Canadian pondweed (*Elodea canadensis*), reed-grass (*Phalaris arundinacea*), shining pondweed (*Potamogeton lucens*), perfoliate pondweed (*P. perfoliatus*), water bistort (*Polygonum amphibium*) and water buttercups (*Ranunculus* subgenus *Batrachium*) together with a well-developed fringeing reedswamp.

Recovery from pollution takes a few years. Haslam (1978) reported that the seriously polluted river Cole (Trent) was cleaned up in 1971–2 but species diversity did not increase dramatically till 1976 when the pollution-tolerant *Potamogeton crispus* and *P. pectinatus* started to decline. A similar time lag occurred on the Strine (Severn). When using water plants to gauge pollution a complicating factor is that small populations of quite sensitive macrophytes may be maintained by downstream inoculation from clear headwaters. Also the contribution of enhanced scouring in urban reaches must be assessed when gauging the general richness of the aquatic flora.

Higher plants seem little affected by the heavy-metal concentrations found in British rivers but an example has been noted of a plant

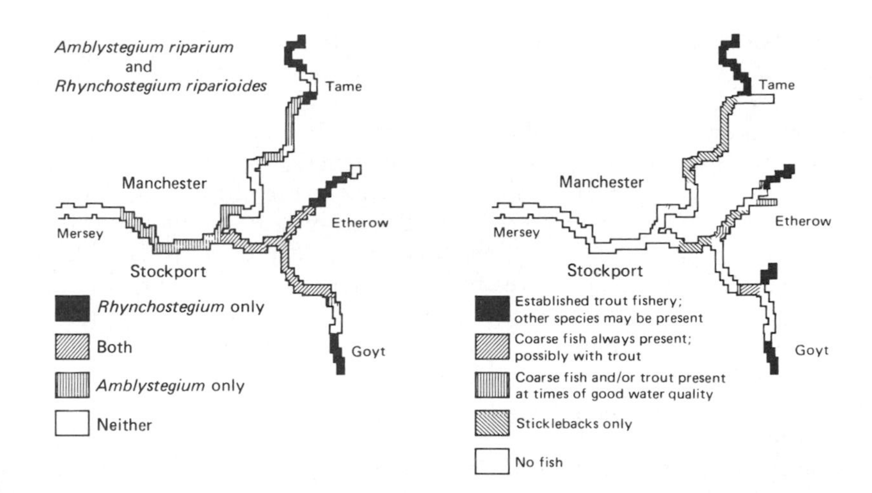

Fig. 15.3. Distribution of two aquatic mosses (left) and fish (right) in the Mersey catchment (Holland and Harding, 1984).

favoured by 'ochre water'. When water rich in iron is pumped out of collieries finely divided ferric hydroxide (ochre) precipitates out and can turn the bed of rivers orange. Where such a discharge enters water-courses the jointed rush (*Juncus articulatus*) is often extremely luxuriant.

Recent work on aquatic bryophytes in the Mersey catchment (Holland and Harding, 1984) has shown that they can form useful biological indicators of pollution. With increasing nutrients the liverwort (*Scapania undulata*) disappears and the alga *Cladophora glomerata* starts to become dominant. Below the first sewage outfall clumps of the moss *Rhynchostegium riparioides* come in and take over as the dominant bryophyte. It is soon joined by *Amblystegium riparium* a reliable indicator of marked nutrient enrichment (Fig. 15.3). Both mosses favour well-illuminated riffles; an increase in organic pollution is suggested by the sequence (1) neither; (2) *Rhynchostegium* only; (3) both present; (4) *Amblystegium* only; (5) neither. *Fontinalis antipyretica* is about as pollution tolerant as *R. riparioides*. Mosses are highly tolerant of heavy-metal pollution while the alga *Cladophora glomerata* is sensitive, so its absence from likely habitats is worth investigating.

Bankside vegetation

This occupies the transition zone between the channel and top flood level; it represents an ecotone along which soil moisture, fertility, disturbance and competition are the main variables. The vegetation is so diverse, with species from many plant communities coming together, that phytosociologists have been unable to classify much of it; the presence of willows is the only constant feature. As this urban habitat has been largely ignored by ecologists, an account is given of the banks of the River Don in Sheffield.

The dominant plant is Japanese knotweed (*Reynoutria japonica*) which forms dense stands along the river commencing where it enters the built-up area; it may occupy all lateral space in the corridor and stems arch far out over the water (Fig. 15.2). Mixed with it at lower levels are patches of another alien, Himalayan balsam (*Impatiens glandulifera*). Japanese knotweed, which never fruits in Sheffield, is a perennial which spreads to new sites by the dispersal of rhizome and root fragments while the balsam, an annual, is more of an opportunist so its detailed distribution changes each year. Between them these two species have, over the last 20 years, eclipsed acres of what were formerly rich tall-herb communities.

Where well established the *Reynoutria* is tolerant of both scour and deposition, forming on the one hand tussocks and on the other scattered shoots from a rhizome system that can grow vertically to adapt to new ground levels. Associated floristic diversity is low; stands above top flood level may only include twining plants such as *Calystegia sepium* owing to the heavy shade and persistent stem litter that accumulates within the stands. Below flood level, however, stem litter gets removed and silt deposition brings in propagules from the pre-vernal community of deciduous woodland, so riverine *Reynoutria* stands in central Sheffield are characterized by permanent populations of early leafing herbs such as *Allium ursinum, Anemone nemorosa, Hyacinthoides non-scripta, H. hispanica* and *Ranunculus ficaria*. All can compete with at least 2 years accumulation of litter through petiole extension. Beds of Himalayan balsam are slightly more diverse and include the annual, nitrophilous scrambling species *Galium aparine*. A third introduced species, the giant hogweed *Heracleum mantegazzianum*, is expected to increase dramatically along the disturbed margins of the river over the next decade. Himalayan balsam is the tallest (2 m) annual in British flora, giant hogweed is the tallest (3.5 m) monocarpic perennial herb and Japanese knotweed the equal tallest (3 m) polycarpic perennial herb which gives some idea of the favourable conditions for growth in this habitat.

Near water level disturbed bars and spreads of silt carry rich communities of annuals, many of which are typical of highly fertile topsoil; their density may reach 40 species in a 2 × 2 quadrat. Regularly present are a good range of yellow flowering Cruciferae belonging to genera such as *Sisymbrium, Erysimum, Sinapis, Brassica, Rorippa* and *Barbarea* together with other nitrophilous annual weeds such as *Capsella bursa-pastoris, Polygonum aviculare, Stellaria media, Matricaria perforata* and *M. matricarioides*; many of these may be dispersed through the storm water drainage system. Other annuals are more characteristic of arable land – *Veronica persica, Viola tricolor, Galeopsis tetrahit, Polygonum persicaria, Anagalis arvensis, Papaver rhoeas, Euphorbia helioscopa.* It is possible that well-illuminated silt banks by rivers are a natural habitat for annuals which today have their main populations in fields of root crops. A further element comprises exotics originating from sewage (tomato, cultivated strawberry) or gardens (honesty, lobelia, opium poppy).

Where aggressive aliens have not yet colonized, the base of the bank is occupied by an irregular strip of reed canary-grass (*Phalaris arundinacea*) above which is a belt of tall nitrophilous herbs typical of border zones, e.g. *Alliaria petiolata, Artemisia vulgaris, Aster novi-belgii, Cirsium arvense, Conium maculatum, Tanacetum vulgare* and *Urtica dioica.* Zoned above this is an urban meadow community dominated by Umbelliferae and grasses, and also containing *Hesperis matronalis, Saponaria officinalis* and *Silene dioica.*

Scattered trees and shrubs occur along the banks but only rarely do they form closed woodland. A female clone of crack willow (*Salix fragilis* var *russelliana*) is the commonest; it establishes from detached stems which get embedded in flood debris. Ash, sycamore, elder, apple,

Fig. 15.4. Wild fig tree (*Ficus carica*) by the River Don. The seed originated from sewage, its subsequent growth being facilitated by heated water discharged from steelworks (I. Smith).

goat-willow, white willow, laburnum and alder (*Alnus glutinosa*) are all widespread and suggest incipient succession to a woodland typical of fertile soils. An unusual member of this woodland is the wild fig (*Ficus carica*), 30 or 40 of which occur by the urban Don (Fig. 15.4). The seeds originate from sewage (Gilbert and Pearman, 1988). The age of the Sheffield figs is difficult to determine but none are young. It is likely that the trees established when manufacturing industry was at its height, at which time river water was used for cooling and the Don ran at a constant 20° C. Under the influence of this special microclimate the figs established in large numbers but following the collapse of heavy industry water temperature returned to normal. If this historical explanation is correct the trees are very much part of the town's industrial heritage.

Animals

Invertebrates

River quality is monitored jointly by chemists, who measure such

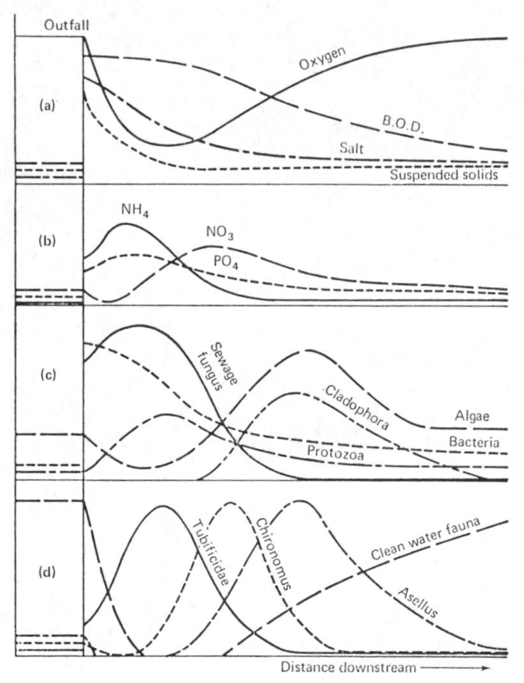

Fig. 15.5. Diagrammatic presentation of the effects of an organic effluent on a river and the changes as one passes downstream from the outfall. (a) and (b) physical and chemical changes; (c) changes in micro-organisms; (d) changes in larger animals (Hynes, 1960).

Table 15.1. Families of freshwater invertebrates arranged into groups showing an increasing sensitivity to water pollution from 1 to 10 (the amended DOE/NWC 'Biological Monitoring Working Party' scores)

	Families	*Score*
Mayflies	Siphlonuridae, Heptageniidae, Leptophlebiidae, Ephemerellidae, Potamanthidae, Ephemeridae	10
Stoneflies	Taeniopterygidae, Leuctridae, Capniidae, Perlodidae, Perlidae, Chloroperlidae	
River bug	Aphelocheiridae	
Caddis	Phryganeidae, Molannidae, Beraeidae, Odontoceridae, Leptoceridae, Goeridae, Lepidostomatidae, Brachycentridae, Sericostomatidae	
Crayfish	Astacidae	8
Dragonflies	Lestidae, Agriidae, Gomphidae, Cordulegasteridae, Aeshnidea, Corduliidae, Libellulidae	
Caddis	Psychomyidae, Philopotamiidae	
Mayflies	Caenidae	7
Stoneflies	Nemouridae	
Caddis	Rhyacophilidae, Polycentropidae, Limnephilidae	
Snails	Neritidae, Viviparidae, Ancylidae	6
Caddis	Hydroptilidae	
Mussels	Unionidae	
Shrimps	Corophiidae, Gammaridae	
Dragonflies	Platycnemididae, Coenagriidae	
Bugs	Mesoveliidae, Hydrometridae, Gerridae, Nepidae, Naucoridae, Notonectidae, Pleidae, Corixidae	5
Beetles	Haliplidae, Hygrobiidae, Dytiscidae, Gyrinidae, Hydrophilidae, Clambidae, Helodidae, Dryopidae, Elminthidae, Chrysomelidae, Curculionidae	
Caddis	Hydropsychidae	
Craneflies	Tipulidae	
Blackflies	Simuliidae	
Flatworms	Planariidae, Dendrocoelidae	
Mayflies	Baetidae	4
Alderflies	Sialidae	
Leeches	Piscicolidae	
Snails	Valvatidae, Hydrobiidae, Lymnaeidae, Physidae, Planorbidae	3
Cockles	Sphaeriidae	
Leeches	Glossiphoniidae, Hirudidae, Erpobdellidae	
Hog louse	Asellidae	
Midges	Chironomidae	2
Worms	Oligochaeta (whole class)	1

aspects as dissolved oxygen, BOD, SS and nitrogen, and by biologists who observe invertebrate communities. The principles of biological monitoring have been clearly laid out by Hynes (1960) from whose book *Polluted Waters* Fig. 15.5 has been taken. This diagram, which deserves close scrutiny, represents the effects of organic effluent on 16 physical, chemical and biological features of a river, though in practice the smooth curves of the graph would be complicated by multiple sewage outfalls, the entry of clean water tributaries and discharges of toxic industrial waste. Most water authorities have developed their own biological index of water quality based on the number of taxa present with a weighting for abundance, though there is a move towards adopting a common one working with Table 15.1 which lists families of freshwater invertebrates in order of increasing tolerance to pollution. Although most of these organisms can be found in urban areas, those with low scores in the table tend to be dominant, while ones scoring in the range 7–10 are scarce or absent.

Under the worst conditions of urban organic pollution (class 4) the river bed is carpeted with masses of a brown, yellow or white-coloured slime known as sewage fungus. This is a mixture of bacteria, fungi and ciliated protozoans amongst which live a few particularly resistant animals such as tubifex worms (*Oligochaeta*), chironomid larvae and rat-tailed maggots, all of which have specialized methods of obtaining oxygen. At intermediate levels of pollution (class 3) the hog-louse (*Asellus aquaticus*) becomes super-abundant; any stone, piece of wood or rubbish pulled out of the water will be crawling with them. Also present will be a range of leeches and snails (Table 15.1). With improving water quality diversity increases, case-forming caddis and freshwater shrimps (*Gammarus pulex*) being distinctive markers for class 2 waters. It should be noted that, owing to the tolerance of the *Baetidae*, the presence of mayfly nymphs is not in itself an indication of pure water (class 1); the entire invertebrate fauna of a riffle or pool needs to be sampled before an assessment is made.

The effects of toxic trade effluent can complicate the interpretation of simple biological scales designed to monitor oxygen levels. Each organism has its own threshold for a particular poison and a wide range of poisons can occur. After a toxin enters the river sensitive animals are eliminated, then gradually reappear, often many kilometres downstream, as its concentration declines. Among the first organisms to be reduced are certain common algae so those which survive as dominants may be unusual, e.g. *Batrachospermum* in the lead and zinc streams of South Wales. Sometimes *Cladophora glomerata* colours the river bed bright green because there are no animals to graze it, and then slowly declines as invertebrates return. Unexplained absences or species of unexpected abundance are always an indication of possible toxic effects.

Holland and Harding (1984) consider that in parts of the Mers... ment zinc is responsible for the scarcity of the freshwater limpet (... *fluviatilis*), wandering snail (*Lymnaea peregra*), leeches (*Erpobdella* sp...), blackflies (*Simuliidae*) and *Asellus*. Elsewhere on the same system, ... textile works discharging chlorine into a clean stream causes an immediate deterioration to a tubifex–chironomid community. Coking plants on the River Rother produce a liquor containing un-ionized ammonia which has the same effect; kick sampling in the river bed for many kilometres below these works produces tubifex worms by the thousand, chironomid larvae by the score and very occasionally *Lymnea peregra* and *Asellus* which have been washed down from the cleaner headwaters. Often there is a policy of not diverting toxic trade wastes to sewage works in case they reduce the sale value of the sludge; this approach has slowed the cleaning up of many rivers.

Saline discharges may be as salty as sea water and are particularly a feature of the Yorkshire coalfield where the River Dearne for example contains *Gammarus duebonii* which is more salt tolerant than the usual *G. pulex*. The major aquatic plants of inland saline waters are *Myriophyllum spicatum* and *Potamogeton pectinatus* which also commonly occur in brackish water ditches near the sea. Sewage works discharge sodium chloride but not in biologically significant amounts, though it is sufficient to make the water in London taste flat.

The ecological effects of the increased torrential flow to which urban watercourses are subject are little known, but a few observations have been made that have a bearing on this. It is frequently reported that severe floods sweep away many animals, and several workers have quantified this; Jones (1951) for example reported that summer floods in the River Towy reduced the invertebrate population from 300–1000 m^{-1} to 40–48 m^{-1}. Another aspect of this phenomenon is that scouring controls the detailed nature of the substratum which in turn controls the density of invertebrates. Percival and Whitehead (1929) have shown that at least in unpolluted water animal density falls off sharply as erosion increases. The unusual patterns of scouring and siltation coupled with pollution combine together in a variety of ways to reduce invertebrate numbers in urban watercourses.

Fish

During the industrial revolution, fish populations in the mid and lower reaches of most major rivers were eliminated by the mixture of oxygen-depleting sewage and toxic industrial effluent discharged into them. As a result of this, no-one can be sure what the natural fish population of most of the rivers that flow through our towns was. Certain stretches remain virtually fishless, particularly on the Aire, Don, Mersey, Rother and Tame (Trent). Where improvements have taken place fish return in the approximate order: eel; three-spined stickleback; gudgeon, roach; perch, chub, dace, pike, minnow; grayling, brown trout, salmon. Being naturally mobile and also liable to reintroduction by angling clubs, fish quickly respond to improvements as the recent history of the River Thames has shown. In the late 1950s the lower 70 km of the Thames was almost fishless, but following the construction of extensive new sewage treatment plants conditions improved so much that by 1968, 41 freshwater and marine fish had returned and by 1975, 86 species were present; in addition large flocks of wildfowl started to feed on the burgeoning fauna of the mud. Around 1979 the goal of 30% minimum dissolved oxygen was met and the passage of migratory fish commenced. Today over 150 species are believed to be present but major fish kills still occur if there is an unfortunate combination of events. For example, in 1986 a hot spell which encouraged algal growth ended with a sudden storm that caused combined sewerage systems to overflow so foul sewage entered the river. The Thames Water Authority's 'bubbler' which is normally towed up and down the river during summer was out of action and this combination of circumstances led to a pronounced oxygen sag which resulted in the death of many thousands of fish.

The extent of damage to fish populations varies considerably from reach to reach; Fig. 15.3 illustrates this for the Mersey catchment. Toxic industrial effluents explain the fishless state of the upper stretches of the Goyt (textile works), Tame (paper mill) and Etherow (zinc); elsewhere deoxygenation associated with sewer overflows is responsible. This is a dynamic situation, and it is expected that fish will move upstream as pollution controls tighten.

Birds

The bird community centred on urban rivers is composed of species closely dependent on aquatic conditions; birds that breed in terrestrial habitats such as patches of woodland or scrub within the river corridor and others that use the river for feeding or resting. None are urban specialists. Mallard and moorhen nest wherever there is cover and can be seen from bridges at all times of the year; their breeding success is limited by floods and rat predation. Nesting kingfishers normally require vertical earth banks at least 1.5 m high, but in towns have taken to nesting in disused pipes which formerly discharged into the river. This facility has enabled them to claim territories along the complete length of most urban rivers; they feed on minnows and sticklebacks.

During winter many more birds use river corridors for feeding or resting. As water quality improves tufted duck and black-headed gulls congregate, herons and cormorants are seen, and common sandpiper, terns and many other species are recorded on autumn passage. Each city has its characteristic bird population but as yet regional patterns have not been worked out.

15.3 CANALS

Canals are amongst the oldest structures to be found in towns, most having originated between 1758 and 1805. They are particularly widespread in the Midlands where the Birmingham conurbation possesses a greater length than Venice. Since 1968, canals have been divided into commercial, cruising and remainder waterways; most fall into the middle category so are available for pleasure-boating, fishing and other recreational uses. The general features of urban canals as wildlife habitats have been described in a survey of the Regent's and Hertford Union Canals, Inner London (Nicholson, 1985). His findings show they provide a wildlife resource of outstanding value, certain lengths rivalling in quality those of any other waterbody in the capital. The associated habitats are also of considerable interest. Seven ecological zones can be recognized across a canal (Fig. 15.6).

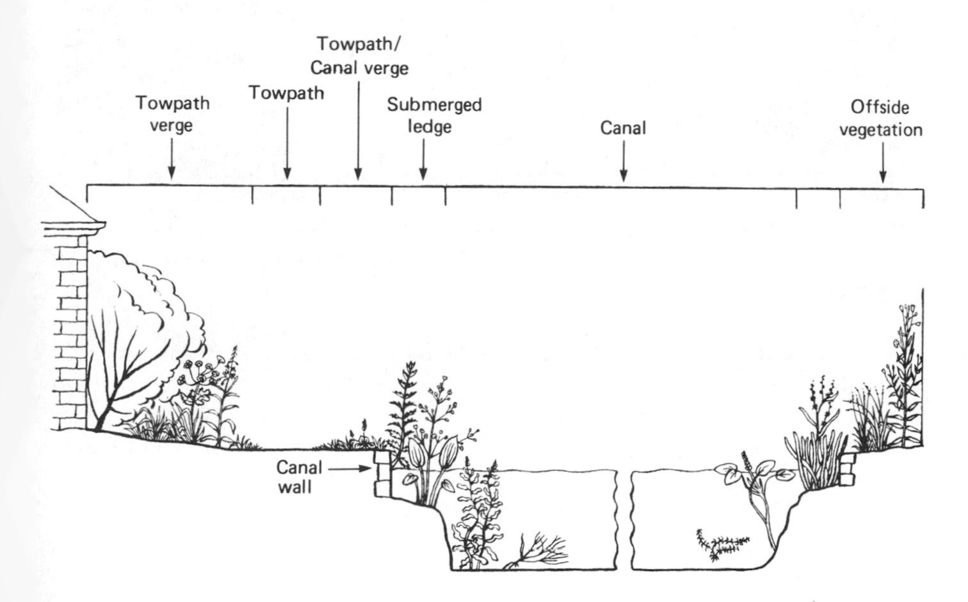

Fig. 15.6. Diagrammatic cross-section of a canal showing seven ecological zones.

Vegetation

Tow path verge

The predominant vegetation in this zone is a mixed rank neutral grassland and tall-herb community of varied composition. Self-sown trees and shrubs such as sycamore, elder, goat willow and buddleia are frequently present. Owing to its age, unexpected species such as dactylorchids may be present while the often commercial surroundings

Fig. 15.7. The canal system has been both the point of introduction and the main route of spread for many plant and animal species. This is helped by commercial surroundings.

facilitate the entry of numerous aliens (Fig. 15.7). Areas of acid grassland may also be present either as relic features or developed on industrial waste. A further element has hedgerow connotations. This varied strip of unmanaged vegetation only 1–3 m wide is of considerable value to butterflies, grasshoppers and diptera, and is a source of food for seed-eating birds.

Tow path

Tow paths are usually paved or surfaced with aggregate; trampling limits diversity.

Tow path/canal verge

This zone is mostly occupied by a trampled *Lolium* grassland containing clumps of tall herbs and species typical of disturbed sites such as wall barley, burdock and mallow. Sedges, which as a group are rare in towns, may be present in considerable diversity, e.g. *Carex flacca, C. hirta, C. nigra, C. otrubae, C. ovalis* and *C. pendula*; their presence reflects the age of the site and the fact that canals are clay lined.

Canal wall

This is usually built of large stone or concrete blocks and carries a luxuriant vegetation of native tall herbs not often seen in urban areas, e.g. angelica, bur-marigolds, greater water dock, gypsywort, hemlock water dropwort, marsh woundwort and skull-cap. Aliens may also be present such as *Angelica archangelica, Aster novi-belgii, Impatiens* spp., *Paspalum paspalodes* or the two introduced bur-marigolds (*Bidens connata, B. frondosa*) that have started to spread in this niche (Burton, 1979). Species with floating seeds spread quite fast as there is usually a slight flow in canals.

Off-side

Thick untrampled strips of marginal vegetation occur along the off-side of canals, they frequently include the ferns *Athyrium filix-femina, Dryopteris filix-mas* and *D. dilatata*. This zone is the main nesting site for water fowl on the canal.

Submerged ledge

This feature typically carries a tall deep-rooted reed bed but with increasing boat traffic it can be destroyed; the bank may then require protecting with metal piling. Sites where it persists are angles associated with locks, bridges and wharfs.

The channel

Factors that determine the nature of channel vegetation are, in order of decreasing importance, the density of boat traffic, water quality, weed control and shading. In a survey of 118 sites Murphy and Eaton (1983) found a mean of 10 taxa/50 m length (range 0–28). Using cluster analysis on the data, four types of site were recognized; each had significantly different boat traffic densities. At highest traffic density a small homogeneous group of submerged aquatics was sparsely present.

They included *Cladophora glomerata* and *Elodea canadensis* which have numerous growing points and little or no dependence on roots, the streamlined *Potamogeton pectinatus* and a fringe of *Glyceria maxima* a strongly-rooted emergent quick to recover after physical damage. Intermediate traffic encouraged a very diverse fringeing belt of emergent species including *Iris pseudacorus, Lythrum salicaria, Sagittaria sagittifolia* and *Spargarium erectum,* with yellow water-lily (*Nuphar lutea)* and *Potamogeton natens.* Floating-leaved species are only present at the canal margin because propeller action breaks them up. As in the previous group, submerged vegetation is only weakly developed as turbidity of the water cuts down light penetration reducing photosynthesis.

With falling boat traffic submerged communities are able to develop fully, particularly *Callitriche stagnalis, Ceratophyllum demersum, Lemna* spp., *Myriophyllum spicatum, Potamogeton crispus, P. berchtoldii, P. perfoliatus, Nitella* spp. and the moss *Fontinalis antipyretica.* Relatively fragile emergents like *Alisma plantago-aquatica, Butomus umbellatus* and the canal specialist *Alisma lanceolata* also increase. *Glyceria maxima* competes very successfully in these canals forming dense almost pure beds.

Overall, submerged communities are significantly depressed by heavy boat traffic while emergent community abundance shows a weaker negative correlation with traffic. The method of suppression is predominantly a turbidity effect on the former, while direct physical damage by boats may be the principal effect on emergent and floating-leaved macrophytes. The limited appeal of urban canals to pleasure boat cruisers often results in relatively light traffic and a consequently rich vegetation.

Though water pollution is not an extensive problem in the 2498 km of canals in England and Wales (class 1 quality, 38%; class 2, 51%; class 3, 10%; class 4, 1%) their quality is generally poorer than that of rivers (Department of the Environment, 1986). Where urban canals that only have light traffic support a low abundance and diversity of submerged macrophytes, water pollution should be considered a possible cause.

It is clear from the accounts of fishermen that much canal vegetation is in a dynamic state, increasing in response to cleaner water and reductions in boat traffic. Interestingly, Haslam (1978) considers that dredging is not a major factor in the ecology of canals, species usually returning after 2 years and vegetation being back to normal within 3.

Special features of the vegetation

Alisma lanceolata, Carex otrubae, Glyceria maxima and *Elodea canadensis* are species for which canals appear to be the optimum habitat in Britain. The spread and subsequent decline of Canadian pondweed is well documented. The first English record was made in 1847 from near Market Harborough, Leicestershire, then an important node on the canal system. From there it spread like wildfire and 10 years later was present throughout southern England and the Midlands, often in prodigious quantities. There are reports of rivers and canals blocked by it, of extra horses needing to be yoked to barges and of ducks unable to swim across ponds. The spread was entirely by fragmentation as only female plants were known then and though male plants were found in 1879, they have never been proved to fruit in this country. By 1880 it was becoming clear that colonization of a fresh site involved a rapid increase to pest proportions which took 3–4 years, by when most other aquatic plants would be eliminated. This population density would be maintained for up to 10 years followed by a gradual decline over the next decade. A small relic population might remain or the plant disappear altogether, possibly returning a few years later but remaining subordinate. This phenomenon has never been satisfactorily explained but can still be observed when Canadian pondweed invades a new water body. In 1966 another North American adventive, *Elodea nuttallii,* was identified in England. It is now widespread and is rapidly replacing *E. canadensis* though it has not yet shown quite the invasive power of that species (Simpson, 1984). The small introduced water fern, *Azolla filiculoides*, which may entirely cover canal backwaters, also shows a cycle of abundance and decline.

The niche provided by warm water discharge has also been exploited by introduced plants. Shaw (1963) and Lousley (1970) recount how in the 'hot lodges' or 'pounds' where water used for cooling in the Lancashire cotton mills was released into canals, exotic aquatics such as *Vallisneria spiralis* used to grow so profusely that it had to be repeatedly dragged out. Other exotics that colonized this niche were *Egeria densa, Lagarosiphon major, Luronium natans, Ludwigia palustris* and *Najas graminea*. Though several of these are probably aquarium rejects the *Najas* is thought to have been introduced with Egyptian cotton along with the leafy pondwood (*Potamogeton epihydrus*) which also occurs in some Yorkshire canals. Forty years after its discovery as a cotton alien, this North American plant was found growing as a native in the Outer Hebrides. In places the water temperature, at 38° C, is too hot for plant life, the zone of exotics only starting where it has fallen below 29° C. After about 1960 the mills started turning to alternative fuel sources, so heated discharges are a decreasing urban habitat.

Among the very few generally accepted alien freshwater algae in the British checklist are *Chara braunii, Pithophora oedogonia* and *Composopogon* sp. which for over 50 years occurred in the Reddish Canal, Manchester, eared soon after their habitat, a point source of artificially ter, ceased (Swale, 1962). The importance of the site was only ter the cotton mills closed in 1959 so no thorough investigation of the algal flora was undertaken.

Invertebrates and fish

Invertebrate richness is largely controlled by water pollution (Table 15.1) and the biomass of aquatic plants. A particularly rich canal is the 8 km Gotts Park to Calverley Bridge stretch of the Leeds–Liverpool which, where it traverses Leeds, is an SSSI. It is one of the finest habitats for freshwater molluscs in Yorkshire with 41 species recorded from the 12 km running west from the city centre. Unusual water crickets, water beetles, caddis and several dragonflies are also present. The reason for this prolific wildlife lies in the origin of the water. The canal is fed with clean alkaline water from reservoirs in the Pennines, little effluent enters the system, and boat traffic is just sufficient to maintain a reasonable flow through the locks.

Where water quality is on the class 2/3 boundary or better, considerable numbers of dragonflies may be present. *Ischnura elegans,* which is tolerant of mild pollution, is the commonest dragonfly on canals in Britain, but *Aeshna cyanea, A. grandis* and *Sympetrum striolatum* may also be present. Two moths frequent by canals are the small china-mark moth and the brown china-mark moth. They have aquatic larvae that make themselves cases of plant material and crawl around just under the water (Fig. 15.8).

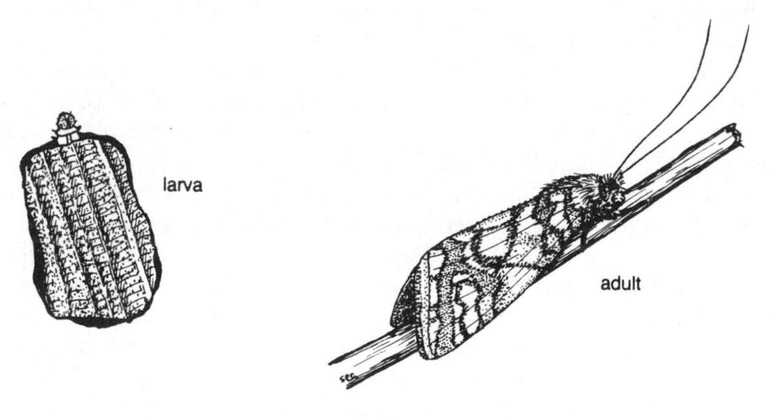

Fig. 15.8. Adult and larva of the brown china-mark moth. The larvae are truly aquatic living in flat cases made from two pieces of their food plant the broad-leaved pondweed (*Potamogeton natans*) (S. Garland).

The canal system has turned most of our rivers into a web of interconnecting waterways which, together with their use as highways has greatly facilitated the spread of freshwater organisms. Not only have native species been able to extend their range, aliens sometimes introduced on timber or other cargoes have found they offer ideal routes for dispersal. Macan (1974) discusses 16 freshwater invertebrates which are probably aliens of recent introduction and suggests that for many the canal system has been both the point of introduction and the main route of spread. An outstanding example is the American amphipod *Crangonyx pseudogracilis*. Between 1937 and 1955 this species, which looks like a small freshwater shrimp, became very widespread in the canal system of central England and is still spreading. Many of the other introductions he discusses show some preference for warm water effluents associated with industry, e.g. the snails *Planorbis dilatatus* and *Physa* spp., an oligochaete *Branchiura sowerbyi* and the amphipod *Corophium curvispinum* var *devium*. The zebra mussel (*Dreissena polymorpha*) was introduced to Britain via the London Docks (1824) from where it spread and increased greatly. Maps of its present-day distribution coincide closely with the canal region. In a number of waterways it appears to be present only as dead shells which indicates a decline since last century.

The fish most commonly found in canals are adaptable natives such as roach, bream, perch and pike, together with introduced common, mirror and crucian carp. Others such as bleak and ruffe were once confined to the eastern rivers of England but canal building enabled them to swim across the watershed. This has obscured a biogeographical distinction of long standing as the richness of the eastern rivers of England is believed to originate from early post-glacial time when they were connected to the Rhine. A stretch of the St Helens canal in Lancashire is unique in containing a large colony of guppies (*Poecilia reticulata*), together with a breeding colony of the cichlid (*Tilapia zillii*) as well as non-breeding angel-fish, catfish, chequered and rosy barbs, gouramies and mollies. They all established in 1963 when the stock of a defunct pet shop was released into the canal. They have been able to persist as the water is warmed by discharge from the adjacent Pilkington Brothers glass factory (Lever, 1977).

In conclusion, the flora and fauna of urban canals is frequently of very high quality in terms of its abundance, diversity, presence of notable species, its value to nature conservation and visual attractiveness. This is due to light boat traffic and improving water quality. They have been the point of introduction and means of spread for numerous alien species through their links with commerce. Heated discharges provide a specialized environment where tropical species have naturalized.

15.4 PONDS, LAKES, RESERVOIRS AND WATER MAINS

It is convenient to regard ponds as sufficiently shallow for rooted vegetation to cover the bottom and with margins that do not experience wave action. Lakes are larger deeper waterbodies, the margins of which are exposed. Both ponds and lakes are commoner in towns than in the countryside. Beebee (1979), for example, found that 15% of the gardens he surveyed contained a pond. In new towns such as Runcorn, Warrington and Milton Keynes there has been a policy of incorporating field ponds into the urban landscape, while ornamental ponds, defined here as containing a fountain or statue, extend aquatic habitat into the heart of our cities. As lakes take up more space, they tend to be found on the fringe of built-up areas so are not subject to such strong urban influences. The artificial lakes surrounding London include supply reservoirs, which, as they contain water for human consumption, are kept reasonably free of pollution, compensation reservoirs built in the last century to keep the canal system supplied with water, e.g. the 51 ha Brent Reservoir now an SSSI, and flooded sand, gravel and clay pits. Balancing lakes to collect urban run-off are now being added to this huge acreage of still water.

Ponds

The ecology of ponds is well known in a general sense but the information is scattered and few studies have been made of small artificial ones. Those in gardens or school grounds are typically very small and shallow, get topped up with tapwater, and have a surface area of less than 6 m^2; most are lined with polythene, fibre glass or concrete. Being artificially created and of recent origin they are usually initially stocked with a variety of plants bought from commercial suppliers to which others collected from the wild are later added. Most garden aquatics are species native to Britain though several cultivars are offered such as the kingcups *Caltha palustris* 'Alba' and 'Plena', yellow flags *Iris pseudacorus* 'Golden Queen', 'Sulphur Queen' and 'Variegata', corkscrew rush *Juncus effusus* 'Spiralis', greater spearwort *Ranunculus lingua* 'Grandiflora'. Establishing vegetation is not a problem – most garden ponds are solid with it.

Once the base of the food chain has been provided, spontaneous colonization commences. Midges lay their eggs, water boatmen, pond skaters and water beetles fly in, while molluscs hatch from eggs introduced on water plants. Garden ponds have turned out to be ideal habitats for several amphibians which are now so anthropocentric that it is not known which native habitats they favoured previously. The common frog, which has been declining in many parts of Britain (Prestt *et al.*, 1974), is thriving in garden ponds. Beebee (1979) found that breeding occurred in half the garden ponds he surveyed from which he calculated that his urban study area of 70 km^2 in the south of England contained 44 000 frogs. Neither age of pond, or volume, or amount of aquatic vegetation were important in determining the suitability of a pond for breeding (Beebee, 1981). Table 15.2 emphasizes the increasingly important role of garden ponds as breeding sites for frogs. A survey in the Sheffield area (Whiteley, 1977) confirmed that gardens are now the main habitat in which frogs are seen: they provided 31% of all records (sample size 655).

The common toad has also been declining (Cooke, 1972) and, as with frogs, garden ponds are an important refuge for them (Table 15.2), but, owing to their limited volume, not quite such a significant one. Toads prefer large ponds where they can form extensive breeding colonies. A further behavioural difference, picked up in the Sheffield survey, was

Table 15.2. Distribution and abundance of amphibian breeding sites in Sussex, England

	Chalk hill sites per km²	*Suburban sites per km²*	*Marsh sites per km²*	*Agricultural sites per km²*
Common frog	0.014	~ 100	1?	0.1
Common toad	0.014	~ 30	0	0
Smooth newt	0.1	~ 40	1?	0.5
Palmate newt	0.04	~ 0	0.1?	0.4
Crested newt	0.04	~ 0	0	0.25

Data from Beebee (1981).

that 27% of the toad records (sample size 338) were traffic casualties compared to about 1% for frogs. Nearly all the dead were sexually mature individuals found during April while migrating to their breeding ponds. However, the significance of suburban gardens for toads should not be underestimated; they provide food, shelter and refuge. The association of the midwife toad with garden ponds is discussed in Chapter 14.

The ideal newt pond according to Beebee (1981) has no fish, a surface area of less than 200 m^2, a 5–50% cover of aquatic vegetation and a depth of 0.5–1.0 m; amphibians in general also like dense vegetation around part of the perimeter. The abundance of such habitats in suburbia has enabled palmate and smooth newts, helped by a fair amount of introduction, to become widespread in urban areas, while at the same time loss of breeding grounds in the countryside has led to a decline nationally. Great crested newts are more particular in their choice of breeding sites; on no account should they be introduced into garden ponds from wild stock; not only is this illegal, few garden ponds are suited to their needs. Newts live on land for long periods and are as likely to be encountered hiding in moist crannies, among moss and around rockeries as in the pond.

The ecology of amenity ponds found in parks and open spaces differs from that of garden pools. Most are kept supplied from a nearby stream or have run-off directed into them. This water is inevitably eutrophicated and silt laden; as a result dense plankton growth and floating masses of filamentous algae develop in early summer and persist. Aquatic vegetation is often limited by the need for frequent dredging, and also by the grazing and trampling of waterfowl. A feature of the macrophytes is that a number have usually originated as aquarium rejects, so alongside Canadian pondweed (*Elodea canadensis*) and *E. nuttallii,* temporary populations of species such as frogbit (*Hydrocharis morsus-ranae*), water soldier (*Stratiotes aloides*), tape-grass (*Vallisneria spiralis*) and the water fern *Azolla filiculoides* may turn up.

The mollusc population is similarly affected by introductions; for example, ponds in public parks in the west of Sheffield are sited on clean soft-water streams yet many contain populations of large thick-shelled snails which nationally show a preference for hard water. They are suspected of being derived from aquaria, and species such as great pond snail (*Lymnaea stagnalis*), ear pond snail (*L. auricularia*) and great ram's horn snail (*Planorbis corneus*) probably rely on constant reintroduction to maintain themselves. Records of the duck and swan mussel (*Anodonta anatina, A. cygnaea*) show a similar patchy and shifting distribution.

Fish populations in amenity lakes are mostly healthy, self-perpetuating and have a species composition controlled by introductions. Best suited to the eutrophicated water and also the needs of anglers are gudgeon, roach, bream, tench, perch, ruffe and carp. Among alien ornamental species the goldfish and golden orfe are extremely hardy and maintain thriving feral populations. Sometimes the former revert to their original wild-type green coloration. Fourteen species of fish occur in the Royal Parks, London. A study of parasites on fish in the Serpentine showed that they were infected with 14 species, none of which adversely affected them (Lee, 1977). The parasite population was that expected in a small eutrophic water body.

Amenity ponds are sufficiently large to accommodate a resident bird population. Mallard, moorhen and, in winter, tufted duck are the typical inhabitants; if there is sufficient cover, mute swan, coot and little grebe may breed. Canada geese sometimes call in for a few days in winter while in summer, house martin, swallow, migrant warblers and bats are attracted by emerging insects. After a day spent on the rubbish tips gulls may call in for a pre-roost bathe before flying off to spend the night on local reservoirs or buildings. Pinioned muscovy and other ornamental ducks are a major public attraction but if populations get too large they degrade the pond through a combination of stirring up the mud and their droppings, both of which reduce water quality and lead to algal problems. Trampling, grazing and defaecating by ducks conflict with high standards of park maintenance, which can result in ornamental plants having to be fenced with netting and grasscrete laid on islands. Further discussion of the avifauna of amenity ponds can be found in Chapter 11 and in Goode (1986).

Lakes and reservoirs

The rich wildlife associated with lakes and reservoirs in the urban fringe is too far removed from the influence of towns to rate as mainstream urban ecology; a few crosslinks occur however. The London reservoirs are extremely important overwintering sites for birds; all the species of duck (18), grebe (five) and diver (three) that normally visit Britain, except for eider duck, are regularly seen on them. In particular, the birds favour the older shallower waterbodies such as those at Barn Elms and the canal reservoirs that have the richest aquatic and bank vegetation; modern deep steep-sided examples are more popular with gulls which need large expanses of open water on which to roost. The nearest reservoirs to central London, those of Stoke Newington, lie within the urban heat island and when the main reservoirs are frozen over they experience a great influx of birds. There is no doubt that the reservoirs bring into London very large numbers of waterfowl which would never ordinarily visit the smaller amenity lakes.

The Berlin Havel lakes lie well within the city and have long been used as storm water collection basins, commercial waterways, and for intensive recreation; this has led to serious problems of eutrophication. Changes in waterside vegetation have been monitored by Markstein and Sukopp (1980) who reported that reed beds (*Phragmites australis*), which surrounded 40% of the margin in 1962, could only be found along 17% of the shoreline in 1977. The cause of the decline has been linked to eutrophication (Bornkamm *et al.*, 1980). Though enrichment could be looked on as having a beneficial fertilizing effect and the reeds do appear particularly vigorous, it eventually hastens their decline as, along with the increase in size, the stems become more liable to breakage. Pot experiments showed that while nitrogen application increases height, leaf area and stem diameter, phosphorus reduces the diameter and cross-sectional area of the outer ring of strengthening sclerenchyma tissue in the stems, so they are less rigid and more liable to break. The final cause of disappearance is mechanical disturbance by wind, boat traffic and bathers which result in the weakened stems lodging. Eutrophication is also believed to be responsible for reducing fish diversity in the Havel from 28 species (1900–1960) to 22 (1979) as the reed beds provided important nurseries for the fry and were the main habitat of crucian carp and chub. One day the balancing lakes incorporated into Britain's new towns may provide similar problems and oppotunities for wildlife.

Water mains

An outline account of the fauna and flora of waterworks is given by Fitter (1945); it is frontline urban ecology. He relates that Krapelin, who investigated the luxuriant fauna of the Hamburg water system, sought, among other things, proof of his theory that the blind and colourless water fauna which inhabits certain cave systems and very deep lakes would either colonize or evolve in the very similar environment of water pipes. He did not find what he was looking for, but a breeding colony of a blind shrimp-like crustacean (*Niphergus* sp.) has been found in culverts under a filter bed in London. The rest of the ecosystem is composed of algae, fungi and bacteria, including iron bacteria of the genus *Gallionella* which coat the inside of water pipes, doing considerable damage. With filtration and purification the community has been seriously depleted since 1912, when 90 tons of the zebra mussel (*Dreissena polymorpha*) were removed from 400 m of water main at Hampton, Middlesex. Water boards are sensitive about the populations of animals that penetrate their water pipes but it is common knowledge that breeding colonies of *Asellus* and *Gammarus* occur, feeding on fungi, algae, bacteria and suspended solids. Oligochaetes are also regularly found together with parthenogenetic species of chironomid larvae which can breed without passing through the adult stage. One of the few molluscs to flourish in mains is *Potamopyrgus jenkinsi* which is ovoviviparous and parthenogenetic so a single specimen can give rise to a population. A problem for these animals is that of attachment, but corroded or furred up pipes offer a range of crannies sheltered from the current and molluscs like the zebra mussel are attached by a holdfast. Successful water main populations are almost always of benthic organisms. The increasing use of eutrophic water and treatment processes designed to maximize throughput are expected to lead to a greater animal infestation of the distribution system.

Chapter 16

WOODLAND

This chapter is mainly concerned with woods surrounded by housing and experiencing a high level of public use; they are criss-crossed by paths, used for adventure play by children, and many receive a considerable management input from the local authority. They form a continuum with the more extensive woods to be found at the city limits, the ecology of which falls within the province of standard works such as Rackham (1976) and Peterken (1981). The North American concept of the 'Suburban Forest' (Waggoner and Ovington, 1962) where every tree in a town is thought of as contributing to a network of woodland which is studied as a whole, often by aboriculturalists, is attractive but only applicable to climates where planting for shade is a priority. The European definition of woodland as an area of more or less closed canopy trees associated with a shade-bearing ground layer is accepted here.

Key factors in the ecology of any woodland are its age and origin which provide a convenient framework within which to examine urban woods. Table 16.1 uses these variables to construct five categories ranging from ancient seminatural woodland with a high wildlife value to recent plantations. The numbers show the estimated proportions of the various categories in Britain in 1978 which also forms a rough guide to their proportional occurrence in urban areas today. Noteworthy features are the unprecedented wave of recent amenity planting in towns and the unexpected scarcity of spontaneous secondary woodland. While most rural woods can be pigeon-holed into such a classification, it can be more difficult with urban ones as they have often experienced more disturbance and consequently show greater complexity. This will be

Table 16.1. The five main types of urban woodland defined in terms of their origin and degree of naturalness. Estimated proportions in Britain as a whole in 1978 are given

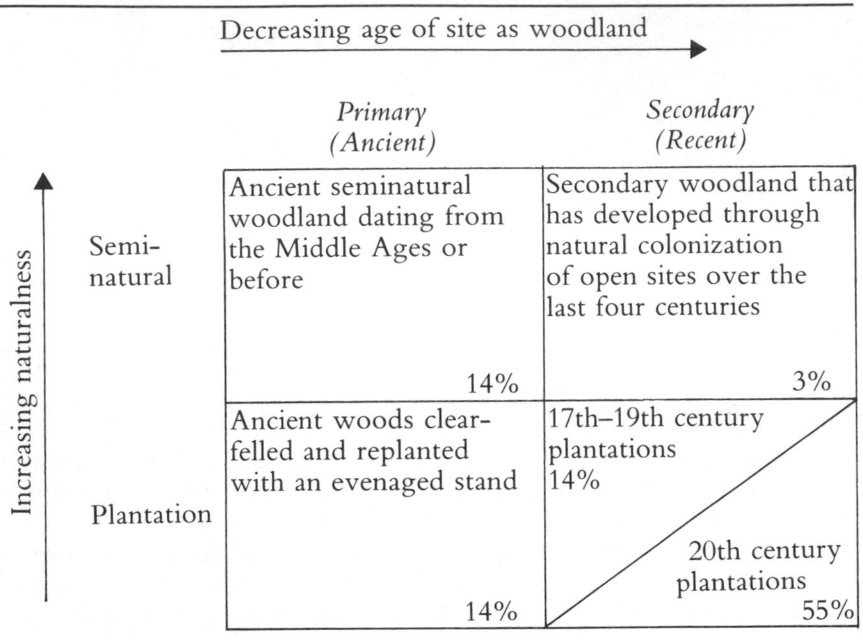

Decreasing age of site as woodland →

Increasing naturalness ↑	*Primary (Ancient)*	*Secondary (Recent)*
Semi-natural	Ancient seminatural woodland dating from the Middle Ages or before 14%	Secondary woodland that has developed through natural colonization of open sites over the last four centuries 3%
Plantation	Ancient woods clear-felled and replanted with an evenaged stand 14%	17th–19th century plantations 14% 20th century plantations 55%

Taken from Peterken (1981).

demonstrated by applying the methods of woodland archaeology (Rackham, 1976, 1980) to a small urban wood in Sheffield.

16.1 THE ARCHAEOLOGY OF A SMALL URBAN WOOD

Little Roe Wood, 3 ha in extent, occupies the sides of a small valley 1.5 km from the centre of Sheffield. It is high-canopy deciduous woodland bounded by housing and a school and contains a close network of public paths. The name Roe Wood is probably a reference to roe deer, it having once been part of a mediaeval deer park, though up to at least 1635 it was commonly referred to in documents as Cockshutt Roe, a reference to woodcock. Neither animal has been seen there within living memory. By 1711, part had been cleared to form fields separating the area into two, known as Great Roe and Little Roe, both described as spring woods. This coppice function continued through the 19th century and a glimpse of its composition is provided from the details of a sale notice of 1817 which mentions 331 oak poles, 14 ash poles and 2 owler (alder) poles. Much of Little Roe Wood disappeared

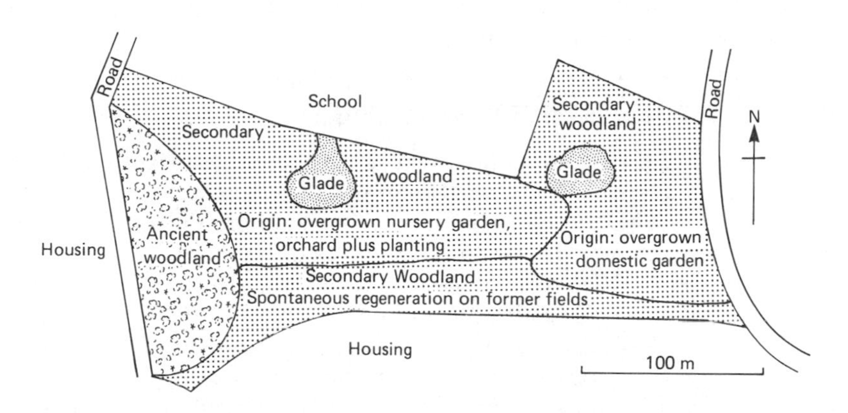

Fig. 16.1. Map showing the history of Little Roe Wood, Sheffield.

under housing and allotments in the late 19th century but in 1897 what was left was presented to Sheffield by its owner, the Duke of Norfolk, to commemorate a visit of Queen Victoria to the city; since then it has been managed by the recreation department as public open space. Little extra information can be gleaned from the shape of the wood and there are no associated earthworks.

The vegetation (Fig. 16.1) provides evidence that the present Little Roe Wood contains a small fragment of ancient woodland preserved as a patch of oak, rowan and hazel standing over infertile soil supporting bracken, bluebells and wood anemones in a matrix of creeping soft-grass (*Holcus mollis*). A steep slope to the south of the stream, shown as fields on mid-19th century maps, today carries a mature ash, elm, sycamore, elder and bramble community which has every appearance of being first-generation secondary woodland which spontaneously colonized the old fields once grazing ceased. The largest part of the wood originated from the now heavily overgrown grounds of the former residence of the Duke of Norfolk's wood warden. To the casual user it appears a normal wood but an ecologist will soon spot the remains of hedges, an orchard and ornamental trees mixed in among the secondary woodland, while two glades containing species that require a fertile soil, i.e. nettle, elder and great hairy willowherb (*Epilobium hirsutum*), mark the site of former buildings. In 1950 the recreation department underplanted a small area with elm.

This example has been quoted at length as it demonstrates two points. Firstly, the methods of woodland archaeology can be applied equally to small urban sites buried in housing as to the large coppice woodlands of East Anglia where the technique was first developed. Secondly, Little Roe Wood contains elements of the five types of woodland defined in Table 16.1 so is unusually complex.

16.2 ANCIENT SEMINATURAL WOODLAND

Survival of ancient woodland through a phase of urbanization is determined by combinations of topography, ownership and chance. Some of the most heavily wooded cities in Britain are those containing slopes too steep for easy development. Sheffield, on the dip slope of the Pennines, possesses 31 parcels of ancient woodland (Jones, 1986), many of which lie on steep slopes in the middle suburbs, while in Newcastle upon Tyne the deep ravines of Jesmond Dene and Scotswood Dene carry high-quality woodland well into the inner suburbs. The same is true of the 'cloughs' in Skelmersdale New Town, Castle Eden Dene at Peterlee and river cliffs elsewhere. By contrast, towns in flatter areas such as Manchester, Birmingham, Liverpool, Leicester and Leeds contain less relic woodland, as intensive farming was followed by intensive building. Certain woods, such as Brier Marsh Wood in the suburbs of Leeds, owe their persistence to a high ground water-table.

Ownership provides the surest protection against development. The most extensive ancient urban woods are those bought by or presented to local authorities to manage for public amenity. This was amply demonstrated by a survey of broad-leaved woodland in Greater London (Greater London Council, 1986) which found that predominantly publicly owned ancient woodland covered 1.7% (2600 ha) of the area. Additionally it reported that since the 1930s only 18% had been lost or seriously damaged which is the lowest rate of loss yet reported for any area in Britain. Owing to a new awareness of the amenity and nature conservation value of ancient woodland, the rate of deliberate destruction, particularly in urban areas, is likely to remain very low. Occasionally, however, a local authority still wishes to build on one of their woodland holdings, an example being Southwark Council who in 1984 announced plans to put council housing on 4 ha of Sydenham Hill Woods, having just leased the remaining 6 ha to the London Wildlife Trust. After a public inquiry during which it was argued that the woodland was valuable in itself and that building on half the area would inevitably damage the rest, the threat was averted.

Tracking down small fragments of ancient woodland within a city provides an insight into the complexities of survival by chance. If undertaking such a survey, it is best to start by consulting large-scale maps that pre-date urbanization from which wooded areas can be transferred to modern 1:10 000 or better 1:2 500 plans. The most likely

Table 16.2. A selection of ancient woodland vascular plant species in central Lincolnshire*: fast-colonizing woodland species and shade-bearing weeds.

Species with a strong affinity for ancient woods, showing little or no ability to colonize secondary woodland and rarely found in other habitats

Anemone nemorosa	*Lysimachia nemorum*
Carex pallescens	*Melica uniflora*
C. pendula	*Milium effusum*
C. remota	*Oxalis acetosella*
Equisetum sylvaticum	*Paris quadrifolia*
Galeobdolon luteum	*Potentilla sterilis*
Galium odoratum	*Primula vulgaris*
Luzula pilosa	*Sorbus torminalis*
L. sylvatica	*Tilia cordata*

Species with a mild affinity for ancient woods. Some occur sparingly in other seminatural habitats

Adoxa moschatellina	*Geum rivale*
Allium ursinum	*Hyacinthoides non-scripta*
Campanula latifolia	*Hypericum hirsutum*
Carex sylvatica	*Mercurialis perennis*
Chrysosplenium oppositifolium	*Ranunculus auricomus*
Conopodium majus	*Stellaria holostea*
Corydalis claviculata	*Veronica montana*
Fragaria vesca	*Viola riviniana*

Fast-colonizing woodland species

Circaea lutetiana	*Glechoma hederacea*
Dryopteris dilatata	*Hedera helix*
D. filix-mas	*Rubus fruticosus*
Epilobium montanum	*Silene dioica*
Geranium robertianum	*Stachys sylvatica*
Geum urbanum	*Urtica dioica*

Shade-bearing weeds

Arcticum lappa	*Heracleum sphondylium*
A. minus	*Hieraceum* spp.
Arrhenatherum elatius	*Holcus lanatus*
Aster novi-belgii	*Lapsana communis*
Cirsium arvense	*Poa annua*
C. lanceolatum	*P. trivialis*
Dactylis glomerata	*Potentilla reptans*
Epilobium angustifolium	*Rumex obtusifolius*
E. ciliatum	*Senecio squalidus*
Galium aparine	*Stellaria media*
	Tussilago farfara

* Taken from Peterken (1981).

refugia are the grounds of what were formerly large private houses or institutions such as schools and hospitals, possibly now redeveloped as flats. Here, in marginal tree belts, awkward corners, on banks, and especially along small streams, fragments of ancient woodland may be found. They can be recognized chiefly by the ground flora which should contain species with a low colonizing ability. Peterken's (1974) carefully compiled list of old woodland indicator species derived from his work in Lincolnshire (Table 16.2) has been widely applied in Lowland Britain and found to be reasonably accurate, despite his advice that it should only be used in the study area. If combinations of species are appraised and a general appearance of naturalness taken into account, few mistakes will be made, though a secondary wood in Leicester is known to have acquired a large population of wood anemone through invasion by rhizome growth from an ancient hedge. Tree and shrub layers provide less useful evidence in urban areas as both have usually been strengthened by planting with exotics.

Fig. 16.2 shows the current status of a group of six small woods that were accurately delimited on Fairbank's 1789 map of the Sheffield area; today they lie 1.5–2.5 km from the city centre. The easternmost wood has almost entirely vanished under housing though a fragment remains in a corner of the Botanical Gardens as a dozen oaks standing over a bluebell – *Holcus mollis* ground layer. Endcliffe Wood, now part of a heavily used public park, has 70–80 year old oaks standing over bare ground or a discontinuous *Deschampsia flexuosa–Holcus mollis* sward. Part of Smith Wood was preserved in the grounds of a convent (now a

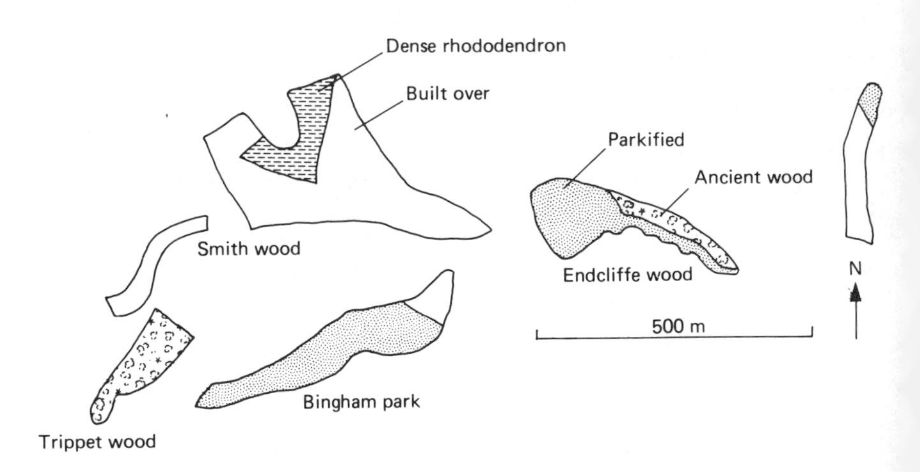

Fig. 16.2. The current status of a group of small woods marked on Fairbank's 1789 map of Sheffield.

school) where, along the course of the Oak Brook and in a marginal tree belt, a few huge oaks survive among planted sycamore, beech and horse chestnut; the ground flora, which includes bluebells, is threatened by encroaching rhododendron. The best piece of Smith Wood was on a steep slope within the private grounds of a large house, but this was destroyed when the property changed hands in 1970 and 160 flats were built. Little survives of the wood in Bingham Park on the opposite side of the valley which has suffered 'parkification' (see later) but Trippet Wood on a very steep slope in the same park is still a heathy oakwood with bilberry, honeysuckle, wood sorrel and hazel on acid soil.

Many chance phenomena operate both to preserve and destroy woodland in urban areas. The survival of odd corners and strips cut off during the construction of railways or roads can be locally important, while a number of woods in northern towns were grubbed up for fuel during the general strike of 1926 and the following years of depression. Patches of persistent species such as wood sage, together with a few multistemmed oak set in acid grassland, are often all that mark the site of what 60 years ago were flourishing woods.

Clearance is not the only hazard facing ancient woodland in towns. The effect of inappropriate management has been dramatically illustrated by Latimer (1984) who studied Highgate Wood (29 ha) and Queens Wood (21 ha) in the London Borough of Haringey. Apart from Kenwood they are the only ancient woods within a 16 km radius of central London. Until fairly recently both were notable for their mature oaks and hornbeams, wild service trees, midland hawthorns and rich ground flora, but here similarities cease. In 1885 the Corporation of London undertook to maintain Highgate Wood as public open space since when it has lost a great deal of its plant and animal interest, while Queens Wood which since 1898 has been the responsibility of Haringey Borough still retains much of its ancient character.

The policy in Highgate Wood has been to open it up to allow unimpeded and safe walking throughout so the shrub layer and bramble areas have been repeatedly cleared, leaves and dead wood burnt, low boughs removed, potentially unsafe trees felled and a close network of tarmac paths laid. Permanent estate workers lived in the wood which now contains nine buildings. This wholesale disturbance together with that caused by trampling and the passage of machinery throughout the wood devastated the plants so that by 1915 (Fitter, 1945) most of the bluebell–wood anemone-dominated ground flora had been lost and natural regeneration was at a standstill. Since the late 1960s underplanting with exotics such as Douglas fir, western hemlock, red oak, Turkey oak and Norway maple has taken place, especially in the more open areas. Queens Wood was also opened up for public access in the 19th century but financial stringency limited management to the removal of dead wood and some drainage works which still continue. As a consequence of this lower level of management, both regeneration of the tree layer and species diversity are much better. Latimer's report makes suggestions for future management, most of which involve the cessation of current practices and the removal of introduced exotics, though disturbance in Highgate, which in the 1960s included disc harrowing and flail mowing, may have gone too far for much recovery.

This 'parkification' of ancient woodland in urban areas is not confined to London and the methods can be more insidious. In 1972 the choicest of Sheffield's ancient amenity woods – Ecclesall Woods – which extends to 117 ha of oak–bluebell communities was surveyed by consultants who made recommendations for its continued management as public open space. Their prescription, which included the clearance of unwanted debris, thinning of dense young growth, introduction on a large scale of exotic trees, drainage of wet areas, control of the weeds bramble and bracken, and the provision of car parks and picnic areas, was unpopular with the public who use the wood. After holding a public meeting the recreation department took the precaution of setting up an Amenity Woodlands Advisory Committee to help avoid future conflicts; meetings are held two or three times a year, often in the field. The committee has worked well and the next few paragraphs set out some of the problems of urban woodland management that have been identified over the last 10 years.

The introduction of non-native species always promotes discussion. A very moderate line is taken, as the prime object of management is usually general amenity, so employing small amounts of larch (*Larix decidua*) or Roble beech (*Nothofagus obliqua*) – specifically to replace dead elms – is acceptable. Only in the most natural woods is there an embargo on exotics and even here the situation is complicated by the presence of long-established hornbeam (*Carpinus betulinus*) and alder buckthorn (*Frangula alnus*) both believed to have been planted originally but now affectionately regarded by local botanists. Large old sweet chestnuts abound and are valuable for hole-nesting birds.

In some instances it has been recognized that rough grassland adjacent to a wood is important both visually and ecologically; steps are then taken to redraw the wood boundary so as to include these ecotones. Further steps then need to be taken to see that they do not get planted up.

Only occasionally is there a polarization of view in the committee and then it is between members with a forestry training who view woods as requiring substantial management and the ecologists/naturalists who believe that amenity woodland should be largely self-maintaining. Foresters commonly speak of the problems of old age, the dangers of insects and fungal diseases, and persist in referring to trees as 'the crop.'

They believe that no trees should be allowed to grow till they fall naturally and have suggested a rotation for amenity woodland about a third longer than for a commercial crop, viz., beech, ash and elm 100–120 years, oak 120–150 years. They particularly favour the taking out of decrepit trees and those in decline. Much is made of a point that natural regeneration often fails because it does not get the care and maintenance essential for survival. The neat, tidy and predictable woodland which results from this approach appeals to the local authority members because they have a legal responsibility to keep the woods safe for the public and are liable to be sued if an accident occurs.

Ecologically trained members see woodland differently, particularly the larger blocks such as Ecclesall Woods. Being aware that the degenerate stages of woodland, such as large old trees and glades, are extremely rare nationally they would prefer to see parts of the woodland grow through to senility and decay. If the long view is taken, woods are self-maintaining, though regeneration is often sporadic and follows catastrophies such as the Sheffield gale of February 1962, which produced so many gaps, now filled with 10 m high regeneration cores, that the future of most of Sheffield's woods is assured for the next 200–300 years. The key to the ecologists' approach is to let natural processes shape the wood.

The artificial thinning of regeneration cores to leave a few main stems is particularly contentious. In the view of ecologists these should be left to self thin to the trees most suited to the site. A recent survey of an unmanaged compartment revealed that in small gaps rowan becomes dominant, larger ones are dominated by birch with an oak understory, the presence of sweet chestnut saplings is related to available seed source, and sycamore and willows were only present on the site of old bonfires. To thin all cores to favour oak reduces complexity and naturalness.

The responsibility of the local authority over safety is difficult to get round despite the fact that no member of the public has been injured in Sheffield's woods. Along tracks, footpaths and in areas where public use is greatest more intensive management may be appropriate in the interests of safety. The main tool offered by ecologists is a zoning concept whereby remoter parts of a wood are zoned as wildlife areas and scheduled to receive a minimum of management. An appearance of naturalness is itself an amenity.

Most ancient woods in urban areas have experienced some 19th and 20th century invasion by or planting of exotic trees such as larch, scots pine, sweet chestnut, beech (in the north) and rhododendron, but contain a preponderance of locally native species. The shrub and ground layers tend to be more resistant to invasion by exotics. Close to gardens bird sown-seeds of cherry laurel (*Prunus laurocerasus*), Oregon grape

(*Mahonia aquifolium*) and holly cultivars may occasionally convert to mature shrubs, while shade-loving species originating from garden refuse, e.g. periwinkle, London pride (*Saxifraga* × *urbium*) and rose of Sharon (*Hypericum calycinum*) may form large patches.

One of the few systematic studies of the ingress of exotics into the ground flora of urban woods is by Moran (1984) who examined the influence of adjacent land use on understory vegetation in New York. Forest stands were chosen where the margin was bounded by a residential area, a road or agriculture. Residential edges exhibited greater species richness, cover, number of herbaceous dicotyledons, number of introduced species and number of annuals. This was interpreted as being due to species exchange with the adjacent gardens through tipping of refuse, dispersal on footwear and possibly the activities of pets. Higher light levels along residential edges may also be important. Thirty metres in from the edge the effects were hardly significant with only *Taraxacum officinale* and *Hieracium aurantiacum* showing an enhanced presence. Typical residential edge species were introduced taxa such as *Chelidonium majus, Glechoma hederacea, Hypericum perforatum, Plantago major* and native *Solidago* spp.

16.3 PLANTATIONS ON THE SITE OF ANCIENT WOODLAND

The flora and fauna of plantations has a well-deserved reputation for being monotonous even when they occupy the site of ancient woodland. This is largely due to heavy shade cast by the even-aged canopy trees which often include a high proportion of beech, sycamore or conifers. The total list from such a wood can be extensive but most species are represented by very small populations in rides or at the extreme edge. In Sheffield, opportunities have arisen for experimental management in Bowden Housesteads which is a mixed highly urban dense even-aged plantation (1898) of oak, beech, sycamore and sweet chestnut on the site of what was until the late 1870s a fine 'spring wood' (Fig. 16.3). In 1979 a 0.5 ha glade was cut in a relatively uniform area to enhance habitat diversity. A great increase in the density of bluebells occurred (Fig. 16.4), though the species is now declining as regeneration grows up. Species from the seed bank, such as bramble, figwort (*Scrophularia nodosa*), St John's-wort (*Hypericum pulchrum*), toadrush, soft rush and woodsedge (*Carex sylvatica*), have appeared while others have blown in, for example goat willow, rosebay willowherb, thistles and coltsfoot. Foxgloves deliberately introduced as seed have established well. Already the flowers are attracting meadow brown and orange tip butterflies deep into the wood.

Though originally created for wildlife the public look on the new glade as a meeting place and children of all ages play there. This has

Fig. 16.3. A glade cut in this heavily used plantation, on the site of an ancient wood, resulted in the appearance of a luxuriant ground layer including bluebells (see. Fig. 16.4).

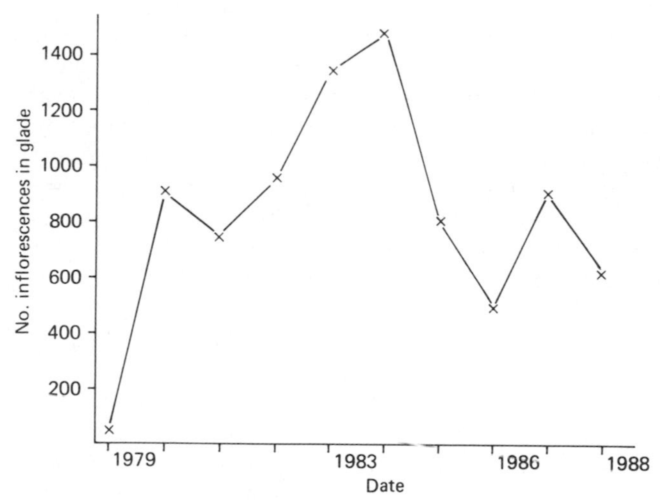

Fig. 16.4. The development of a bluebell (*Hyacinthoides non-scripta*) population following the creation of a 0.5 ha glade in an urban plantation on the site of ancient woodland. After 7 years competition is starting to reduce the density of bluebells.

necessitated a complete reversal of thinking; instead of regarding the glade as a nature conservation area, it is more realistic to view it as an activity glade where children play 'cowboys and indians' around sprouting stools, light fires, sit on logs and generally romp about. Adults also make a habit of visiting the glade so a new path network is developing. The effect of this activity on the botany has not been excessive; dense tree regeneration is occurring and species previously suppressed by the low light intensity such as honeysuckle have started to flower. Glade creation offers a method of increasing the amenity value of ancient woods which have been converted to plantations.

16.4 SPONTANEOUS RECENT WOODLAND

Secondary woodland that has developed from natural regeneration on bare ground over the last few centuries or so is not particularly common in towns where new woodland, if required, is usually planted. The best examples occur on neglected sites such as abandoned allotments, railway land, old cemeteries (Fig. 13.4), strips beside rivers, and on ground found to be too steep or wet for any positive land use once farming ceased. The characteristics and successional relationships of the woodland occurring in these habitats is considered in the relevant chapters, so general principles will be considered by examining a small (0.5 ha) even-aged wood that has colonized a disused railway yard.

Over the 35 years since the site was abandoned, a closed canopy woodland dominated by goat willow (*Salix caprea*) has grown up. This species is the pioneer woody colonizer of most urban sites in South Yorkshire. Within the wood, in which it is possible to move about quite freely under the 7 m-high canopy, saplings of sycamore and ash are starting to grow up and will one day take over as dominants while young elder and hawthorn will probably remain subordinate and non-flowering except where there is side light. Ash, sycamore and elder form a series with an increasing ability to survive and grow in deep shade, their distribution and performance within the young wood illustrates this.

The ground flora contains two elements. The first are shade-bearing weeds which have persisted from the early 'open ground' or 'meadow stage' of the succession. They include the 20 species given in Table 16.2, most of which are not woodland plants; their populations are assumed to be slowly declining. Some, such as *Senecio squalidus* and *Epilobium ciliatum*, probably rely for their persistence on reintroduction from outside as they flower and set seed only poorly at low light intensities. The other smaller group (Table 16.2) are fast-colonizing woodland herbs. Apart from the male fern and broad buckler-fern which have wind-dispersed spores, and ivy with its berries, the others possess no obvious mechanisms that fit them for travelling rapidly about urban areas. These lists are quite similar to those given by Peterken (1981) for secondary woodland in rural Lincolnshire, though, as might be expected, his maturer woods contained more truly woodland species. It must be a matter of speculation as to how long it would take twayblade, wood sanicle and dewberry (*Rubus caesius*) to colonize this type of woodland, but I have seen 80% of his list in recently created urban sites. This suggests that woods separated by large expanses of arable land may have similarities to ones isolated in a sea of housing; the ground flora of both have links with a hedgerow flora. The town facies is distinctive due to its regular complement of aliens, e.g. Oxford ragwort, Michaelmas daisy and a high proportion of nitrophiles.

The development of the ground flora of secondary urban woods is a continuous process, as it is unlikely to be left undisturbed sufficiently long for any kind of stability to be attained. In this 35-year-old wood shade-bearing weeds are gradually being replaced by woodland species with expanding populations, and, as sycamore and ash take over from goat willow as the main canopy formers, further changes can be expected. Species often show a patchy distribution around the original site of colonization which imposes a low species diversity on individual sample plots, even though a list from the entire wood might be quite long. This contrasts with the ground vegetation of ancient woods which tend to be an intimate mixture of species, the distribution of which is controlled rather precisely by soil factors, past management and the light climate.

Many people imagine that spontaneous secondary woodland in urban areas is dominated by sycamore. This is not so; though stands of dense young sycamores can be found, there is not much evidence that they convert to sycamore woods, most mature examples of which were originally planted. In its native haunts in the mountains of central and southern Europe the species occupies moist loamy hollows and narrow valley floors; it will not tolerate waterlogging or grow on shallow limestone soils. Continental writers agree that it requires a deep moist freely drained soil of relatively high base status. These demands tend to confine the habitats where it naturalizes readily in urban areas to places where humus is decaying rapidly, the soil is fertile and there is a certain level of disturbance; it is to some extent a nitrophile.

In woodlands within its native area, which approaches as close as north-east France and southern Germany, the species rarely attains more than subdominance and is very often associated with ash. Jones (1945), who has monographed sycamore, reported that in Britain unplanted mature stands are possibly never found, but in the pole stage small groups are not infrequent, since it is very quick growing when young.

As a wood ages the proportion of sycamore decreases and it gets suppressed. Jones was referring to sycamore in the countryside but the same is largely true of towns. The tree is a pioneer species with a somewhat weedy character, springing up readily in odd corners, on walls, waste ground, recently turned earth and the like, but few of the myriads of seedlings convert to saplings and even fewer of the saplings turn into mature trees. An exception is where there is a lack of seed parent trees of other species, if there are no other trees around, it can form large pure stands but even these are intolerant of competition from grass.

Having pointed out the rarity of spontaneously regenerated sycamore woods, it must be acknowledged that the sycamore is so well established that nothing can stop it taking its place as an 'honorary native' in fertile valley woodland where mixed stands of sycamore and ash are likely to occupy the niche recently left vacant by elm. The unpopularity of the species with some conservationists is largely misplaced. The physical effects of the moderately persistent leaf litter are not great; only bryophytes and certain grasses such as rough-stalked meadow-grass (*Poa trivialis*) are unable to maintain themselves against it. The spear shoots of bluebell and ramsons pierce through, yellow archangel crawls out, the 'penknife' shoots of wood anemone and dog's mercury insinuate, creeping soft-grass (*Holcus mollis*) survives with the help of its rhizome store, while other species such as celandine and violet (*Viola riviniana*) manage to persist through petiole elongation. Tree seedlings are also good at coping with the litter which tends to aid their survival (Sydes and Grime, 1981 a,b). In addition to exerting a physical effect there is some evidence that, during decay, litter releases toxic components capable of inhibiting seed germination and/or plant growth. After a series of investigations Sydes and Grime (1981b) concluded that sycamore litter, which combines moderate toxicity with an intermediate degree of persistence, exerts its main influence through physical rather than chemical causes. An example of the smothering effect of sycamore litter, which can be observed in the field, is the way it controls the distribution of the woodland moss *Mnium hornum* by restricting it to tree stumps, fallen logs, tree bases, the tops of hummocks and other sites where litter gets blown away.

Another criticism levelled at sycamore is that, being an introduced species, it has only a low number of associated insect species; but with 43, twenty of which are lepidoptera, it is comparable with the native field maple (*Acer campestris*) which has 51 insect spp., 24 being lepidoptera. If insect biomass is taken into account, sycamore, with its very high density of the sycamore aphid, compares favourably with most native trees. The aphids form a step in an important food chain (Fig. 9.6) which includes the carnivorous larvae of hoverflies, lacewings and ladybirds, and large flocks of foraging blue-tits in the autumn. Additionally the neutral bark of old sycamores supports the same range of epiphytic lichens and bryophytes as native elm and ash trees.

Only in urban woods of SSSI status is the presence of sycamore likely to conflict with objectives of management, and, even there, its control can be counter-productive as the tree thrives on disturbance. Not only does it quickly seed up gaps, but 'stagnating' seedlings, kept in check by low light intensities, are likely to be present on the floor of the wood; these respond quickly to any clearance. After creating a gap in the canopy of fertile woodland containing sycamore the only way to keep it free from the species is hand-pulling each year. It should be recalled that seed flights of up to 4 km have been recorded (Jones, 1945).

It is natural for urban woods to contain urban species; to try to maintain them as examples of native high forest is contrary to the ecology of communities which receive a substantial seed rain of exotics. As their populations increase, many alien species such as Norway maple (*Acer platanoides*), holm oak (*Quercus ilex*), Turkey oak (*Q. cerris*), horse chestnut, acacia (*Robinia pseudoacacia*), larch, Norway spruce and several pines are likely to build up considerable spontaneously regenerating populations in our woods in the way sycamore and sweet chestnut have in the past. Nicholson and Hare (1986) have labelled the unquestioning removal of sycamore the 'John Wayne syndrome'; cut down first, ask questions later. Sheffield City Council recently upset an entire neighbourhood by removing sycamore from a popular suburban wood. Ironically disturbance of the wood floor during extraction has probably guaranteed the long-term survival of this species on that site.

16.5 PLANTATIONS

18th and 19th century

Plantations from the 18th and 19th century, now forming fine mature woods, are often present in well-defined areas within a town; they mark the site of former estates. Originally planted for shelter, screening, boundary definition and sometimes to encourage game, they are today poor botanically but often good for birds. The canopy-forming trees include mixtures of beech, sycamore, pine, larch, lime and horse chestnut with a scattering of more exotic species. This suggests that timber return and ornament were also objectives of planting. Below the canopy there is usually an understory in which evergreens such as holly, yew and rhododendron are prominent, mixed with hawthorn, elder, snowberry and patches of saplings. These may be either planted or self sown. The ground flora comprises those plants typical of secondary woodland (Table 16.2) supplemented by plantings of persistent shade-tolerant species of horticultural origin. What distinguishes urban estate planting from that in the countryside is that over the decades the former have had their ground flora augmented by species such as daffodil, snowdrop, winter aconite, lily of the valley, Spannish bluebell, lungwort (*Pulmonaria officinalis*), London pride, sweet violet (*Viola odorata*), stinking iris (*Iris foetidissima*), hellebores and blue-eyed Mary (*Omphalodes verna*), as attempts were made to transform them into wild Victorian gardens. Woody species employed to this end include butcher's broom, periwinkle, ivy cultivars, *Hypericum* spp. and *Mahonia* spp. The thick cover and lack of disturbance encourages a wide variety of birds and enables them to breed successfully.

Naturalistic plantations

Over the last few years a type of woodland planting called 'naturalistic' has been tried in many towns. Taking its philosophy originally from the Dutch, the idea is to use mostly indigenous species mixes and employ management techniques that encourage ecological diversity. Examples over 10 years old can be seen in Warrington New Town (Tregay, 1985) where belts and clumps composed of ash, cherry (*Prunus avium*) and oak, with some rowan and field maple and occasional pine are being established, nursed by birch and alder. Understory shrubs are represented by hazel, holly and elder, while a narrow, variable and discontinuous edge mix of dog rose, bramble and gorse is employed.

A diverse natural character is achieved not by leaving all to nature but through intensive and demanding management involving herbicides, slow release fertilizers, seeding, mowing, raking, thinning, trimming, strimming, pruning, coppicing and, after 5 years, the planting of hand-raised ground flora species such as primrose, red campion, celandine and bluebell. Edge treatments designed to foster an ecotone are particularly intensive. The resulting nature-like communities are an impressive example of what can be achieved by gardening with wild species (Fig. 16·5) but have departed in concept from the Dutch philosophy.

The Dutch do not take a totally purist approach towards plant species selection; so long as the planting remains random and informal and the total effect created is natural they are satisfied. If no native species are available to fulfil a specific requirement demanded by site use, then an exotic is chosen. This frequently happens with regard to evergreen species and certain rose cultivars produce a more dense free-flowering bush which is less susceptible to vandalism than native ones. This technique recognizes the principle that urban planting is nearly always dual purpose and has an amenity role to play well beyond the fostering of wildlife. Aping the countryside is rarely a valid objective. The influx

Fig. 16.5. Wood anemone (*Anemone nemorosa*) introduced into secondary woodland at Highgate Cemetery.

of animal life into these naturalistic plantations has yet to be studied in detail.

Kennedy and Southwood's (1984) estimates of the number of phytophagous insects and mites associated with 28 British tree species or genera are often quoted as one of the chief justifications for planting native species. Their work shows how the number of herbivorous insect species associated with a tree is determined largely by the abundance of the tree and the length of time it has been present in Britain. As a result of these two factors, introduced trees, particularly if they are not common, are likely to support a poorer insect fauna than widespread native trees. Of this there is no doubt, though there are exceptions such as the unexpected richness of Norway spruce. Where uncertainty creeps in is over the appropriateness of transferring data obtained from studying high-quality natural vegetation to the urban scene. For example, how many of the 141 alder insects are likely to be found in a belt of alders beside an urban road when most of the sites investigated were riverain woodland? Similarly of the 98 beech insects, how many occur only within its native area in south-east England? These questions can be answered by research, but it has not yet been undertaken.

16.6 BIRDS

The typical small urban wood linked to public open space consists of a group of mature trees standing over a grassy fieldlayer; it is particularly unfavourable to bird life. The only species likely to be found there are woodpigeon, magpie and carrion crow which exploit the high canopy and do not require cover. If nesting holes are present, starlings, which actually prefer houses, may also occur. If a shrub layer and freedom from unacceptable levels of disturbance are introduced, the extra cover brings in a range of songbirds such as robin, blue-tit, great-tit, missel thrush, song thrush, blackbird, wren, dunnock and chaffinch which can breed successfully in quite tiny woods (0.1 ha) provided the structure is right. If sited near gardens, greenfinch and long-tailed tits may be present. The presence of a few trees approaching senility and decay may bring in the occasional tawny owl and possibly a pair of great spotted woodpeckers. The presence of conifers can add considerably to the variety of birdlife by favouring goldcrest and coal-tit. A scrub margin may bring in blackcap, willow warbler and chiffchaff.

A disturbing feature of well-structured small woods in public places is that despite their attractiveness to birds, breeding success is very low. In the corner of a Sheffield park which is managed particularly for the benefit of nesting birds, eggs usually disappear within a week of being laid; the respective roles of boys, squirrels and magpies in this has yet to be determined. The net result is that over 4 years only one young blackbird and one juvenile thrush were seen in this 2 ha plantation, despite something like a maximum of 13 song thrush nests and 21 blackbird nests being built each season. Small species nesting in closer cover such as dunnock, chaffinch, robin and wren have been more successful (Miller, 1981).

Classic native deciduous woodland of the type only present in towns as encapsulated countryside may support in addition breeding populations of tree-creeper, hawfinch, nuthatch, jay, wood warbler, woodcock, spotted flycatcher, sparrow hawk and greater spotted and lesser spotted woodpeckers. Out of the breeding season these may forage more widely and be seen by a considerable number of people. The central 14 ha of one such wood in Sheffield has been managed as a bird sanctuary for the last 40 years with access restricted and little or no woodland management undertaken. An unexpected result of this is that it has grown rather uniform, so the sanctuary contains fewer species, often present at a lower density, than the surrounding public, more heavily managed, parts. Other results of regular recording have shown

Table 16.3. The food of tawny owls in London

	Holland Park	*Hamp-stead Heath*	*Morden*	*Rich-mond Park*	*Esher*	*Book-ham Common*
Distance from city centre (km)	6.5	8	16	16	24	32
No. of mammal species	3	9	4	6	7	11
Mammals (%)	7	48	55	64	34	90
Birds (%)	93	36	45	33	48	10
Frogs/fish (%)	–	16	0	3	18	–

Data from Beven (1982).

that nest boxes get badly vandalized, so these are no longer provided on the grounds that it is wrong to attract birds into them when there is little chance of successful breeding compared with the outcome of using natural holes. Surveys also revealed that areas where beech was dominant or codominant were particularly unfavourable for breeding birds when compared to areas with a fairly well-developed shrub layer.

A great deal of information exists on the distribution of woodland birds in urban areas but far fewer investigations have been made of their ecology. The insight such work can provide is demonstrated by studies that have been carried out on the food of tawny owls (*Strix aluco*) in London (Beven, 1965, 1982). Pellets were examined from six sites ranging from the central area (Holland Park) to the outer suburbs (Bookham Common). The results showed that at the central site where there are very few mammals the owls feed largely on birds (93%) which also make up a good part of their diet in other open spaces within the city (Table 16.3). The proportions are reversed at the city outskirts where mammals usually form more than 50% (up to 90%) of their diet. A site-specific element heavily overlies the results, with owls feeding extensively on rabbits and dung beetles at Richmond Park, while frogs form a significant part of the diet at Hampstead Heath and Esher where eight goldfish were also taken. Grey squirrels were rarely eaten.

Following a major study of breeding-bird communities in 32 urban woods in Massachusetts, Tilghman (1987) identified the following characteristics of the habitat as affecting their success. The size of the woodland explained 79% of the variation in total species richness. Half the species were more commonly found in the larger woods (43–69 ha) while only eight were more abundant in the smallest ones (1–5 ha); these included the American robin, northern mocking bird, warbling vireo and house sparrow. Other significant factors were the density of buildings in adjacent areas (negative) and the density of shrubs within the wood (positive), while close proximity to trails and the presence of taller trees in the canopy both resulted in a lower diversity. Woods with scattered patches of conifers tended to have a greater number of birds. These results support the idea that 'habitat islands' in an urban setting follow some of the basic concepts of island biogeography (MacArthur and Wilson, 1963, 1967), especially the species–area relationship.

The impact of recreation (disturbance) on the density of breeding birds in urban woodland has been studied in Holland by Van der Zande *et al.* (1984). Of the 13 species studied in detail, eight were found to have a density negatively correlated with recreation intensity. In increasing order of sensitivity they are chaffinch, blackcap, woodpigeon, chiff-chaff, *Turdus philomelos*, willow warbler, garden warbler and, shyest of all, turtle dove. The order of susceptibility differed slightly in coniferous woodland due to cover there being predominantly in the tree tops rather than a shrub layer. The recreation effect was thought to be determined more by weekday activity rather than weekend recreation intensity. The great-tit showed a positive correlation with recreation because the location of most territories was determined by the presence of nesting boxes which are usually hung near houses.

Chapter 17

LIVING WITH WILDLIFE

Nature is often untidy and many people brought up to expect a well-kept city landscape find it difficult to accept the informal appearance of thriving wildlife habitats in a town. In fact, so culture bound are our aesthetic feelings that even the most ardent urban ecologists occasionally find such sites jarring to their sensibilities. So at best our attitude to urban wildlife sites as visual landscape is ambivalent. From early May till the end of October, they are colourful places alive with flowers, insects and the chirping of grasshoppers; in this condition they provide a useful foil to more formal areas of landscape. For the rest of the year, however, dead plant material is one of their chief characteristics.

Scruffiness is not the only source of criticism levelled at urban wildlife sites. They also tend to attract tipping on a scale varying from the occasional boot load of garden refuse, through domestic rubbish such as furniture, carpets and mattresses which are too large for the dustbin, to regular lorry loads of builders' rubble or retail debris. This occasionally gets so bad that the margins of isolated sites become a health hazard, with populations of rats visible; the local authority then feels obliged to clean them up. A further cause of alarm is the belief that undesirables such as alcoholics, glue sniffers, petty criminals and muggers are attracted to these sites. This is unwarranted, as such people mostly prefer unsuccessful, therefore underused, tended open space of which there is plenty.

The most powerful social argument against ecological landscapes in towns is that they do not fit in with the expectations of the majority of citizens who are ignorant of their objectives. Low-level management is mistaken for neglect, the sites are regarded as a disgrace rather than an amenity, and people feel threatened by change. If they are to win support, ecologists need to be quite clear about the conflicts involved in their approach and of how to counter them. A further, often overlooked, factor is that parks department staff have not been trained to manage sites for wildlife, so difficulties may arise due to their inexperience. It has been found for example that if a special mowing regime is proposed for an area of grass so that it falls outside the regular programme of work, it is likely to be left uncut all year.

17.1 THE AESTHETIC CONFLICT

When attempting to persuade a local authority or other landowner to let natural succession take its course on a patch of urban land so that it can become a haven for wildlife, it is helpful to have data on the phenology of the species present. Fig. 17.1 shows the flowering periods of

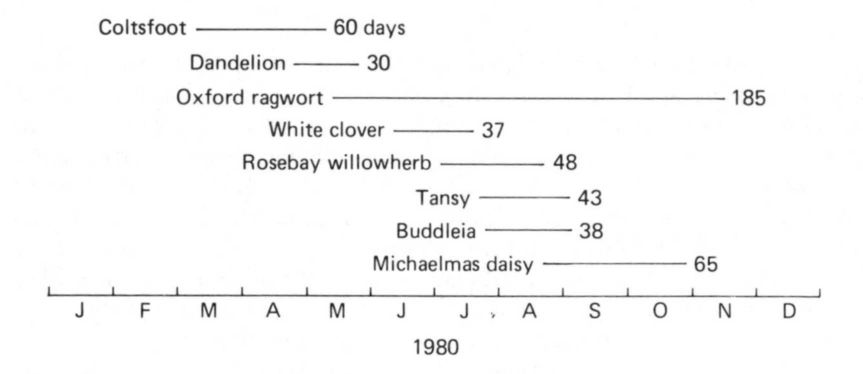

Fig. 17.1. Chart illustrating the main flowering periods of the dominant species on an urban common in Sheffield.

dominants on an urban common in Sheffield. It demonstrates that plants with attractive flowers are in bloom from early March till late autumn, reaching a peak in the July–August period. As a number of the species are garden escapes with particularly spectacular blooms, the description flowery meadows is appropriate for the community.

While information on flowering periods is useful, even more valuable is a knowledge of total phenology. Fig. 17.2 gives complete phenological information on a range of common urban species. Many turn sites green with clumps of dense leafy shoots well before the flowers come out, while others like rosebay willowherb and Japanese knotweed display striking autumn colours. The data also highlights a visual disadvantage of many tall herbs common on urban sites: their dead

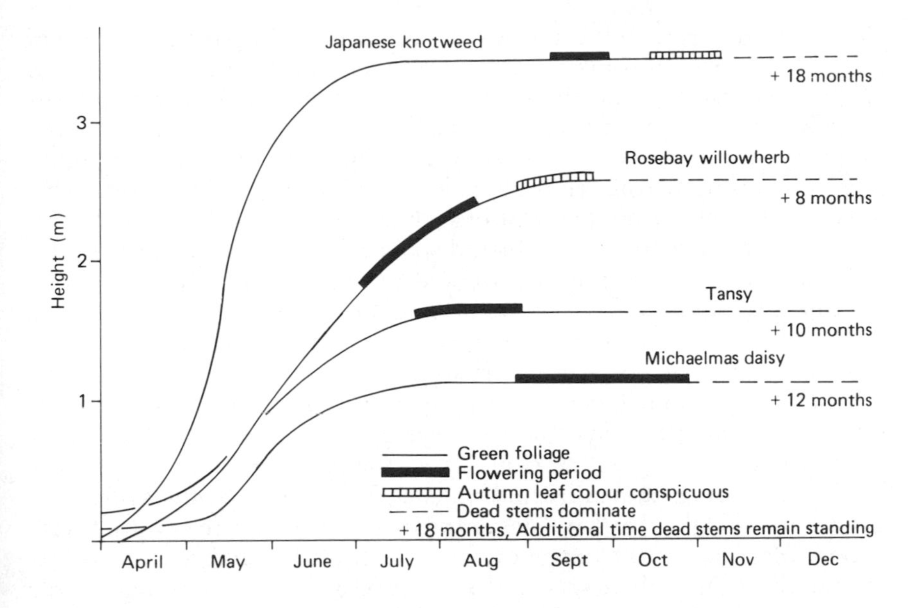

Fig. 17.2. Phenological data collected from individual plants of a range of common urban species; Sheffield 1980. ———, green foliage; ▬▬, flowering period; □□□, autumn leaf colour conspicuous; – – –, dead stems dominate; + x months, additional time dead stems remain standing.

stems persist for over a year before falling, so they dominate the vegetation from late autumn through till the early summer flush of new growth. Why the stems should be so persistent, 2½ years in the case of *Reynontria*, is not known, though it has been suggested that in their natural habitats they may help protect the rootstock against grazing or provide shelter from the wind.

Observing the appearance of 'natural sites' in towns during the winter has confirmed that the dead stems of grasses and herbs are very conspicuous, though their starkness is reduced by the presence of winter green species such as Yorkshire-fog, Oxford ragwort, meadow grasses (*Poa* spp.), clover and moss. The silver foliage of wormwood can also be important at this time of year. The aversion of most town dwellers to dead plant material is partly due to the preference that landscape architects have for specifying evergreen material and partly due to the energies of gardeners who spend much of the autumn removing it from sites in their care. Thus people are largely conditioned to equate the annual cycle of growth and decay with neglect. It appears acceptable in a few situations, such as railway embankments which have never been managed with aesthetic objectives but as soon as management is relaxed along a road verge, in a cemetery or the corner of a park the criticism starts. This is clearly a matter of expectations and these can be changed.

The disposal of rubbish on to urban wildlife sites occurs almost entirely at the margins so can be solved by appropriate edge treatments. In many northern towns the standard edge treatment is a roughly formed 1 m-high bank of rubble designed to keep gypsies off. This looks untidy and encourages tipping while the alternative of a ditch is an even greater invitation for rubbish disposal. A row of kerbstones set on end in concrete is not much of a deterrent either; a more formal treatment is required. Two of the best I have seen are a timber-rail fence and a split chestnut paling supported on frequent posts; both are easily erected by volunteers. An alternative, now adopted as routine by Sheffield City Council, is to flatten any gypsy bank, topsoil a 3 m-wide strip around the site, sow this with rye-grass, then mow it five or six times a year. This gives the site a cared-for appearance, ties it into the local surroundings and almost eliminates fly tipping. At the same time any informal paths that have developed across the site are surfaced with a loose material, such as shale or ash, which plants can grow in so that the width of the path adapts to the use it receives. Encouraging people on to sites like this helps with acceptability and promotes self-policing.

17.2 SOCIAL CONSIDERATIONS

The fostering of wildlife areas in cities is too complex an operation to be left in the hands of ecologists. In order to gain maximum benefit from the concept, or sometimes just acceptance, it is necessary to know how people use their environment and what they want from it. This is a young but vital field of study. For example, social scientists and environmental psychologists are now beginning to discover what form the often quoted 'popular demand for nature in cities' takes and just how difficult it is to provide for.

Work to date has shown that aspirations differ greatly according to income group, age, sex and cultural background, with women, ethnic minorities and low-income groups showing a particular reluctance to use areas of more informal landscape. Work on Merseyside by P. Fitzpatrick has given a glimpse of how this might be overcome with the discovery that the use public open space receives correlates strongly with the level of management the vegetation is given; length of surfaced footpath and number of access points also play positive, but less important roles. The middle classes appear to appreciate aesthetic aspects of public open space more than the lower-income groups who look on it mainly as background for enjoying the company of others – a social rather than an environmental experience. Women on their own are apt to fear natural sites, though this applies less to those accompanied

Table 17.1. Oxleas Wood, London; feelings expressed during group discussions with users of this urban wood

Positive associations		
General comments		
Nice (38)	Favourite landscape (6)	Good (2)
Like it (37)	Pretty (5)	Very attractive (2)
Lovely (20)	Happy there (3)	Love woods (2)
Beautiful (14)	Gorgeous (3)	Fantastic (2)
Feel at home (8)	Looks good (3)	Relaxing (2)
Naturalness		
Natural (16)	Alive (4)	Like it in autumn (3)
Green (14)	Wild (4)	Back to nature (3)
Nice smells (7)	Cool in summer (4)	Fresh air (2)
Variety		
Interesting (13)	Adventurous (3)	Colourful (2)
Lots to see (8)	Variety (2)	
Solitude/escape		
Peaceful (16)	Quiet (12)	Tranquil (3)
Romantic (2)		
Negative associations		
General comments	Don't like woods	
Naturalness	I would prefer a path	
Variety	(No negative comments)	
Fear/danger		
Not safe on your own (12)		Fear of getting lost (5)
Fear because woods are frightening (9)		Dangerous (2)
Too dark (9)		
Rapists/male strangers/dirty old men (3)		Wouldn't go there (2)

Information provided by C. Harrison and J. Burgess (unpublished).

by a dog, while Asian women may have a dread of snakes or racial attack.

Having a few key well-managed sites in a city – urban wildlife on show – will never fulfil the demand, though they can prove useful in providing examples of what an ecological site looks like and thus allay fears. Many people want natural sites close to their homes to experience on an everyday basis. Small areas are seen as best because large expanses can be threatening. The pleasures derived include colour, the unexpected event, contact with birds and animals, quietness, collecting and the variety. As a bonus, social interaction appears to be enhanced by the informal countryside appearance. Frequently expressed points of dissatisfaction which provide a challenge for managers include too overgrown, untidy, messy, too rough, not safe on your own, dangerous. The very wide range of feelings expressed during group discussions about a wood in London is shown in Table 17.1, from which it can be seen that an appreciation of its naturalness and variety was tempered by images of fear and danger.

17.3 DESIGN AND MANAGEMENT

Ecological areas need to be designed into their setting. Some of the main problems are achieving the correct balance between formal and informal landscape treatments for a given locality, manipulating space using structure planting, and designing the interface with adjacent land uses. Frequently, areas rich in wildlife already exist in a locality and it is the job of a landscape architect to obtain maximum benefit from these through appropriate site planning whether a road verge, cemetery, urban common, neglected garden, disused canal or some odd corner is involved.

Close to buildings, access points and main circulation routes, it is most suitable to design an area transitional to more formal heavily maintained landscape. For example, public open space adjacent to a housing development might be roughly zoned into ornamental horticulture, mown grass and trees, naturalistic planting with a variety of mowing regimes, and furthest away from the buildings a wildlife zone with minimum maintenance. Within such a framework a skilled designer should be able to accommodate a wide range of existing and/or newly created wildlife habitats. Apart from manipulating the trees and shrubs, preferably to include a sheltered sunny south-facing ecotone, the wildlife zone is often best left for natural succession to take its course. The local seed rain and the history of the site, working through their influence on environmental conditions, will together provide something unique, robust and totally appropriate to the locality. If required the field layer can be manipulated by seeding into disturbed open ground or introducing pot-grown material. This might be permissible on an educational site or a centrally located 'showpiece' plot but as a general principle the less obvious the hand of man the more satisfactory the site is liable to become; nature is the best provider of incident, variety and local character.

With regard to management the first question to ask is whether any is necessary. Non-intervention is a positive option and will almost certainly be the appropriate one over a large part of any site. Most sites will, however, benefit from a small management input which should be sufficiently simple to be carried out by unsupervised contract labour. It is unreasonable to expect that groups like the British Trust for Conservation Volunteers or local parties working at weekends can ever be responsible for maintaining more than a small fraction of wildlife sites. Jobs such as mowing the margins and fence repair, footpath surfacing, the removal of tipped material and litter that has blown in, and especially keeping access points tidy make the biggest impact. Locally, cutting down the dead stems in tall-herb communities may be appreciated, though this should not be carried out on a large scale as they form valuable overwintering sites for many organisms. On one site where, with the best of intentions, a flail mower was used for this job it also cut down many spontaneously regenerating young saplings putting the plant succession back several years.

17.4 BENEFITS FOR ALL

To forestall the accusation that a minority group is imposing a new pattern of urban landscape on a reluctant population, some of the advantages of managing areas in towns for wildlife will be recalled. Firstly they provide a home or feeding ground for a wide range of plants and animals which many people find attractive. Interest in wildlife is maintained at a high level by television programmes, newspaper articles, popular journals and a very wide range of books so that easy access to nature is frequently regarded by citizens as important as access to parks, leisure centres and other amenities. Owing to popular demand, local authorities have a responsibility to promote and provide for wildlife throughout their area. Set-piece ecological parks are useful for pointing the way but will never fulfil the requirement.

Part of the attractiveness of 'natural sites' is their informality and dynamic nature (Fig. 17.3). Every site is different not only from its neighbours but from season to season, and also, as succession proceeds, from year to year. Flexibility on this scale is unusual in urban areas where gardeners spend much of their time fossilizing the landscape. With the horticultural approach sites tend to look rather similar while the ecological style emphasizes differences and provides local character.

Fig. 17.3. Part of the attractiveness of natural sites is their informality and dynamic nature.

As this is a feature that is disappearing fast, its restoration through favouring spontaneous vegetation is appreciated.

Urban nature areas should rarely be looked on as other than public open space as they are dual purpose. Zoning schemes and the manipulation of circulation may be employed to control use but it is as unsatisfactory to have areas in busy towns set aside exclusively for wildlife as it is to have landscapes that cater for people but preclude wildlife. The vegetation will be robust as it is a result of the environment selecting from the available propagules only those species that will grow there. It is consequently capable of self-maintenance and adapting to change so many wildlife sites can double up as recreation areas. There is some evidence that such multipurpose sites help divert vandalism from tended open space as they allow complete involvement with the environment (Gilbert, 1981).

At a pragmatic level, being low cost to create and maintain, ecologically based landscapes appeal increasingly to land owners who have to manage on shrinking budgets. Though often arising accidentally through neglect, they come to be appreciated on a much wider basis and do not attract the expected adverse comment. An example of this occurred in the early 1970s when a directive was issued by the Ministry of Transport to cease the regular mowing of motorway verges. This primarily financial decision produced what are now highly valued tall grasslands and brought kestrels to the notice of everyone.

The ecological approach adds a further option to the choices available to landscape architects. It is a particularly suitable method of keeping land in store for an unpredictable future and providing areas of complexity in the urban scene. It also caters for a neglected age group (children). The concept has similarities to the philosophy behind the London commons (Hampstead Heath, Wimbledon) particularly with regard to low management input and a natural appearance. However, the most satisfying examples are not relic countryside, nor do they attempt to mimic it; they are composed of quite different species, ones that have been selected by the urban environment to produce resilient communities of plants and animals typical only of towns. It is our destiny to live side by side in mutual tolerance with synanthropic animals and introduced plants.

REFERENCES

Allen, D.E. (1965) The flora of Hyde Park and Kensington Gardens, 1958–1962. *Proc. bot. Soc. Brit. Isl.,* **6,** 1–20.

Andre, H.M. and Lebrun, P. (1982) Effects of air pollution on corticolous microarthropods in the urban district of Charleroi, Belgium. In *Urban Ecology: 2nd European Ecological Symposium* (eds R. Bornkamm, J.A. Lee and M.R.D. Seward), Blackwell Scientific Publications, Oxford and Edinburgh, pp. 191–207.

Appleton, J. (1975) *The Experience of Landscape*, John Wiley, Chichester.

Atkins, D.P., Trueman, I.C., Clark, C.B. and Bradshaw, A.D. (1982) The evolution of lead tolerance by *Festuca rubra* on a motorway verge. *Environ. Pollut. (Ser. A),* **27,** 233–41.

Avery, B.W. (1980) *Soil Classification for England and Wales. (Higher Categories),* Soil Survey Technical Monograph No. 14., Harpenden.

Bangerter, E.B. and Kent, D.H. (1957) *Veronica filiformis* Sm. in the British Isles. *Proc. bot. Soc. Brit. Isl.,* **2,** 197–217.

Barkman, J.J. (1958) *Phytosociology and Ecology of Cryptogamic Epiphytes,* Van Gorcum, Assen, Netherlands (reissued 1969).

Barnes, H.F. (1949) The slugs in our gardens. *New Biol.*, **6,** 29–49.

Bateson, S.H. and Dripps, J.S. (1972) Long-term survival of cockroaches out of doors. *Environ. Hlth.*, **80,** 340–1.

Batten, L.A. (1972) Breeding bird species diversity in relation to increasing urbanization. *Bird Study,* **19,** 157–66.

Beebee, T. (1979) Habitats of the British amphibians: 2. Suburban parks and gardens. *Biol. Conserv.,* **15,** 241–97.

Beebee, T. (1981) Habitats of the British amphibians: 4. Agricultural lowlands and general discussion of requirements. *Biol. Conserv.,* **21,** 127–39.

Beven, G.(1965) The food of tawny owls in London. *Lond. Bird Rep.,* **29,** 56–72.

Beven, G. (1982) Further observations on the food of tawny owls in London. *Lond. Nat.*, **61,** 88–94.

Bilham, E.G. (1938) *The Climate of the British Isles,* Macmillan and Co., London.

Birdsall, C.W., Grue, C.E. and Anderson, A. (1986) Lead concentrations in bullfrog (*Rana catesbeiana*) and green frog (*Rana clamitans*) tadpoles inhabiting highway drainages. *Environ. Pollut. (Ser. A),* **40,** 233–47.

Blockheim, J.G. (1974) Nature and properties of highly disturbed urban soils, Philadelphia, Pennsylvania. Paper presented before Division S–5, Soil Sci. Soc. Illinois, Chicago.

Bloomfield, H.E., Handley, J.F. and Bradshaw, W.D. (1981) Topsoil quality. *Landscape Design,* **135,** 32–4.

Blum, A. (1925) Beitrage zur Kenntnis der annuellen Pflanzen. *Bot. Arch.,* **9,** 3–36.

Bond, T.E.T. (1972) Observations on the macro-fungi of an apple orchard in relation to cover crops and fertilizers. *Trans. Brit. Mycol. Soc.,* **58,** 403–416.

Bond, T.E.T. (1981) Macro-fungi on a garden lawn, 1971–78, *Bull. Brit. Mycol. Soc.,* **15,** 99–138.

Bonte, L. (1930) Beitrage zur adventiveflora des rheinisch-westfalischen industriegebietes. *Verh. Naturhist. Vereines Preuss. Rheinl. Westfalens,* **86,** 141–255.

Bornkamm, R., Raghi-Atri, F. and Koch, M. (1980) Effects of entrophication on *Phragmites australis. Garten + Landschaft,* **80,** 15–19.

Bradshaw, A.D. (1980) Mineral nutrition. In *Amenity Grassland: An Ecological Perspective* (eds I.H. Rorison and R. Hunt), John Wiley, Chichester, pp. 101–18.

Bradshaw, A.D. (1983) Topsoil quality – proposals for a new system. *Landscape Design,* **141,** 32–4.

Bradshaw, A.D. and Chadwick, M.J. (1980) *The Restoration of Land. Studies in Ecology,* Vol 6, Blackwell Scientific Publications, Oxford and Edinburgh.

Braun, S. and Fluckiger, W. (1984a) Increased population of the aphid *Aphis pomi* at a motorway; Part 1 – Field evaluation. *Eviron. Pollut. (Ser. A),* **33,** 107–20.

Braun, S. and Fluckiger, W. (1984b) Increased population of the aphid *Aphis pomi* at a motorway: Part 2 – the effect of drought and de-icing salt. *Environ. Pollut. (Ser. A),* **36,** 261–70.

Braun, S. and Fluckiger, W. (1985) Increased population of the aphid *Aphis pomi* at a motoway: Part 3 – the effect of exhaust gases. *Environ. Pollut. (Ser. A),* **39,** 183–92.

Briggs, D.J. (1972) Population differentiation in *Marchantia polymorpha* L. in various lead pollution levels. *Nature, Lond.,* **238,** 166–7.

Briggs, D. (1976) Genecological studies of lead tolerance in groundsel (*Senecio vulgaris* L.). *New Phytol.,* **77,** 173–86.

Brown, V.K. (1985) Insect herbivores and plant succession. *Oikos,* **44,** 17–22.

Brown, V.K. and Hyman, P.S. (1986) Successional communities of plants and phytophagous coleoptera. *J. Ecol.,* **74,** 963–75.

Bultitude, J. (1984) *Apples: a Guide to the Identification of International Varieties,* Macmillan Press, London.

Burges, R.C.L. (1946) Adventive flora of Burton-on-Trent. *Rep. Bot. Soc. Exch. Club. Brit. Isl.,* **13,** 815–19.

Burgess, E.W. (1925) The growth of the city; an introduction to a research project. In *The City* (eds R.E. Park, E.W. Burgess and R.D. McKenzie), University of Chicago Press, Chicago, pp. 47–62.

Burton, R.M. (1979) *Bidens* in Britain. *London Nat.,* **58,** 9–14.

Burton, R.M. (1983) *Flora of the London Area,* London Natural History Society, London.

Busvine, J.R. (1980) *Insects and Hygiene,* 3rd edn., Chapman and Hall, London and New York.

Campbell, C.A. and Dagg, I.N. (1975) Bird populations in downtown and suburban Kitchener-Waterloo, Ontario. *Ont. Field Biol.,* **30,** 1–22.

Carson, R. (1962) *Silent Spring,* Riverside Press, Massachusetts.

Cass, R. (1983) Liverpool 1984: The International Garden Festival. *Landscape Design,* **145,** 17–20.

Chandler, T.J. (1965) *The Climate of London,* Hutchinson, London.

Changnon, S.A. (1976) Inadvertent weather modification. *Water Resourc. Bull.,* **12,** 695–718.

Clapham, A.R. (1953) Human factors contributing to a change in our flora: the former ecological status of certain hedgerow species. In *The Changing Flora of Britain* (ed. J.E. Lousley), Botanical Society of the British Isles, pp. 29–36.

Clapham, A.R., Tutin, T.G. and Warburg, E.F. (1962) *Flora of the British Isles,* Cambridge University Press, Cambridge.

Clarke, D.R. (1979) Lead concentrations: bats vs terrestrial small mammals collected near a major highway. *Environ. Sci. Technol.,* **13,** 338–41.

Clarkson, K. and Birkhead, T. (1987) Magpies in Sheffield – recipe for success. *Brit. Trust Ornithol. News,* **151,** 8–9.

Clements, F.E. (1916) *Plant Succession. An Analysis of the Development of Vegetation,* Carnegie Institute, Washington.

Clifford, H.J. (1956) Seed dispersal on footwear. *Proc. bot. Soc. Brit. Isl.,* **2,** 129–31.

Clinging, V. and Whiteley, D. (1985) Mammals. In *The Natural History of the Sheffield Area* (ed. D. Whiteley), Sorby Natural History Society, Sheffield, pp. 84–104.

Conolly, A.P. (1977) The distribution and history in the British Isles of seven alien species of *Polygonum* and *Reynontria. Watsonia,* **11,** 291–311.

Cooke, A.S. (1972) Indications of recent changes in status in the British Isles of the frog (*Rana temporaria*) and the toad (*Bufo bufo*). *J. Zool, Lond.,* **167,** 161–78.

Coombs, C.F.B., Isaacson, A.J., Murton, R.K., Thearle, R.J.P. and Westwood, N.J. (1981) Collared doves (*Streptopelia decaocta*) in urban habitats. *J. Appl. Ecol.,* **18,** 41–62.

Coppins, B.J. (1973) The 'drought hypothesis'. In *Air Pollution and Lichens* (eds B.W. Ferry, M.S. Baddeley and D.L. Hawksworth) Athlone Press of the University of London, London, pp. 124–42.

Cornwell, P.B. (1968) *The Cockroach,* Vol. 1. Hutchinson, London.

Cramp, S. and Gooders, J. (1967) The return of the house martin. *London Bird Rep.,* **31,** 93–8.

Craul, P.J. (1985) A description of urban soils and their desired characteristics. *J. Arboric.*, **11,** 330–9.

Craul, P.J. and Klein, C.F. (1980) Characterization of streetside soils in Syracuse, New York. *METRIA*, **3,** 88–101.

Creed, E.R. (1974) Two-spot ladybirds as indicators of intense local air pollution. *Nature, Lond.*, **249,** 390–2.

Curl, J.S. (1972) *The Victorian Celebration of Death,* David and Charles, Newton Abbot.

Curl, J.S. (1980) *A Celebration of Death,* Constable, London.

Curtis, R. (1931) Adventive flora of Burton-on-Trent. *Rep. bot. Soc. Exch. Club. Brit. Isl.*, **9,** 465–9.

Czechowski, W. (1980) The ants *Lasius niger* as indicators of the degree of environmental pollution in a city. *Przeglad Zoologicnzy*, **24,** 113–22.

Darlington, A. (1969) *Ecology of Refuse Tips,* Heinmann Educational Books, London.

Davis, B.N.K. (1976) Wildlife, urbanization and industry. *Bio. Conserv.*, **10,** 249–91.

Davis, B.N.K. (1978) Urbanization and the diversity of insects. In *Diversity of Insect Faunas* (eds L.A. Mound and N. Waloff), Blackwell, Oxford and Edinburgh, pp. 126–38.

Davis, B.N.K. (1979) The ground arthropods of London gardens. *Lond. Nat.*, **58,** 15–24.

Davis, B.N.K. (1982) Habitat diversity and invertebrates in urban areas. In *Urban Ecology: 2nd European Ecological Symposium* (eds R. Bornkamm, J.A. Lee and M.R.D. Seaward), Blackwell Scientific Publications, Oxford and Ediburgh, pp. 49–63.

Davison, A.W. (1970) The ecology of *Hordeum murinum* L.I. Analysis of the distribution in Britain. *J. Ecol.*, **58,** 453–66.

Davison, A.W. (1971) The ecology of *Hordeum murinum* L.2. The ruderal habitat. *J. Ecol.*, **59,** 493–506.

Davison, A.W. (1977) The ecology of *Hordeum murinum* L.3. Some effects of adverse climate. *J. Ecol.*, **65,** 523–30.

Dawe, G. and McGlashan, S. (1987) The ecology of urban hoverflies in relation to spontaneous and managed vegetation. *Brit. Ecol. Soc. Bull.*, **8,** 168–71.

Dawson, F.H. and Haslam, S.M. (1983) The management of river vegetation with particular reference to shading effects of marginal vegetation. *Landscape Planning*, **10,** 147–69.

Department of the Environment (1986) *River Quality in England and Wales 1985,* HMSO, London.

Department of the Environment: Department of Transport (1977) *Design Bulletin 23. Residential Roads and Footpaths: Layout Considerations,* HMSO, London.

Detwyler, T.R. (1972) Vegetation of the city. In *Urbanization and Environment* (eds T.E. Detwyler and M.G. Marcus), Duxbury Press, California, pp. 230–59.

Dighton, J. (1978) Effects of synthetic lime aphid honeydew on populations of soil organisms. *Soil Biol. Biochem.*, **10,** 369–76.

Disney, R.H.L. (1972) Some flies associated with dog dung in an English city. *Entomol. Mon. Mag.*, **108,** 93–4.

Dixon, A.F.G. (1973) *Biology of Aphids,* Edward Arnold, London.

Dony, J.G. (1953a) *Flora of Bedfordshire,* The Corporation of Luton Museum and Art Gallery, Luton.

Dony, J.G. (1953b) Wool aliens in Bedfordshire. In *The Changing Flora of Britain* (ed. J.E. Lousley), Botanical Society of the British Isles, Oxford, pp. 160–3.

Dony, J.G. (1955) Notes on the Bedfordshire railway flora. *Bedfordshire Nat.,* **9,** 12–16.

Dorney, R.S. (1979) The ecology and management of disturbed urban land. *Landscape Architecture,* **69,** 268–72, 320.

Douglas, I. (1983) *The Urban Environment,* Edward Arnold, London.

Druce, G.C. (1927) *The Flora of Oxfordshire,* 2nd edn, Clarendon Press, Oxford.

Durwen, R. (1978) *Kartierung der Relativen Warmeverhaltnisse Munsters mittels Phlanologischer Spektren,* Diplom-arbeit, Munster, 88pp.

Dutton, R.A. and Bradshaw, A.D. (1982) *Land reclamation in Cities: a Guide to Methods of Establishment of Vegetation on Urban Waste Land,* HMSO, London.

Ecological Parks Trust (1982) *Ecological Parks Trust: 2nd Report 1981–1982,* Burlington House, London.

Edelsten, H.M. (1940) The insect fauna of the waste areas of Tilbury dock. *Proc. R. Entomol. Soc. Lond. (A),* **15,** 1–11.

Ellenberg, H. (1954) Naturgemasse Anbauplanung, Melioration und Landespflege, *Landw. Pflanzen,* **3,** 1–109.

Ellenberg, H. (1974) Zeigerwerte der Gefasspflanzen Mitteleuropas. *Scr. Geogot,* **9,** E. Goltze-Verlag, Gottingen.

Elton, C.S. (1966) *The Pattern of Animal Communities,* Methuen, London.

Ely, W.A. (1985) The midwife toad in South Yorkshire. *Sorby Record,* **23,** 29–30.

Erzinclioglu, Y.Z. (1981) On the diptera associated with dog dung. *Lond. Nat.,* **60,** 45–6.

Evenson, N. (1979) *Paris: A Century of Change 1878–1978,* Yale University Press, New Haven.

Falinski, J.B. (1971) Synanthropization of plant cover. II. Synanthropic flora and vegetation of towns connected with their natural conditions, history and function. *Mater. Zakl. Fitosco. Stos. U.W. Warszawa-Bialowieza,* **27,** 1–317 (Polish, English summary).

Fitter, R.S.R. (1945) *London's Natural History,* Collins, London.

Franken, E. (1955) Der beginn der Forsythienblute in Hamburg 1955. *Met. Rdsch.,* **8,** 113–14.

Garland, S. (1985) Butterflies and moths (Lepidoptera). In *The Natural History of the Sheffield Area* (ed. D. Whiteley), Sorby Natural History Society, Sheffield, pp. 135–51.

Gaudefroy, E. and Movillefarine, E. (1871) Note sur des plantes meridionales observees aux environs des Paris (Florula Obsidionalis). *Bull. Soc. Bot. France,* **18,** 246–52.

Gemmell, R.P. (1982) The origin and botanical importance of industrial habitats. In *Urban Ecology: 2nd European Ecological Symposium* (eds

R. Bornkamm, J.A. Lee and M.R.D. Seaward), Blackwell Scientific Publications, Oxford and New York.
Gilbert, O.L. (1968a) *Biological indicators of air pollution,* Ph.D. Thesis, University of Newcastle upon Tyne.
Gilbert, O.L. (1968b) Bryophytes as indicators of air pollution in the Tyne Valley. *New Phytol.,* **67,** 15–30.
Gilbert, O.L. (1970a) Urban bryophyte communities in north-east England. *Trans. Brit. Bryol. Soc.,* **6,** 306–16.
Gilbert, O.L. (1970b) A biological scale for the estimation of sulphur dioxide pollution. *New Phytol.,* **69,** 629–34.
Gilbert, O.L. (1971) Some indirect effects of air pollution on bark-living invertebrates. *J. Appl. Ecol.,* **8,** 77–84.
Gilbert, O.L. (1974) An air pollution survey by school children. *Environ. Pollut.,* **6,** 175–80.
Gilbert, O.L. (1976) A lichen-arthropod community. *Lichenologist,* **8,** 86.
Gilbert, O.L. (1981) Plant communities in an urban environment. *Landscape Res.* **6,** 5–7.
Gilbert, O.L. (1983a) The wildlife of Britain's wasteland. *New Sci.,* **67,** 824–9.
Gilbert, O.L. (1983b) Chatsworth, the Capability Brown lawn and its management. *Lanscape Design,* **146,** 8.
Gilbert, O.L. (1983c) The ancient lawns of Chatsworth, Derbyshire. *J. R. hort. Soc.,* **108,** 471–4.
Gilbert, O.L. (1985) A wildflower mix with a short life. *Landscape Design,* **157,** 47–9.
Gilbert, O.L. (1988) Urban demolition sites: a neglected habitat. *Bull. Brit. Lich. Soc.,* **62,** 1–3.
Gilbert, O.L. and Pearman, M.C. (1988) Wild figs by the Don. *Sorby Record,* **25,** 31–3.
Glue, D.E. (1973) Adaptations by predators to urban life. *Bird Study.* **66,** 411.
Glue, D. (1982) *The Garden Bird Book,* Macmillan, London.
Godde, M. and Wittig, R. (1983) A preliminary attempt at a thermal division of the town of Munster, West Germany on a floral and vegetational basis. *Urban Ecol.,* **7,** 255–62.
Goldstein, E.L., Gross, M. and DeGraaf, R.M. (1986) Breeding birds and vegetation: a quantitative assessment. *Urban Ecol.,* **9,** 377–85.
Goode, D. (1986) *Wild in London,* Michael Joseph, London.
Goodwin, D. (1952) The colour varieties of feral pigeons. *Lond. Bird. Rep.*, **16,** 35–6.
Goodwin, D. (1960) Comparative ecology of pigeons in inner London. *Brit. Birds,* **55,** 201–12.
Gorer, R. (1985) Mystery of the London plane. *Country Life,* **177,** 500–1.
Graham, G.G. (1988) *The Flora and Vegetation of County Durham,* The Durham Flora Committee and the Durham County Conservation Trust, Durham.
Greater London Council (1986) *A Nature Conservation Strategy for London: Woodland, Wasteland, the Tidal Thames, Ecology Handbook No. 4.*, Greater London Ecology Unit (Formerly Ecology Section of the GLC), London.
Greenoak, F. (1985) *God's Acre*, Orbis/W.I. Books, London.

Greenwood, F.E. and Gemmell, R.P. (1978) Derelict industrial land as a habitat for rare plants in S. Lancs, (V.C. 59) and W. Lancs (V.C. 60). *Watsonia,* **12,** 33–40.

Grenfell, A.L. (1983) Aliens and adventives. *Bot. Soc. Brit. Isl. News,* **35,** 10–11.

Grenfell, A.L. (1985) Aliens and adventives. *Bot. Soc. Brit. Isl. News,* **39,** 6–7.

Grenfell, A.L. (1986) Adventive News 33: Avonmouth Docks. *Bot. Soc. Brit. Isl. News,* **42,** 16.

Grime, J.P. (1979) *Plant Strategies and Vegetation Processes,* John Wiley, Chicester.

Grime, J.P., Hodgson, J.G. and Hunt, R. (1988) *Comparative Plant Ecology,* Unwin Hyman, London.

Grime, J.P. and Hunt, R. (1975) Relative growth rate, its range and adaptive significance in a local flora. *J. Ecol.,* **63,** 521–34.

Grime, J.P. and Lloyd, P.S. (1973) *An Ecological Atlas of Grassland Plants,* Edward Arnold, London.

Grodzinska, K. (1982) Plant contamination caused by urban and industrial emissions in the region of Cracow city, Poland. In *Urban Ecology: 2nd European Ecological Symposium* (eds R. Bornkamm, J.A. Lee and M.R.D. Seaward), Blackwell Scientific Publications, Oxford and Edinburgh, pp. 149–60.

Grue, C.E., Hoffman, D.J., Beyer, N.W. and Franson, L.P. (1986) Lead concentrations and reproductive success in european starlings *Sturnus vulgaris* nesting on roadside verges. *Environ. Pollut. (Ser. A),* **42,** 157–82.

Grue, C.E., O'Shea, T.J. and Hoffman, D.J. (1984) Lead exposure and reproduction in highway-nesting barn swallows. *Condor,* **86,** 383–9.

Gupta, P.L. and Rorison, I.H. (1975) Seasonal variation in the availability of nutrients down a podzolic profile. *J. Ecol.,* **63,** 521–34.

Hadden, R.M. (1978) Wild flowers of London W1. *Lond. Nat.,* **57,** 26–33.

Hall, M.J. (1984) *Urban Hydrology,* Elsevier Applied Science Publishers, London and New York.

Handley, J.F. and Bulmer, P.G. (1988) *Making the Most of Green Space. A report to the Department of the Environment,* The Groundwork Trust, St. Helens (in press).

Hanson, C.G. and Mason, J.L. (1985) Bird seed aliens in Britain. *Watsonia,* **15,** 237–52.

Harris, S. (1986) *Urban Foxes,* Whittet Books, London.

Haslam, S.M. (1978) *River Plants,* Cambridge University Press, Cambridge.

Haupler, H. (1974) Statistische answertung von punktrasterkarten der gefa pflanzenflora sud-Niedersachsens, *Scripta Geobot,* **8,** Gottingen.

Hawksworth, D.L. and Rose, F. (1970) Qualitative scale for estimating sulphur dioxide air pollution in England and Wales using epiphytic lichens. *Nature, Lond.,* **227,** 145–8.

Hayward, I.M. and Druce, G.C. (1919) *The Adventive Flora of Tweedside,* T. Buncle, Arbroath.

Herter, K. (1965) *Hedgehogs: A Comprehensive Survey,* Phoenix House, London.

Hiller, H.G. (1983) *Hillier's Manual of Trees and Shrubs,* 5th edn., Van Nostrand Reinhold Company, London.

Hodgson, J.G. (1986) Commonness and rarity in plants with special reference to the Sheffield flora. Parts I–IV. *Biol. Conserv.*, **36,** 199–252, 253–74, 275–96, 297–314.

Hogg, J. (1867) On the ballast-flora of the coast of Durham and Northumberland. *Annals and Magazine of Natural History, Ser. 3.* **19,** 38–42.

Holland, D.G. and Harding, J.P.C. (1984) Mersey. In *Ecology of European Rivers* (ed. B.A. Whitton), Blackwell Scientific Publications, Oxford and Edinburgh, pp.113–44.

Holler, A. (1883) Die eisenbahn als verbreitungsmittel von pflanzen. *Flora, Jena,* **66,** 197–205.

Hollis, G.E. (1977) Water yield changes after the urbanization of the Canon's Brook catchment, Harlow, England. *Hydrol. Sci. Bull.,* **22,** 61–75.

Holm, J. (1987) *Squirrels,* Oxford University Press and Whittet Books, London.

Holyoak, D. (1971) Movements and mortality of Corvidae. *Bird Study,* **18,** 97–106.

Horbert, M. (1978) Klimatische und Lufthygienische Aspekte der Stadt- und Landschafts planung. *Natur und Heimat,* **38,** 34–49.

Hounsome, M. (1979) Bird life in the City. In *Nature in Cities,* (ed. I.C. Laurie), John Wiley, Chicester, pp.179–201.

Howard, E. (1898) *Tommorow: A Peaceful Path to Real Reform,* Swan Sonnenschein, London.

Howes, C.A. (1976) Notes on suburban hedgehogs. *Naturalist, Hull,* **101,** 147–8.

Howes, C.A. (1985) Feral cat. In *Yorkshire Mammals* (eds C.A. Howes and M.J. Delany) Bradford University, Bradford, pp. 175–6.

Hudson, H.W. (1898) *Birds in London,* Longman, London.

Hynes, H.B.N. (1960) *The Biology of Polluted Waters,* Liverpool University Press, Liverpool.

Jackowiak, B. (1982) Occurrence of *Puccinellia distans* in the city of Poznan. *Badania Fizjograficzne nad Polska Zachodnia,* **33,** 129–42. (English summary).

Jackson, P.W. and Skeffington, M.C. (1984) *Flora of Inner Dublin,* Royal Dublin Society, Dublin.

Johnson, M.S. (1978) Land reclamation and the botanical significance of some former mining and manufacturing sites in Britain. *Environ. Conserv.,* **2,** 223–8.

Jones, E.W. (1945) Biological Flora of the British Isles, *Acer,* L. *J. Ecol.,* **32,** 215–52.

Jones, J.R.E. (1951) An ecological study of the river Towy. *J. Anim. Ecol.,* **20,** 68–86.

Jones, M. (1986) Ancient woods in the Sheffield area: The documentary evidence. *Sorby Record,* **24,** 7–18.

Kelcey, J.G. (1975) Industrial development and wildlife conservation. *Environ. Conserv.,* **2,** 99–108.

Kelcey, J.G. (1985) Nature conservation, water, and urban areas in Britain. *Urban Ecol.,* **9,** 99–142.

Kellett, J.E. (1982) The private garden in England and Wales. *Landscape Planning,* **9,** 105–23.

Kennedy, C.E.J. and Southwood, J.R.E. (1984) The number of species of insect associated with British trees; a re-analysis. *J. Anim. Ecol.,* **53,** 455–78.

Kent, D.H. (1950) Notes on the flora of Kensington Gardens and Hyde Park. *Watsonia,* **1,** 296–300.
Kent, D.H. (1955) Scottish records of *Senecio squalidus* L. *Proc. bot. Soc. Brit. Isl.,* **1,** 312–13.
Kent, D.H. (1956) *Senecio squalidus* L. in the British Isles – 1, early records (to 1877). *Proc. bot. Soc. Brit. Isl.,* **2,** 115–18.
Kent, D.H. (1960) *Senecio squalidus* L. in the British Isles – 2, the spread from Oxford (1879–1939). *Proc. bot. Soc. Brit. Isl.,* **3,** 375–9.
Kent, D.H. (1964a) *Senecio squalidus* L. in the British Isles – 4, Southern England (1940→) *Proc. bot. Soc. Brit. Isl.,* **5,** 210–13.
Kent, D.H. (1964b) *Senecio squalidus* L. in the British Isles – 5, The Midlands (1940→) *Proc. bot. Soc. Brit. Isl.,* **5,** 214–16.
Kent, D.H. (1964c) *Senecio squalidus* L. in the British Isles – 6, Northern England (1940→) *Proc. bot. Soc. Brit. Isl.,* **5,** 217–19.
Kettlewell, H.B.D. (1973) *The Evolution of Industrial Melanism,* Clarendon Press, Oxford.
Kienle, H. and Luz, H. (1977) *Dachbegrunung, Luxus oder Notwendigkeit,* Frankische Rohrwerke: Optima: Zinco.
King, D.L. and Ball, R.C. (1967) Comparative energetics of a polluted stream. *Limnol. oceanogr.,* **12,** 27–33.
Kirby, P. (1984) Heteroptera colonising demolition sites in Derby. *Entomol. Mon. Mag.,* **120,** 253–8.
Kirkham, J.F. (1982) *Design Guide for Residential Roads,* South Yorkshire County Council, Wakefield.
Knebel, G. (1936) Monographie der algenreihe der Prasiolales, insbesondere von *Prasiola crispa. Hedwigia,* **75,** 1–120.
Kreuzpointner, J.B. (1876) Notizen zur flora Munchens. *Flora, Jena,* **59,** 77–80.
Kuhnelt, W. (1982) Free-living invertebrates within the major ecosystems of Vienna. In *Urban Ecology: 2nd European Ecological Symposium* (eds R. Bornkamm, J.A. Lee and M.R.D. Seaward), Blackwell Scientific Publications, Oxford and Edinburgh, pp. 83–7.
Landsberg, H.E. (1981) *The Urban Climate,* Academic Press, New York and London.
Latham, J.B. (1984) A survey of the flora of Kensal Green and St. Mary's cemeteries 1981–1983. *Lond. Nat.,* **63,** 53–67.
Latimer, W.L. (1984) *Woodland Contrasts: Wildwood or City Park,* London Wildlife Trust, London.
Laundon, J.R. (1967) A study of the lichen flora of London. *Lichenologist,* **3,** 277–327.
Lawrynowicz, M. (1982) Macro-fungal flora of Lodz. In *Urban Ecology: 2nd European Ecological Symposium* (eds R. Bornkamm, J.A. Lee and M.R.D. Seaward), Blackwell Scientific Publications, Oxford and Edinburgh, pp. 41–47.
Lazenby, A. (1983) Ground beetles (Carabidae) and other coleoptera on demolition sites in Sheffield. *Sorby Record,* **23,** 39–51.
Lazenby, A. (1988) Urban beetles in Sheffield. *Sorby Record,* **25,** 22–31.
Lazro, R.J. (1979) *Urban Hydrology,* Ann Arbor Science Publishers, Michigan.
Lee, J.A. and Greenwood, B. (1976) The colonisation by plants of calcareous

wastes from the salt and alkali industry in Cheshire, England. *Biol. Conserv.*, **10,** 131–49.

Lee, R.L.G. (1977) The Serpentine fish and their parasites. *Lond. Nat.*, **56,** 57–70.

Lees, F.A. (1941) *A Supplement to the Yorkshre Floras* (eds C.A. Cheetham and W.A. Sledge), A Brown, London.

Lehmann, E. (1895) Flora von Polnisch-Livland. *Arch. Naturk. Liv-Ehst-Kurlands (Ser. 2), Biol. Naturk,* **11,** 1–422.

Lever, C. (1977) *The Naturalised Animals of the British Isles,* Hutchinson, London.

Lewis, G. and Williams, G. (1984) *Rivers and Wildlife Handbook,* Royal Society for the Protection of Birds, Sandy, Bedfordshire.

Loudon, J.C. (1843) *On the Laying Out, Planting and Managing of Cemeteries, and on the Improvement of Churchyards,* Longman, London (Facsim, Ivelet Books, Redhill, 1981).

Lousley, J.E. (1948) *Ficus carica* L. in Britain. *Rep. bot. Soc. Exch. Club. Brit. Isl. 1948*, 330–3.

Lousley, J.E. (1953) The recent influx of aliens into the British flora. In *The Changing Flora of Britain* (ed. J.E Lousley), Botanical Society of the British Isles, Oxford, pp. 140–59.

Lousely, J.E. (1961) A census list of wool aliens found in Britàin, 1946–1960. *Proc. bot. Soc. Brit. Isl.*, **4,** 221–47.

Lousely, J.E. (1970) The influence of transport on a changing flora. In *The Flora of a Changing Britain* (ed. F. Perring), Classey, Faringdon, pp. 73–83.

Lund, A.C. (1974) Analysis of urban plant communities of Atlanta, Georgia. Ph.D. Thesis, Emory University, Atlanta, Georgia, USA.

Luniak, M. (1983) The avifauna of urban green areas in Poland and possibilities of managing it. *Acta Ornithol. Warsz.*, **19,** 1–56.

Lusis, J.J. (1961) On the biological meaning of colour polymorphism of lady beetle *Adalia bipunctata* L. *Latv. Entomol.*, **4,** 3–29.

Macan, T.T. (1974) Freshwater invertebrates. In *The Changing Flora and Fauna of Britain* (ed. D.L. Hawksworth), Academic Press, London and New York, pp. 143–55.

MacArthur, R.H. and Wilson, E.O. (1963) An equilibrium theory of insular zoogeography. *Evolution,* **17,** 373–87.

MacArthur, R.H. and Wilson, E.O. (1967) *The Theory of Island Biogeography (Monographs on Population Biology,* No. 1), Princeton University Press, Princeton, NJ.

Macdonald, D.W. (1987) *Running with the Fox,* Unwin Hyman, London and Sydney.

Mannerkorpi, P. (1944–45) Uhtuan taistelurintamalle saapuneista tulokaskasveista. *Ann. bot. Soc. Zool – Bot. Fenn. 'Vanamo', (Not. Bot.* 15), **20,** 39–51.

Markstein, B. and Sukopp, H. (1980) The waterside vegetation of the Berlin Havel (1962–77), *Garten + Landschaft,* **80,** 30–36.

Massey, C.I. (1972) A study of hedgehog road mortality in the Scarborough district. *Naturalist, Hull,* **97,** 103–5.

Matheson, C. (1944) The domestic cat as a factor in urban ecology. *J. Anim. Ecol.*, **13,** 130–3.

Matthews, P. and Davison, A.W. (1976) Maritime species on roadside verges. *Watsonia,* **11,** 146–7.

McMillan, R.C. (1954) Effects of smoke pollution on plant life. *Public Works and Municipal Services Congress, 1954.*

McNab, A. and Price, S. (1985) Lineside Landscape. *Landscape Design,* **158,** 14–15.

McNeill, S. and Southwood, T.R.E. (1978) The role of nitrogen in the development of insect/plant relationships. In *Biochemical Aspects of Plant and Animal Coevolution* (ed. J.B. Harborne), Academic Press, London, pp. 77–98.

Meijer, J. (1974) A comparative study of the immigration of carabids (Coleoptera: Carabidae) into a new polder. *Oecologia,* **161,** 185–208.

Merrett, C. (1666) *Pinax rerum naturalium Britannicarum,* Roycroft, London.

Messenger, K.G. (1968) A railway flora of Rutland. *Proc. bot. Soc. Brit. Isl.,* **7,** 325–44.

Metcalf, C.R. (1953) Effects of atmospheric pollution on vegetation, *Nature, Lond.,* **172,** 659–61.

Miles, J. (1979) *Vegetation Dynamics,* Chapman and Hall, London.

Miller, A. (1981) Birds of the Sorby Plantation 1977–1980. *Sorby Record,* **19,** 25–30.

Moran, M.A. (1984) Influence of adjacent land-use on understory vegetation of New York forests. *Urban Ecol.,* **8,** 329–40.

Morgan, R.A. and Glue, D.E. (1981) Breeding survey of black redstarts in Britain 1977. *Bird Study,* **28,** 163–8.

Morris, P. (1966) The hedgehog in London. *Lond. Nat.,* **45,** 43–9.

Morris, P. (1983) *Hedgehogs,* Whittet Books, Weybridge.

Muggleton, J., Lonsdale, D. and Benham, B.R. (1975) Melanism in *Adalia bipunctata* L. and its relationship to atmospheric pollution. *J. Appl. Ecol.,* **12,** 451–64.

Muhlenbach, V. (1979) Contributions to the synanthropic (adventive) flora of the railroads in St. Louis, Missouri, U.S.A. *Ann. Mo. bot. Gdn,* **66,** 1–108.

Murphy, K.J. and Eaton, J.W. (1983) Effects of pleasure-boat traffic on macrophyte growth in canals. *J. Appl. Ecol.,* **20,** 713–29.

Myerscough, P.J. (1980) Biological flora of the British Isles: *Epilobium angustifolium* L. *J. Ecol.,* **68,** 1047–74.

Newbold, C., Purseglove, J. and Holmes, N. (1983) *Nature Conservation and River Engineering,* Nature Conservancy Council, Attingham Park, Shrewsbury.

Nicholson, B. (1985) *The Wildlife Habitats of the Regent's and Hertford Union Canals,* London Wildlife Trust, London.

Nicholson, B. and Hare, A. (1986) The management of urban woodland, *Ecos.,* **7,** 38–43.

Nicholson, E.M. (1951) *Birds and Men,* Collins, London.

Nicholson, E.M. (1987) *The New Environmental Age,* Cambridge University Press, Cambridge.

Numata, M. (1982) Changes in ecosystem structure and function in Tokyo. In *Urban Ecology: 2nd European Ecological Symposium* (eds R. Bornkamm, J.A. Lee and M.R.D. Seaward), Blackwell Scientific Publications, Oxford and Edinburgh, pp. 139–47.

Odum, E.P. (1969) The strategy of ecosystem development. *Science,* **164,** 262–70.

Olkowski, W., Olkowski, H. and Van den Bosch, R. (1982) Linden aphid (*Eucallipterus tiliae*) parasite establishment. *Environ. Entomol.,* **11,** 1023–5.

Owen, D.F. (1971) Species diversity in butterflies in a tropical garden. *Biol. Conserv.,* **3,** 191–8.

Owen, D.F. (1978) *Towns and Gardens,* Hodder and Stoughton, London.

Owen, D.F. and Duthie, D.J. (1982) Britain's successful moth colonist; Blair's shoulder-knot (*Lithophane leautieri*). *Biol. Conserv.,* **23,** 285–90.

Owen, D.F. and Whiteway, W.R. (1980) *Buddleia davidii* in Britain: history and development of an associated fauna. *Biol. Conserv.,* **17,** 149–55.

Owen, J. (1983) *Garden Life,* Chatto and Windus/The Hogarth Press, London.

Owens, M. (1970) Nutrient balances in rivers. *Wat. Treatment and Exam.,* **19,** 239–52.

Palmer, J.R. (1984) The naturalization of oil-milling adventive plant in the Thames estuary. *Lond. Nat.,* **63,** 68–70.

Parslow, J.L.F. (1967) Changes in status among breeding birds in Britain and Ireland. *Brit. Birds,* **60,** 51–9.

Patterson, J.C. (1976) Soil compaction and its effects upon urban vegetation. Better trees for Metropolitan Landscapes Symposium. *Proc. USDA Forest Serv. Gen. Tech. Report NE–22.*

Payne, R.M. (1978) The flora of walls in south-eastern Essex. *Watsonia,* **12,** 41–6.

Peace, J.R. (1962) *Pathology of Trees and Shrubs,* Clarendon Press, Oxford.

Pentelow, F.T.K. (1965) The lake in St. James's Park. *Lond. Nat.,* **44,** 128–38.

Percival, E. and Whitehead, H. (1929) A quantitative study of the fauna of some types of stream-bed. *J. Ecol.,* **17,** 282–314.

Perring, F.H. and Farrell, L. (1977) *British Red Data Books: 1. Vascular plants,* Society for the Promotion on Nature Conservation, Lincoln.

Peterken, G.F. (1974) A method for assessing woodland flora for conservation using indicator species, *Biol. Conserv.,* **6,** 239–45.

Peterken, G.F. (1981) *Woodland Conservation and Management,* Chapman and Hall, London and New York.

Pike, G.V. (1979) Kestrels in the Birmingham area. *West Midland Bird Club Annual Report,* **45,** 23–5.

Plant, C.W. (1979) The status of the hedgehog *Erinaceus europaeus* in the London Boroughs of Barking, Newham, Redbridge and Waltham Forest. *Lond. Nat.,* **58,** 27–37.

Prestt, I., Cooke, A.S. and Corbett, K.F. (1974) British amphibians and reptiles. In *The Changing Flora and Fauna of Britain* (ed. D.L. Hawksworth), Academic Press, London and New York, pp.229–54.

Przybylski, Z. (1979) The effect of automobile exhaust gasses on the arthropods of cultivated plants, meadows and orchards. *Environ. Pollut.,* **19,** 157–61.

Pugsley, H.W. (1948) A prodromus of the British *Hieracia*. *J. Linn. Soc., Bot.,* **54,** 1–356.

Purves, D. (1966) Contamination of urban garden soils with copper and boron. *Nature, Lond.,* **210,** 1077–8.

Purves, D. (1972) Consequences of trace-element contamination of soils. *Environ. Pollut.*, **3,** 17–24.
Purves, D. and Mackenzie, E.J. (1969) Trace-element contamination of parklands in urban areas. *J. Soil. Sci.*, **20,** 288–90.
Rabinowitch, E.I. (1956) *Photosynthesis and Related Processes.* Vol. 2 part 2, Interscience Publishers, New York.
Rackham, O. (1976) *Trees and Woodlands in the British Landscape,* Dent, London.
Rackham, O. (1980) *Ancient Woodland,* Arnold, London.
Rackham, O. (1986) *The History of the Countryside,* Dent, London.
Randhawa, H.S., Clayton, Y.M. and Riddle, R.W. (1965) Isolation of *Cryptococcus neoformans* from pigeon habitats in London. *Nature, Lond.*, **208,** 801.
Ratcliffe, D.A. (1977) *A Nature Conservation Review,* 2 Vols., Cambridge University Press, Cambridge.
Read, H.J., Wheater, C.P. and Martin, M.H. (1987) Aspects of the ecology of Carabidae (Coleoptera) from woodlands polluted by heavy metals. *Environ. Pollut. (Ser. A),* **48,** 61–76.
Richard, W. (1987) Parks as urban elements. *Garten Landschaft,* **97,** 19–25.
Rippon, J.W. (1982) *Medical Mycology,* 2nd edn, W.B. Saunders, London.
Robbins, R.W. (1939) Lepidoptera of a London garden fifty years ago. *Lond. Nat.,* 40–41.
Roberts, D.J. and Roberts, M.J. (1985) Spiders. In *The Natural History of the Sheffield Area* (ed. D. Whiteley), Sorby Natural History Society, Sheffield, pp. 210–21.
Roberts, P. (1977) Magpies in Sheffield. *The Magpie,* **1,** 4–6.
Rorison, I.H. (1967) A seedling bioassay on some soils in the Sheffield area. *J. Ecol.,* **55,** 725–41.
Rothschuh, B. (1968) Die Sicherung des raumbedarfs für den Strabenverkehr. *Strabenbau und Strabenverkehrstechnik,* **66,** 1–75.
Ruff, A. (1979) *Holland and the Ecological Landscapes,* Deanwater, London.
Salisbury, E.J. (1943) The flora of bombed areas. *Nature, Lond.,* **151,** 462–6.
Salisbury, E.J. (1953) A changing flora as shown in the study of weeds of arable land and waste places. In *The Changing of Flora of Britain* (ed. J.E. Lousley), Botanical Society of the British Isles, Oxford, pp. 130–9.
Salisbury, E.J. (1954) Air pollution and plant life. *Proceedings 1954 Conference of the National Smoke Abatement Society,* 105–13.
Salisbury, E.J. (1961) *Weeds and Aliens,* Collins, London.
Sargent, C. (1984) *Britain's Railway Vegetation,* Institute of Terrestrial Ecology, Cambridge.
Saunders, P.J.W. (1966) The toxicity of sulphur dioxide to *Diplocarpon rosae* causing blackspot of roses. *Ann. appl. Biol.,* **58,** 103–14.
Savidge, J.P., Heywood, V.H. and Gordon, V. (1963) *Travis's Flora of South Lancashire,* Liverpool Botanical Society, Liverpool.
Schmid, J.A. (1975) *Urban Vegetation: a Review and Chicago Case Study, Research Paper No. 161,* Department of Geography, The University of Chicago, Illinois.
Schreiber, K.F. (1977) Warmegliederung der Schweiz 1: 200,000 mit Erlauterungen. *Grundlagen für die Raumplanung,* Bern.
Scott, N.E. and Davison, A.W. (1982) De-icing salt and the invasion of road verges by maritime plants. *Watsonia,* **14,** 41–52.

Scott, N.E. (1985) The updated distribution of maritime species on British roadsides. *Watsonia,* **15,** 381–6.
Scurfield, G. (1955) Atmospheric pollution considered in relation to horticulture. *J. R. hort. Soc.,* **80,** 93–101.
Scurfield, G. (1960) Air pollution and tree growth. *For. Abstr.,* **21,** 339–47, 517–28.
Seaward, M.R., (1982) Lichen ecology of changing urban environments. In *Urban Ecology: 2nd European Ecological Symposium* (eds R. Bornkamm, J.A. Lee and M.R.D. Seaward), Blackwell Scientific Publications, Oxford and Edinburgh, pp. 181–9.
Segal S. (1969) *Ecological Notes on Wall Vegetation,* W. Junk, The Hague, Netherlands.
Shaw, C.E. (1963) Canals. In *Travis's Flora of South Lancashire* (eds J.P. Savidge, V.H. Haywood and V. Gordon), Liverpool Botanical Society, Liverpool, pp. 71–3.
Shildrick, J. (1984) *Turfgrass Manual,* National Turfgrass Council, Bingley.
Shildrick, J.P. and Marshall, E.J.P. (1985) *Growth Retardants for Amenity Grasslands. Workshop Report no. 7,* National Turfgrass Council, Bingley.
Simms, E. (1975) *Birds of Town and Suburb,* Collins, London.
Simpson, D.A. (1984) A short history of the introduction and spread of *Elodea* Michx in the British Isles. *Watsonia,* **15,** 1–9.
Sinker, C.A., Packham, J.R., Trueman, I.C., Oswald, P.H., Perring, F.H. and Prestwood, W.V. (1985) *Ecological Flora of the Shropshire Region.* Shropshire Trust for Nature Conservation, Shrewsbury.
Smith, A.H.V. (1985) Birds. In *The Natural History of the Sheffield Area* (ed. D. Whiteley), Sorby Natural History Society, Sheffield, pp. 105–21.
Smith, K.G.V. (1973) Pests and the zoonoses. *Biologist,* **20,** 142–3, 209.
Smyth, R. (1987) *City Wildspace,* Hilary Shipman, London.
Sokal, R.R. and Sneath, P.H.A. (1963) *Principles of Numerical Taxonomy,* Freeman, San Franscisco.
Solomon, M.E. (1965) The ecology of pests in stores and houses. In *Ecology and the Industrial Society* (eds G.T. Goodman, R.W. Edwards and J.M. Lambert), Blackwell Scientific Publications, Oxford and Edinburgh, pp. 345–66.
Spirn, A.W. (1984) *The Granite Garden*, Basic Books Inc., New York.
Sporne, K.R. (1980) A re-investigation of character correlations among dicotyledons. *New Phytol.,* **85,** 419–49.
Stiles, D. (1979) The common wall lizard (*Podarcis muralis*) in Middlesex. *Lond. Nat.,* **58,** 25–6.
Stoddart, D.M. (1980) Notes from the mammal society, No. 40. *J. Zool., Lond.,* **191,** 403–33.
Storrie, J. (1886) *The Flora of Cardiff,* Cardiff.
Stubbs, A.E. and Falk, S.J. (1983) *British Hoverflies*, British Entomological and Natural History Society, London.
Sukopp, H., Blume, H.P. and Kunick, W. (1979) The soil, flora and vegetation of Berlin's wastelands. In *Nature in Cities* (ed. I.C. Laurie), John Wiley, Chichester, pp. 115–32.
Sukopp, H. and Weiler, S. (1986) Biotype mapping in urban areas of the Federal Republic of Germany. *Landschaft und Stadt,* **18,** 25–8.

Sukopp, H. and Werner, P. (1983) Urban environments and vegetation. In *Man's Impact on Vegetation* (eds W. Holzner, M.J.A. Werger and I. Ikusima), W. Junk, The Hague, Netherlands, pp. 247–60.

Summers-Smith, J.D. (1963) *The House Sparrow*, Collins, London.

Sutton, S. (1980) *Woodlice*, Pergamon Press, Oxford.

Swale, E.M.F. (1962) Notes on some algae from the Reddish Canal. *Br. phycol. Bull.*, **2,** 174–6.

Sydes, C. and Grime, J.P. (1981a) Effects of tree leaf litter on herbaceous vegetation in deciduous woodland. I. Field investigations. *J. Ecol.*, **69,** 237–48.

Sydes, C. and Grime, J.P. (1981b) Effects of tree leaf litter on herbaceous vegetation in deciduous woodland. II. An experimental investigation. *J. Ecol.*, **69,** 249–62.

Tabor, R. (1983) *Wild life of the Domestic Cat,* Arrow Books, London.

Tatner, P. (1982) Factors influencing the distribution of magpies *Pica pica* in an urban environment. *Bird Study,* **29,** 227–34.

Taylor, L.R., French, R.A. and Woiwod, I.P. (1978) The Rothamsted insect survey and the urbanization of land in Great Britain. In *Perspectives in Urban Entomology* (eds G.W. Frankie and C.S. Koehler), Academic Press, London and New York, pp.31–65.

Teagle, W.G. (1978) *The Endless Village,* Nature Conservancy Council, Shrewsbury.

Ten Houten, J.G. (1966) Bezwaren van luchtverontreiniging voor de landbouw *Landbonwk. Tijdschr.*, **78,** 2–13.

Thellung, A. (1905) Einteilung der Ruderal und Adventivflora in genetische Gruppen. *Vierteljahrsschr. Naturf. Ges. Zurich,* **50,** 232–305.

Thellung, A. (1917) Stratiobotanik. *Vierteliahrsschr. Naturf. Ges. Zurich,* **62,** 327–35.

Thompson, J.R. and Rutter, A.J. (1982) Planning motorway planting in relation to de-icing salt. In *Urban Ecology: 2nd European Ecological Symposium* (eds R. Bornkamm, J.A. Lee and M.R.D. Seaward), Blackwell Scientific Publications, Oxford and Edinburgh, pp. 332–3.

Thompson, J.R., Rutter, A.J., Ridout, P.S. and Glover, M. (1979) The implications of the use of de-icing salt for motorway plantings in the U.K. In *The Impact of Road Traffic on Plants,* Transport and Road Research Laboratory, Supplementary Report, 513, Crowthorne, Berks.

Thorpe, H. (1969) *Departmental Committee of Inquiry into Allotments,* Cmnd 4166, HMSO, London, pp. 1–460.

Thorpe, H. (1975) The homely allotment, from rural dole to urban amenity, a neglected aspect of land use. *Geography,* **60,** 169–83.

Thorpe, H., Galloway, E.B. and Evans, L.M. (1976) *From Allotments to Leisure Gardens – A Case Study of Birmingham,* Department of Geography, University of Birmingham, pp. 1–73.

Thorpe, H., Galloway, E.B. and Evans, L.M. (1977) *The Rationalisation of Urban Allotment Systems – A Case Study of Birmingham,* Department of Geography, The University of Birmingham, pp. 1–164.

Tilghman, N.S. (1987) Characteristics of urban woodlands affecting breeding bird diversity and abundance. *Landscape and Urban Planning,* **14,** 481–95.

Timson, J. (1959) Variegated foliage. *New Biol.*, **30,** 31–46.

Tregay, R. (1985) A sense of nature. *Landscape Design,* **156,** 34–8.

Tudor-Walters, J. (1918) *Rep. Committee Considering Questions of Building Construction in Connection with the Provision of Dwellings for the Working Classes in England and Wales and Scotland,* Cmnd 9191, HMSO, London.

Tutin, T.G. (1973) Weeds of a Leicester garden. *Watsonia,* **9,** 263–7.

Unwin, R, (1909) *Town Planning in Practice: An Introduction to the Art of Designing Cities and Suburbs,* Unwin, London.

Van der Zande, A.N., Berkhuizen, J.C., Van Latesteijn, H.C., Keurs, W.J. and Poppelaars, A.J. (1984) Impact of outdoor recreation on the diversity of a number of breeding bird species in woods adjacent to urban residential areas. *Biol. Conserv.*, **30,** 1–39.

Van Emden, H.F. (1972) Aphids as phytochemists. In *Phytochemical Ecology* (ed. J.B. Harborne), Academic Press, London and New York, pp. 25–43.

Vick, C.M. and Bevan, R.J. (1976) Lichens and tar-spot (*Rhytisma acerinum*) as indicators of sulphur dioxide pollution on Merseyside. *Environ. Pollut.*, **11,** 203–16.

Visse, G. and Van Wingerden, W.K.R.E. (1982) Aerial dispersal of spiders in a city. In *Urban Ecology: 2nd European Ecological Symposium* (eds R. Bornkamm, J.A. Lee and M.R.D. Seaward), Blackwell Scientific Publications, Oxford and Edinburgh, p. 344.

Waggoner, P.E. and Ovington, J.D. (1962) *Proceedings of the Lockwood Conference on the Suburban Forest and Ecology, Bulletin 652,* Thc Connecticut Agricultural Experimental Station, Connecticut.

Walters, S.M. (1970) The next twenty-five years. In *The Flora of a Changing Britain* (ed. F. Perring), Classey, Hampton, pp. 136–141.

Warren, J.B.L. (1871) The flora of Hyde Park and Kensington Gardens. *J. Bot. Lond.*, **9,** 227–38.

Warren, J.B.L. (1875) Kensington Garden's plants. *J. Bot. Lond.*, **13,** 336.

Wathern, P. (1976) The ecology of development sites, PhD Thesis, University of Sheffield.

Wathern, P. and Gilbert, O.L. (1978) Artificial diversification of grassland with native herbs. *J. Environ. Manag.*, **7,** 29–42.

Watt, A.S. (1947) Pattern and process in the plant community. *J. Ecol.*, **35,** 1–22.

Weigmann, G. (1982). The colonisation of ruderal biotypes in the city of Berlin by arthropods. In *Urban Ecology: 2nd European Ecological Symposium* (eds R. Bornkamm, J.A. Lee and M.R.D. Seaward), Blackwell Scientific Publications, Oxford and Edinburgh, pp. 75–82.

Wells, T.C.E. (1987) The establishment of floral grasslands. *Acta Hortic.*, **195,** 59–69.

Wheeler, A. (1978) Fish in an urban environment. In *Nature in Cities* (ed. I.C. Laurie), John Wiley, Chichester, pp. 159–77.

Whiteley, D. (1977) Amphibian fauna of Sheffield. *Sorby Record,* **15,** 36–48.

Whiteley, D. (1984) The pipistrelle bat: its status and distribution in suburban Sheffield. *Sorby Record,* **22,** 29–37

Whitney, G.G. and Adams, S.D. (1980) Man as maker of new plant communities. *J. Appl. Ecol.*, **17,** 431–48.

Whitton, B.A. (1966) Algae in St. James Park Lake. *Lond. Nat.*, **45,** 26–8.

Whitton, B.A. (1984) *Ecology of European Rivers,* Blackwell Scientific Publications, Oxford and Edinburgh.

Widgery, J.P. (1978) Roesel's bush-cricket *Metrioptera roeselii* in Regent's Park. *Lond. Nat.*, **57,** 57–8.

Wilkinson, R. (1956) The quality of urban rainfall run-off water from a housing estate. *J. Instn. Pub. Health. Engrs,* **55,** 70–84.

Williams, R.J.H., Lloyd, M.M. and Ricks, G.R. (1971) Effects of atmospheric pollution on deciduous woodland 1; Some effects on leaves of *Quercus patraea* (Mattuschka) Leibl., *Environ. Pollut.*, **2,** 57–68.

Williamson, K. (1975) Birds and climatic change. *Bird Study,* **22,** 143–64.

Willis, A.J. (1972) Long-term ecological changes in sward composition following application of maleic hydrazide and 2,4-D. In *Proceedings of 11th British Weed Control Conference,* pp. 360–7.

Willis, A.J. (1985) Long-term effects of maleic hydrazide on roadside vegetation; experiments at Bilbury. In *Growth Retardants for Amenity Grassland. Workshop Report No.* 7. (eds J.P. Shildrick and E.J.P. Marshall), National Turfgrass Council, Bingley, pp. 52–7.

Willis, S.J. (1954) Observations on the weed problem of allotments based on a survey. In *Proceedings of the British Weed Control Conference 1954,* pp. 71–4.

Willstatter, R. and Stoll, A. (1918) *Untersuchungen über die Assimilation der Kohlensaure,* Springer, Berlin.

Wittig, R. and Durwen K.J. (1982) Ecological indicator-value spectra of spontaneous urban floras. In *Urban Ecology: 2nd European Ecological Symposium* (eds R. Bornkamm, J.A. Lee and M.R.D. Seaward), Blackwell Scientific Publications, Oxford and Edinburgh, pp. 23–31.

Woodroffe, G.E. (1955) The Hemiptera-Heteroptera of some cinder-covered wasteland at Slough, Buckinghamshire. *Entomologist,* **88,** 10–17.

Wu, L. and Antonovics, J. (1976) Experimental ecological genetics in *Plantago*. II. Lead tolerance in *Plantago lanceolata* and *Cynodon dactylon* from a roadside. *Ecology,* **57,** 205–8.

Yaldon, D.W. (1980) Notes on the diet of urban kestrels. *Bird Study,* **27,** 235–8.

Zacharias, F. (1972) Initiation of street trees flowering phase (esp. *Tilia × euchlora*) and temperature distribution in West Berlin, Dissertation, Free University of Berlin, pp. 1–309 (in German).

SPECIES AND SUBJECT INDEX

Pages numbers in italics refer to tables or illustrations. Textual material may also occur.